```
TK2851 .K444 2007
Keljik, Jeff.
Electric motors and
 motor controls /
Northeast Lakeview Colleg
33784000120253
```

ELECTRIC MOTORS
and
MOTOR CONTROLS

ELECTRIC MOTORS and MOTOR CONTROLS

SECOND EDITION

JEFF KELJIK

Australia Brazil Canada Mexico Singapore Spain United Kingdom United States

Electric Motors & Motor Controls, 2nd Edition
Jeff Keljik

Vice President, Technology and Trades ABU:
David Garza

Director of Learning Solutions:
Sandy Clark

Senior Acquisitions Editor:
Stephen Helba

Development:
Dawn Daugherty

Marketing Director:
Deborah S. Yarnell

Channel Manager:
Dennis Williams

Marketing Coordinator:
Stacey Wiktorek

Senior Production Manager:
Larry Main

Production Editor:
Benj Gleeksman

Art & Design Coordinator:
Nicole Stagg

COPYRIGHT © 2007 Thomson Delmar Learning, a division of Thomson Learning Inc. All rights reserved. The Thomson Learning Inc. Logo is a registered trademark used herein under license.

Printed in the United States of America
1 2 3 4 5 XX 06 05 04

For more information contact
Thomson Delmar Learning
Executive Woods
5 Maxwell Drive, PO Box 8007,
Clifton Park, NY 12065-8007
Or find us on the World Wide Web at
www.delmarlearning.com

ALL RIGHTS RESERVED. No part of this work covered by the copyright hereon may be reproduced in any form or by any means—graphic, electronic, or mechanical, including photocopying, recording, taping, Web distribution, or information storage and retrieval systems—without the written permission of the publisher.

For permission to use material from the text or product, contact us by
Tel.   (800) 730-2214
Fax   (800) 730-2215
www.thomsonrights.com

Library of Congress Cataloging-in-Publication Data

Keljik, Jeff.
  Electric motors and motor controls / Jeff Keljik. -- 2nd ed.
      p. cm.
  Includes index.
  ISBN 1-4018-9841-6
  1. Electric controllers. 2. Electric motors. I. Title.
  TK2851.K444 2006
  621.46--dc22   2006002546

NOTICE TO THE READER

Publisher does not warrant or guarantee any of the products described herein or perform any independent analysis in connection with any of the product information contained herein. Publisher does not assume, and expressly disclaims, any obligation to obtain and include information other than that provided to it by the manufacturer.

The reader is expressly warned to consider and adopt all safety precautions that might be indicated by the activities herein and to avoid all potential hazards. By following the instructions contained herein, the reader willingly assumes all risks in connection with such instructions.

The publisher makes no representation or warranties of any kind, including but not limited to, the warranties of fitness for particular purpose or merchantability, nor are any such representations implied with respect to the material set forth herein, and the publisher takes no responsibility with respect to such material. The publisher shall not be liable for any special, consequential, or exemplary damages resulting, in whole or part, from the readers' use of, or reliance upon, this material.

Dedication

I would like to dedicate this book to my wife, Susan. She has encouraged me to do what I love to do—help people learn. She has prodded me to stay on the project as other interests compete for my time. She attended events without me, as she knew I had an obligation to complete some writing or was meeting with a colleague.

CONTENTS

Preface xiii

Acknowledgments xv

Introduction xvii

1 Single-Phase Motors **1**

    Key Terms in This Chapter 1
    Introduction 2
    Nameplate Information 2
    Single-Phase Motors 6
    Rotating Magnetic Field 10
    Squirrel Cage Rotor Principles 13
    Split-Phase Motors and Slip 15
    Starting Winding Switches 18
    Capacitor-Start, Induction-Run Motors 21
    Other Capacitor Motors 24
    Shaded Pole Motors 25
    Universal Motors 27
    Summary 28
    Questions 28

2 Three-Phase Motors and Generators **31**

    Key Terms in This Chapter 31
    Introduction 32

Three-Phase Power Generation 32
Frequency 34
Output Voltage and Current Control 35
Generator Output Power 36
Exciters 37
Parallel Operation of Alternators 38
Loss of Alternator Output 42
Standby and Cogeneration Systems 43
Three-Phase Motor Theory 49
Motor Rotation Checkers 52
Three-Phase Motor Connections 52
Multispeed Motors 53
Motor Design Letters 56
Power Factor of Induction Motors 57
Motor Efficiency 58
Troubleshooting 58
Summary 59
Questions 59

3 Basic Motor Control 61

Key Terms in This Chapter 61
Introduction 62
Manual Controls—Controller Design 62
Overload Protection 63
Two-Wire Automatic Control 67
Three-Wire Automatic Control 69
Jog and Reversing Controls 73
Overload Selection 75
Fuse Protection 80
NEC® and Single-Phase Motor Applications 82
Summary 83
Questions 84

4 Controllers, Relays, and Timers 85

Key Terms in This Chapter 85
Introduction 86

Nema Contactors and Starters 86
IEC Standards 93
Pilot Devices 94
Photoelectric Controls 96
Proximity Sensors 102
Enclosures 104
Hazardous Areas and Safety   104
Electromechanical Controls, Relays, and Timers 109
Electronic Timers 121
Solid-State Motor Control 125
Contact Maintenance 127
Summary 128
Questions 130

5 Three-Phase Motors, Controls, and Full-Voltage Starting 131

Key Terms in This Chapter 132
Introduction 132
Manual Control 132
Magnetic Control 134
Reversing Three-Phase Motor Starters 140
Sequence Control/Compelling Control 144
Multispeed Controllers 146
Programmable Controllers 150
Electronic Speed Control 151
Wye–Delta Starting 159
Open/Closed Transition 161
Part Winding Starters 163
Phase Failure Relay 166
Motor Control Centers 166
Summary 166
Questions 167

6 Motor Acceleration and Deceleration 169

Key Terms 169
Introduction 170
Reduced Voltage Starting 170

Primary Resistor Starting 171
Autotransformer Starters 179
Solid-State Reduced Voltage Starting 179
Synchronous Motor Starting 182
Electrical Brakes 186
Plugging Controls 186
DC Braking 188
Mechanical Brakes 191
Summary 193
Questions 193

7 Motor Maintenance and Installation 195

Key Terms in This Chapter 195
Introduction 196
Replacement Motors 196
Bearings—Types and Maintenance 196
Megohmmeters 201
Growlers 202
Lead Identification 203
Delta-Connected Motors—Lead Identification 209
Horsepower Testing 211
Motor Efficiency 214
NEC® and Size of Feeders, Disconnects, and Controllers 215
Disconnects, Controllers, Fuses, Branch Circuit
    Conductors, Overloads 215
Summary 218
Questions 218

8 Special Motors 223

Key Terms in This Chapter 223
Introduction 224
Wound Rotor Motor 224
Repulsion Motors 228
Synchronous Motors 231
Permanent Magnet Motors 235
Stepper Motors 235

Hybrid Motors 238
Torque Motors 238
Summary 239
Questions 239

9 Power Distribution and Monitoring Systems . . . 241

Key Terms in This Chapter 241
Introduction 242
Transformers 242
Voltage Transformation 244
Transformer Design 245
Transformer Operation 249
Transformer Polarity 250
Transformer Losses and Regulation 252
Multivoltage Transformers 253
Transformer Connections for Voltage 253
Autotransformers 255
Control Transformers 256
Instrument Transformers 256
Three-Phase Transformers 261
High Leg or Wild Leg 267
Transformer Nameplate 270
NEC® Requirements 273
Single- and Three-Phase Line Drop 274
Power Factor Correction 277
Demand Meters 277
Energy Management 278
Summary 280
Questions 280

10 DC Motors, Generators, and Controls 283

Key Terms in This Chapter 283
Introduction 284
DC Generation 284
Armatures 286
Series Generator 288

Shunt-Wound Generators 289
Compound Field Generators 290
Armature Reaction 292
Parallel Operation of DC Compound Generators 293
DC Motors 294
Speed and Direction Control 296
Shunt Motors 296
Permanent Magnet Motors 299
Series Motors 299
Brushless DC Motors 301
Electronic Speed Control 302
Manual Control of DC Motors 303
DC Motors and NEC® 303
Summary 305
Questions 305

Appendix A   Wiring Diagrams **307**

Appendix B   Formulas **313**

Appendix C   Symbols **317**

Appendix D   Troubleshooting **323**

Glossary **335**

Index **375**

PREFACE

INTENDED USE

Electric Motors and Motor Controls, 2nd edition, is intended to help student electricians, and practicing electricians, learn how motors operate and how to install them. The book helps students understand how controls are designed and installed to provide either simple or complex control schemes. The book is written for people who already have an understanding of basic electrical theory, including DC and AC theory, electromagnetism, reactance, and capacitance. The math level of the book is intended to be straightforward algebra so that the reader can understand the intent of the computations without having to spend time understanding advanced mathematics.

An effort has been made to keep the book practical, yet provide some of the theory about why a motor and an associated control system work. The concept is if you know how and why an electrical system works, the more likely that you can quickly diagnose and correct problems effectively. The book explains many different systems, both older and newer, because the electrician "on the job" may encounter any of the many styles and vintages of systems.

NEW TO THIS EDITION

This edition of the book has been updated to reflect the current practices and adherence to the 2005 National Electrical Code (NEC®). Some older systems that are becoming rare to work on have been deleted. There are new sections on more advanced motor control systems and more information on types and methods of electronic speed control, including DC and AC drives. As the industry moves to more worldwide utilization of equipment, more emphasis has been place on IEC information in conjunction with the U.S. standard NEMA information. The information needed to understand and apply electronic timers and controls along with electromechanical times has been expanded.

SUPPLEMENTAL PACKAGE

The accompanying *Instructor Guide* is intended to help the teacher with suggestions for class and lab. The answers to the end of chapter questions are provided to help ensure the instructors can verify information presented in the book. A PowerPoint presentation is provided with many of the diagram and schematics so the instructor can present the material to the students and elaborate on any particular point that needs further analysis or explanation.

ABOUT THE AUTHOR

Jeff J. Keljik has been involved with the electrical industry since 1972 and has been involved with teaching in the electrical industry since 1978. Jeff holds a Journeyman electrician as well as a Master electrician license from the state of Minnesota. Jeff is also licensed by Minnesota state colleges and universities to teach post-secondary education. He is approved by the Minnesota Board of Electricity to teach continuing education classes to electricians to qualify for relicensure credit.

Jeff has been a teacher, a department chair, and a director at Dunwoody College of Technology in Minneapolis. He is the Master Electrician of record for the college campus and oversees all the electrical construction and maintenance. While at Dunwoody, Jeff has directed projects for electrical education overseas and has coordinated programs for foreign corporations at Dunwoody. He has taught electrical classes for large corporations at locations throughout the United States.

Jeff is involved with industry groups. He serves as education chair on the executive committee for the North Central Electrical League. He is involved with the Minnesota Electrical Association (MEA) on the training committee. He writes curriculum for MEA classes in apprenticeship and for other classes. Jeff contributes to the Minneapolis newspaper answering electrical questions from readers.

For the last twenty-eight years, Jeff has been involved in educating people about electricity. It is his goal to provide the electrical industry with the best educated, most efficient, effective, and safe electrical workers possible.

ACKNOWLEDGMENTS

I want to thank my colleagues, who helped me with ideas and gave me encouragement. I also would like to thank all the manufacturers and manufacturers' representatives who have helped with photographs and artwork. They supplied literature to provide the electrical industry with the most current information.

Tim Keefe and Viking Electric Supply in Minneapolis for advice and assistance.

Doug Schleisman with Electric Motor Repair in Minneapolis for access to the motor shop for photos.

The electrical staff at Dunwoody College of Technology in Minneapolis for assistance and encouragement.

The author and Delmar Publishers would like to extend their thanks to those who provided detailed reviews of the manuscript. The contributions and suggestions of the following individuals are greatly appreciated.

Ray Lepore
Edison Community College
Piqua, OH

Jeff Woodson
Columbus State Community College
Columbus, OH

Randy Bedington
Catawba Valley Community College
Hickory, NC

Dan Green
Sinclair Community College
Kettering, OH

Chuck Kelley
Northern Maine Community College
Presque Isle, ME

Jeff Chase
Dunwoody Institute of Technology
Minneapolis, MN

Bob Romano
Cincinnati State Community College
Cincinnati, OH

INTRODUCTION

ELECTRICAL SAFETY—YOU ARE RESPONSIBLE FOR YOU!

There are many safeguards that are designed to keep you safe while working in the electrical professions. The most important safeguards are your knowledge of safe work practices and your dedication to doing the job safely!

Even though you may work with electricity daily, never lose the respect for electricity's power to kill or injure. There is enough power delivered to a 100-watt light bulb to throw you across the room, or to burn you severely.

Remember that your body is controlled by electrical signals that operate your heart and muscles. Take all precautions to prevent live power from entering your body and disrupting these signals or burning your body tissue. As little as one-tenth ampere through your heart can cause muscular paralysis and stop your heart. As little as two-tenths of an ampere through your body tissue can cause severe tissue damage.

Electrical safety has been organized by the Occupational Safety and Health Administration (OSHA) into different documents. The Code of Federal Regulation (CFR) 29-1910 has different subparts depending on whether you are doing construction or maintenance, and what voltages are encountered. The National Electrical Code (NEC®) refers to the National Fire Protection Association (NFPA) document 70E in Fine Print Notes (FPN). NFPA 70E provides standards for Electrical Safety in the Workplace and these standards are very similar to the OSHA requirements. 70E also provides guidelines on safe work practices, information on how to determine flash protection, and the types of personal protective equipment (PPE) that are needed for protection from electrical accidents. In addition it refers to the practices of Lock Out and Tag Out (LOTO).

If at all possible, disconnect live power from the circuit on which you are attempting to work. Do as much of the work as you can with the power deenergized and the power locked off. OSHA has extensive rules for working with live circuits. The requirements state that you must have documented training in order to work on live equipment. You must be able to distinguish between live and dead circuits and must be knowledgeable of OSHA safely rules. Remember to test circuits for live voltage using a meter with which you are familiar and are certain is functional.

OSHA—LOCK OUT/TAG OUT

As is always the case when working with electrical systems, safety is of utmost importance. OSHA has rules regarding electrical safety. Some of the most critical rules regard the requirements to lock out (lock in the off position) and tag (with a warning tag and your name) the electrical systems on which you are working. You must be qualified to work on electrical systems in the workplace. According to an OSHA synopsis, "qualified" means that you must be able to distinguish live circuits from deenergized circuits. You must be competent and knowledgeable in reading the correct meters. You must have had documented training in working on electrical circuits and be knowledgeable of the hazards.

If you are able to work on a power system without the power turned on, *do so!* If you don't need the electrical system live to perform your job, turn it off and lock the disconnect in the off position with the controller enclosure door closed. You must use your own lock and place a tag on the lock to indicate who you are, and why the equipment is turned off. It is a violation to willfully cut or remove a lock from a panel if the identified electrician on the tag is available. Protect yourself! Do not rely on a co-worker to put a lock on, to protect you. Multiple lockouts are available. Figure A illustrates how multiple locks (one for each crew member) should be installed to protect you from inadvertent energization.

Even after the power is off, test with a meter that you know is functioning. Make sure all sources of energy are disconnected. Many electricians are hurt because other sources of energy also need to be turned off, locked, or blocked open. For example, perhaps hydraulic pressure is still present, or steam heaters may still be hot on the equipment you are servicing. Check for mechanical energy systems (such as springs, weights, and flywheels) as you put your hands into equipment requiring electrical adjustments.

If you have to work on electrical equipment while it is "hot" (energized), then do so with respect and care. Be sure you are not grounded to earth ground. Be careful of metal screwdrivers. OSHA states that you will use OSHA-approved electrically insulated tools when working in live panels. Figure B shows

FIGURE A Multiple lockout for group lockout.

special insulated tools approved for use up to 600 V.

Also, be careful of your personal attire. Tape chains, metal rings, metal watches, metal pens, or metal-frame eyeglasses are all serious potential hazards when working on live equipment.

Be sure you know the environment in which you are working. Is there potential danger from dust, gas, or flyings? See NEC® Article 500 to determine if a spark or an arc may ignite the surrounding atmosphere while you work on the equipment.

Again, know what you are doing, why you are doing it, and how to do it safely to prevent electrical accidents!

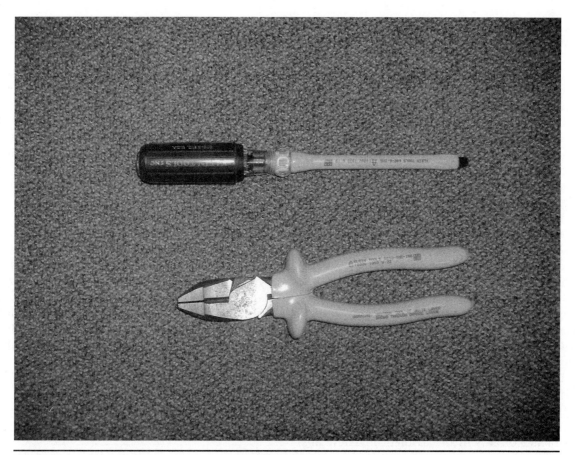

FIGURE B Insulated tools to be used with live circuits.

Be safe in the clothing you wear. Try to wear cotton or wool clothing, as these materials do not burn as easily as synthetic clothing will, when ignited. Do not wear oily or greasy clothes because of ease of ignition. Wear good-quality shoes that do not expose your feet to grounded equipment or to earth ground. Do not allow your steel-toed shoes to become worn so that the steel is exposed. This creates a conducting path from your feet to earth-grounded metal frames through your heart to your hands in the electrical circuit. Wear approved safety glasses to prevent foreign objects from entering your eyes or from molten metal from electrical arcs. Wear hard hats that are approved for electrical work. These hats are dielectrically tested to withstand specific voltages. *Do not* wear earrings, metal watchbands, rings, or metal-frame glasses and be aware of key rings when working with live power.

Do not work alone when working with live electrical circuits. Have another person with you who is familiar with safety procedures and how to rescue you from a live circuit if you are hung up.

Working on live circuits requires that you use insulated tools approved for the purpose. Tool manufacturers have specifically designed tools with voltage insulation ratings. Standard tools with vinyl "comfort grip" handles are not insulated tools for use on live circuits.

You should also be safe with your use of electric tools. When using tools with metal housings, always make sure the tool is grounded through the three-wire cord to a known effectively grounded source. Make sure the three-wire cord is in good condition and that the outlet is properly wired with the hot and neutral in place and the ground securely connected. Receptacles should be tested with a receptacle tester to verify the connections. Some tools are marked "double insulated" to prevent electrical continuity to the user. These tools do not use a three-wire cord and therefore do not use the third U-shaped safety ground connection.

When working in damp or wet locations or at grade level, a ground fault circuit interrupter (GFCI) breaker or receptacle should be used. This device measures the difference between the current flowing to the tool and the current returning from the tool. If there is a specific difference, the breaker or receptacle will disconnect power to the tool. Always inspect cords and cables used in your work to check for breaks or cuts in the insulation. These are potentially lethal points that must be repaired or replaced before connecting power.

You should be familiar with fire safety and fire extinguishers. Be sure to check the surrounding atmosphere where you are working. Are there flammable gases stored or used? Is there explosive dust nearby that could be ignited by an arc or a spark? Are there oily rags or stored flammable materials that could ignite? Be aware of the surroundings and be sure you know how and where to exit when working with electrical systems.

If you are involved in a fire, evacuate all personnel, call the fire department, then try to contain the fire if possible. Know where the fire extinguishers are and which type to use. Type A fires are combustibles (such as wood, paper, and cloth). Type B fires are flammable liquids and greases. Type C fires are electrical fires. *Do not* use water extinguishers on live electrical parts. Type D fires are combustible metals.

- Be careful!
- Pay attention!
- Work safely!

CHAPTER 1

SINGLE-PHASE MOTORS

OBJECTIVES

After completing this chapter and the chapter questions, you should be able to

- Explain all the information found on a motor nameplate
- Identify the types of single-phase motors
- Identify the parts of a single-phase motor
- Explain why an induction motor rotates
- Calculate the synchronous speed of a motor
- Connect motors for different voltages
- Reverse single-phase motor rotation
- Begin troubleshooting single-phase motor problems
- Begin using the National Electrical Code® to help determine motor application

KEY TERMS IN THIS CHAPTER

Armature: Generally, the armature is the part of the electomagnetic circuit that has the voltage induced into it. For example, the rotor of an induction motor is referred to as the armature because it relies on induced voltage. The stator of a synchronous alternator is referred to as an armature because it has voltage induced into it. The rotating member of a DC motor or generator also is referred to as an armature.

Conduction Motor: A motor where electrical connections are brought to the rotor to deliver the current flow needed to cause a magnetic field.

Induction Motor: A motor that is caused to rotate through the principles of induction. A magnetic field is induced into the rotor rather than having electrical connections to the rotor.

Lenz's Law Paraphrased: When a voltage is applied to a coil and a current flows, the current creates a magnetic field, which will induce a voltage into the same coil. This induced voltage is in opposition to the applied voltage. In other words, the effects of induction oppose the cause of induction.

Rotor: The rotating portion of the motor.

Slip: The difference in speed between the rotating magnetic field's synchronous speed and the actual speed of the rotor. This usually is expressed as a percentage.

Speed Regulation: The ability of a motor to hold the design full-load speed. This is expressed as a percentage.

Split Phase: A method of separating a single sine wave to create two voltage waveforms.

Stator: The part of the motor that remains stationary.

Synchronous Speed: The speed of the rotating magnetic field around the stator.

INTRODUCTION

Single-phase motors are found in virtually every home and business in the United States. Single-phase motors are typically small horsepower but provide a wide variety of functions. Fans, blowers, compressors, washers, and dryers generally use single-phase motors for household duty. This chapter will introduce you to the types of single-phase motors and the application of each.

NAMEPLATE INFORMATION

Much of the information needed to make the proper selection for use of a motor is found in the nameplate information. According to the National Electrical Code® (NEC) Article 430, the nameplate of a motor must contain the following information. (See Figure 1-1.)

Manufacturer's Name: This tells the customer who manufactures the motor. It is not essential to replace the motor with a motor from the same manufacturer. Other information on the nameplate will help you to order the proper replacement.

Type: Type letters on the nameplate refer to the way the motor is designed to operate. These letters are not industry-wide nomenclature. The type is a manufacturer identification to track the type of motor for replacement or repair. For instance, a single-phase motor may be listed as type C, which means it is *C*apacitor start or *C*apacitor start, capacitor run. A type P refers to a *P*ermanent split capacitor style of motor. Three-phase motors are listed as type T.

HP: This is the rated mechanical horsepower output of the motor. It is measured at the rated voltage and current and at the proper applied frequency.

Frame: If the motor was manufactured under National Electrical Manufacturers Association (NEMA), the frame number gives all the critical measurements of the motor. (See Figure 1-2.)

Volts: This value is the voltage to be delivered to the motor terminals. If it is a dual-voltage motor, either voltage may be applied, but internal connections must be changed.

Phase: Either single- or three-phase will be designated.

Note: A three-wire service to a home does not constitute three-phase power.

Amperage: This is normal operating current when running at rated HP and rated voltage. If two currents are listed, the higher current is associated with the lower voltage connection and the lower current is associated with the higher voltage.

Note: Running currents for single-phase motors are given in NEC® Table 430.248. You will notice that these currents may not

Single-Phase Motors

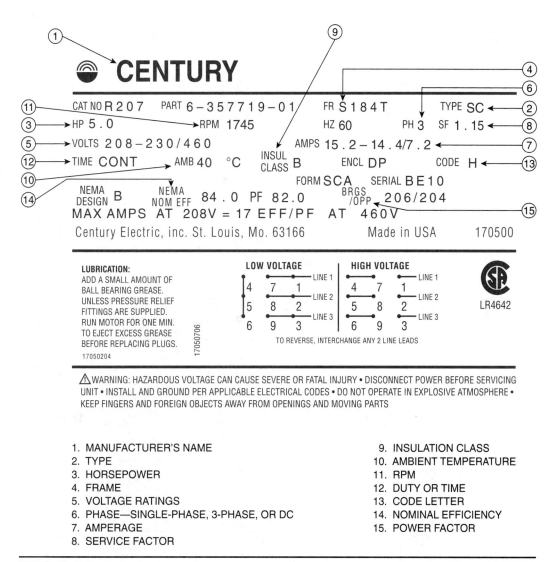

FIGURE 1-1 Typical motor nameplate. *(Courtesy of Wesco)*

correspond to the actual motor nameplate ratings. The code book values are used to size several components of the motor system and are used as a guide for generic motor calculations.

Service Factor (SF): Service factor of the motor is a multiplier applied to the motor HP. It indicates how much the motor can run overloaded on a short-term basis without damaging the motor winding insulation. For example, a 1.15 SF multiplied to a 1-HP motor means it can safely run at 1.15 HP without damage at listed ambient temperature or lower.

Insulation Class: The class of insulation used on the motor windings indicates the maximum operating temperature of the coil windings in the motor. (See Figure 1-3.) The maximum temperature of the winding is the sum of the ambient temperature and the added

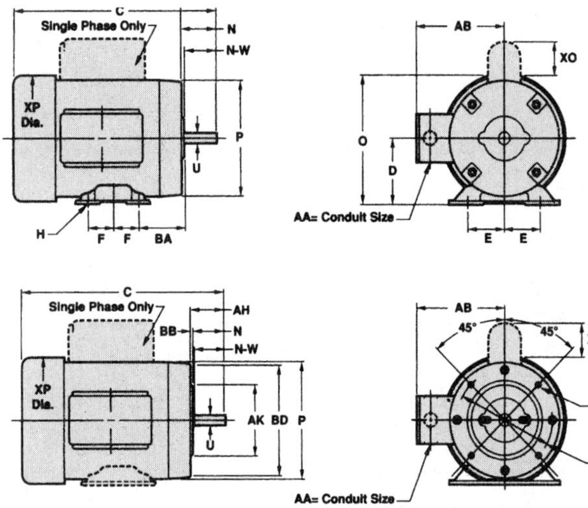

These condensed dimensions are for general reference only. For construction purposes, certified drawings are available.

NEMA Frame Size▲	D	E	F	H	N	O	P	U	N-W	AA	AB	AH	AJ	AK	BA	BB	BD	XO	XP	TAP **	KEY
42	2⅝	1¾	27/32	9/32 Slot	1¼	5 1/16	4⅞	⅜	1⅛	⅜	4½	1 5/16	3¾	3	2 1/16	⅛	4⅞	1⅝	5⅛	¼-20	3/64 Flat
48	3	2⅛	1⅜	11/32 Slot	1 9/16	5 13/16	5 19/32	½	1½	½	4⅞	1 11/16	3¾	3	2½	⅛	5	2¼	5⅞	¼-20	3/64 Flat
S56	3½	2 7/16	1½	11/32 Slot	1 15/16	6 5/16	5 19/32	⅝	1⅞	½	4⅞	2 1/16	5⅞	4½	2¾	⅛	6½	2¼	5⅞	⅜-16	3/16
56						6 13/16	6 19/32				5 5/16								7 5/32		
143T	3½	2¾	2	11/32	2⅜	6 13/16	6 19/32	⅞	2¼	¾	5 5/16	2⅛	5⅞	4½	*2¼	⅛	6½	2¼	7 5/32	⅜-16	3/16
145T			2½																		
182T	4½	3¾	2¼	13/32	2⅞	8¾	8 15/32	1⅛	2¾	¾	6⅝	2⅝	7¼	8½	*2¾	¼	8⅞	2¼	9 3/32	½-13	¼
184T			2¾																		
S213T	5¼	4¼	2¾	13/32	3½	9 15/16	8 15/32	1⅜	3⅜	¾	6⅜	3⅛	7¼	8½	*3½	¼	8⅞	2¼	9 3/32	½-13	5/16
213T			2¾		—	10 11/16	10 13/16			1	8 5/16						9		11 3/32		
215T			3½																		
254T	6¼	5	4⅛	17/32	—	12 15/16	13¼	1⅝	4	1¼	11⅝	3¾	7¼	8½	*4¼	¼	9⅝	—	12⅞	½-13	⅜
256T			5																		
284TS	7	5½	4¾	17/32	—	14½	14¾	1⅝	3¼	1½	11¾	3	9	10½	4¾	¼	11	—	14½	½-13	⅜
284T								1⅞	4⅝			4⅜									½
286TS			5½					1⅝	3¼			3									⅜
286T								1⅞	4⅝			4⅜									½
324TS	8	6¼	5¼	21/32	—	15¾	15¾	1⅞	3¾	2	13½	3½	11	12½	5¼	¼	13⅜	—	15¾	⅝-11	½
324T								2⅛	5¼			5									
326TS			6					1⅞	3¾			3½									
326T								2⅛	5¼			5									
364TS	9	7	5⅝	21/32	—	17 13/16	17⅜	1⅞	3¾	3	15 7/16	3½	11	12½	5⅞	¼	14	—	17¾	⅝-11	½
364T								2⅜	5⅞			5⅜									⅝
365TS			6⅛					1⅞	3¾			3½									½
365T								2⅜	5⅞			5⅜									⅝
404TS	10	8	6⅛	13/16	—	19 9/16	19⅛	2⅛	4¼	3	16 9/16	4	11	12½	6⅝	¼	15½	—	19⅜	⅝-11	½
404T								2⅞	7¼			7									¾
405TS			6⅞					2⅛	4¼			4									½
405T								2⅞	7¼			7									¾
444TS	11	9	7¼	13/16	—	22¼	22	2⅜	4¾	3	21 11/16	8¼	14	16	7½	¼	18	—	19⅜	⅝-11	⅝
444T			7¼					3⅜	8½												⅞
445T			8¼						8½												⅞
447TZ			10						10⅛												⅞
5007C	12½	10	11	15/16	—	30⅝	30⅛	3⅞	11⅝	4	25 19/32	11⅞	—	—	8½	¼	—	—	30⅜	—	1
5009C	12½	10	14	15/16	—	30⅝	30⅛	3⅞	11⅝	4	25 19/32	11⅞	—	—	8½	¼	—	—	30⅜	—	1

* 143-5TC NEMA C Face BA Dimension is 2¾
* 182-4TC NEMA C Face BA Dimension is 3½
* 213-5TC NEMA C Face BA Dimension is 4¼
▲ Shading in this column denotes dimensions established by NEMA standard MG1.
* 254-6TC NEMA C Face BA Dimension is 4¾
** 326TC and smaller have 4 mounting holes in NEMA C Face, 364TC and larger have 8 mounting holes.
\# "C" Dimensions may vary; use for reference only.

ELECTRICAL APPARATUS SERVICE ASSOCIATION · EASA · Associate Member

LEESON ELECTRIC CORPORATION
GRAFTON, WI 53024-0241 U.S.A. TELEPHONE (414) 377-8810 FAX (414) 377-9025

LEESON CANADA
TELEPHONE (416) 670-4770 FAX (416) 670-4378
MISSISSAUGA (TORONTO), ONTARIO L5T 2J3

POWER TRANSMISSION DISTRIBUTORS ASSOCIATION

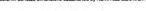

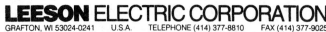

FIGURE 1-2 Frame size and dimension data. *(Courtesy of Leeson Electric Corporation)*

```
Temperature °C = 5/9 × (Temperature °F − 32)
Temperature °F = (9/5 × Temperature °C) + 32
    CLASS A:    105 °C – 221 °F
    CLASS B:    130 °C   266 °F
    CLASS F:    155 °C   311 °F
    CLASS H:    180 °C   356 °F
    CLASS H+:   200 °C   392 °F
```

FIGURE 1-3 Motor wire insulation class temperature rating.

heat of the motor watt losses. The standard insulation classes are B, F, H.

Ambient Temperature: If listed, the ambient temperature is the maximum temperature in degrees centigrade of the surrounding air that will allow rated HP without damage.

RPM: The revolutions per minute is the rated speed of the motor operating at full load or rated HP. Typically, lighter loads will run at higher speeds.

Duty (or time rating): Duty refers to the length of time the motor can run at full load without overheating. Continuous duty means that the motor could run 24 hours per day if all other factors are within specifications.

Code Letter: Locked rotor Kilovolt Amps (KVA) per HP, if 1/2 HP, or more, is listed in the NEC®. The code letter refers to the starting or locked rotor characteristics of the motor. It is listed as Kilovolt Amps per HP, with locked rotor. NEC® Article 430.7(B) is a reference.

Efficiency: Efficiency is sometimes placed on the motor nameplate. It is expressed as a percent of output watts (horsepower watts) compared to input watts at full-load conditions. For example, 85 percent efficient means that 85 percent of the input electrical watts are converted to mechanical output watts or horsepower.

Power Factor: The ratio of true power used in watts and the apparent power delivered. Expressed as a percentage.

Altitude: Motors are typically designed for altitudes below 3300 feet. Air cooling becomes a factor at higher altitudes and derating of the motors' capabilities would be necessary. (Altitude is not normally indicated on the motor nameplate.)

Thermal Protection: If a motor has integral (internal to the motor) protection, it will be marked "thermally protected." If 100 W or less, the nameplate may state TP.

Impedance Protected: If motors correspond to the impedance protected definition NEC® Article 430.32(B)(4) ratings, they may be marked impedance protected or ZP or IMP. prot.

Design Letter: A letter assigned to a particular motor to indicate its operating characteristics such as the starting torque and starting current. Design A motors have steady speed from no load to full load. They have high starting torque and rapid acceleration, but high starting current. Breakdown torque is also high. Design B motors are the general-purpose motors with lower starting current and torque than design A. Comparisons are made to design B as a standard design motor. Design C has higher start torque, but standard starting current. Breakdown torque is the same. Design D motors have high starting torque with low starting current. See pages 000–000 for detailed curves and design details. See Figure 1-4 for an example.

By understanding nameplate information, suitable replacement motors or original design motors can be more precisely selected. For instance, if a motor is not functioning as originally specified, the motor application may be wrong. Your knowledge of motor nameplates will allow you to choose the proper motor for the job.

As you will learn, different types of single-phase motors have different applications. Not all motors will function satisfactorily in all situations. For example, if a split-phase motor continues to fail in the operation of a small refrigeration compressor, the motor type may be incorrect. The proper motor may be a

Design	% Starting torque	% Starting current	% Slip	% Breakdown torque	Applications
A	250–275	High	< 3	> 275	Applications that require constant speed under varying loads
B	100–200	Average	< 5	200–250	Fans, blowers, pumps, etc. with low starting torque needs
C	200–250	Average	< 5	200–250	Conveyors, crushers, etc. starting under load
D	275	Low	> 5	275	High inertia loads such as flywheels and presses

FIGURE 1-4 Motor design letter indicating characteristics.

capacitor start motor that has more starting torque. If a code letter M motor always trips the protection during starting, a replacement with a code letter of K or A may be better. These latter motors have lower inrush current per HP.

Other parameters of the motor also give an indication of misuse or misapplication. Is the service factor (SF) critical for exact replacement of a defective motor? Does an SF of 1.15 fit the needs for replacement of a motor with an SF of 1.0? Normally, this would be acceptable if all other factors are considered, including the sizing of the overcurrent protection. Can a motor with higher insulation class replace one of lower class? Yes, this is acceptable; however, costs are a factor. Higher insulation class usually costs more for the same HP motor. Can duty rating be substituted? Can an intermittent duty motor replace a continuous duty motor? Check the application to see if the motor does, or could, run continuously.

Frames of the motor are also important to note. Open-frame motors need good air circulation to maintain operations without damage. If a motor is in a poorly ventilated area, replacement with the exact type of motor will not prevent future failure. Often considerations need to be made, such as drip-proof motor housing, totally enclosed fan cooled, or explosion-proof motors. Make sure you know which motor frame style you need to satisfy the installation.

SINGLE-PHASE MOTORS

Motors are grouped into two general groups: **conduction motors** and **induction motors**. Conduction motors refer to the method that transfers power to the rotating member. If the power is transferred by physical connections (such as carbon brushes, slip rings, or commutators), the motor is a conduction motor. (See, for example, Figure 10-3.) If the power is transferred electromagnetically, the motor is an induction motor because the power is induced into the rotor. The first types of single-phase motors to be discussed are induction motors.

Induction Motors

The stationary part of the induction motor is the **stator**. The stator contains the wire that has been wound into coils and inserted in the iron slots that make up the inside of the stator core. This wire is called magnet wire because it forms

Single-Phase Motors

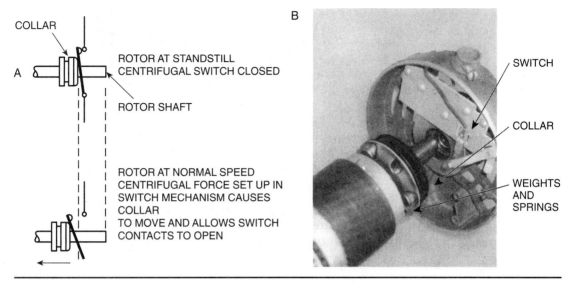

FIGURE 1-5 (A) Operation of centrifugal switch. (B) Cast aluminum squirrel cage rotor.

the electromagnetic fields that cause the motor to spin. The number of coils, the size of the wire, and the number of turns of coil wire help determine the operating characteristics of the motor.

The rotating component of the induction motor is the **rotor**. Some people refer to this as the **armature**. Usually this is true, but it is not always accurate for every motor. The armature is the component of the magnetic system that has voltage induced into it; sometimes this is the rotor and sometimes it is the stator. The rotating member is referred to as the armature on *conduction* motors.

Typically the rotor is a squirrel cage design containing short-circuited bars within an iron shell. As the stator becomes energized, voltage is induced into the rotor. The induced voltage in the rotor causes a current flow, creating a magnetic field within the rotor. This rotor field follows the magnetic field of the stator without using direct electrical connections to the rotor windings. This concept is explained in more detail in the section "Squirrel Cage Rotor Principles."

A third major component of single-phase motors may include a switch used to disconnect part of the stator windings after the motor

has almost reached operating speed. Many motors use a centrifugal switch as shown in Figure 1-5.

Two parts, the rotor, and centrifugal switch, are shown in Figure 1-6A and B.

Design and Operating Principles of Split-Phase Motors

Split-phase motors are rugged, reliable, relatively cheap, and work well in applications where there is not a requirement for high starting torque.

Because the motor has only a single AC sine wave applied to it, an artificial means of creating two voltages out of phase with each other is needed. The motor is designed to split a single sine wave voltage into two waveforms and use the displaced voltage to create a rotating magnetic field in the stator. To do this, the stator is wound with coils of two different sizes of wire and different number of turns or loops in the stator coils.

One coil (the running, or main winding) is a physically larger diameter wire than the other coil (the starting, or auxiliary winding). The running winding is placed into the stator first

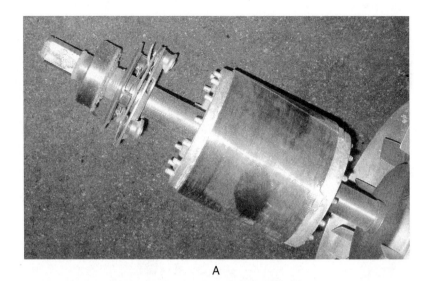

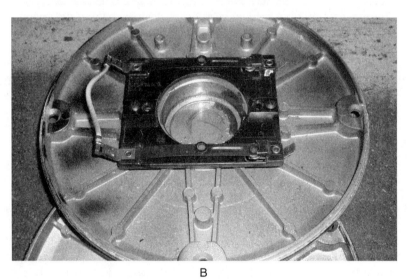

FIGURE 1-6 (A) Rotor with centrifugal switch. (B) Stationary section of switch on frame.

so that it is deeper in the slots. This has two effects on the coil compared to the starting winding. The running winding has less resistance because it is a large diameter wire. Also, because it is seated deeper in the stator iron, it has more inductive reactance. This causes the current in the running winding to lag the applied voltage by approximately 45 degrees electrically.

The second coil (starting winding) of magnet wire is wound with smaller wire so the resistance is higher than the running winding. It also has fewer turns and is placed at the top of the iron slots (inside edge) of the stator. (See Figure 1-7.)

Higher-resistance wire, along with less inductive reactance (less iron permeability and

Single-Phase Motors

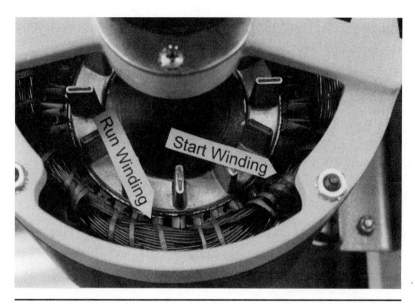

FIGURE 1-7 Stator coil windings.

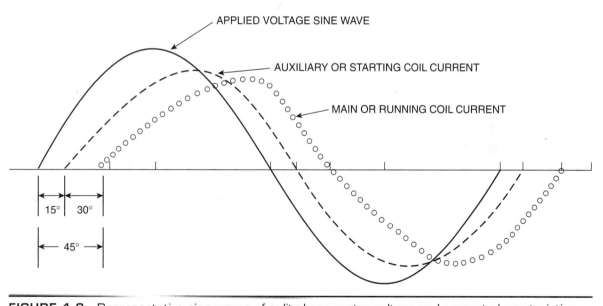

FIGURE 1-8 Representative sine waves of split-phase motor voltage and current characteristics.

fewer turns), causes the current to lag the voltage by approximately 15 degrees electrically. Now the coil currents are split in the stator coils by approximately 30 degrees electrically. This is the difference between the 45-degree lag of current in the running winding and the 15-degree lag of current in the starting winding. (See Figure 1-8.) The relative 30-degree lag in current between the coils will produce the magnetic effects to create a rotating magnetic field.

Magnetic Fields

A brief review of electromagnets and magnetic fields is necessary to understand the effects that occur in the motor's magnetic fields. The magnetic fields that are used in most motors are created by electromagnetic means. As current flows through a wire, a magnetic field is set up around the wire. If the wire is coiled in a circular fashion, the effects of the wire's magnetic field are additive. Many turns of wire in a coil produce a strong magnetic field. The strength of the magnetic field can also be changed by the amount of current flow in the coil. The higher the current flow (more amps), the stronger the magnetic field. The direction of the magnetic field can be determined by using the left-hand rule for a coil: If you grasp the coil of wire with your left hand and your fingers are pointed in the direction of the electron flow, your thumb will point to the magnetic north pole of the coil. The resultant magnetic fields react the same way as permanent magnetic fields. (See Figure 1-9.)

There are two basic laws about magnets: (1) Like magnetic poles repel each other and unlike magnetic poles attract each other; (2) the strength of the magnetic attraction or repulsion is inversely proportional to the square of the distance between the pole faces. Paraphrased, this means that the closer the magnetic poles, the stronger the magnetic attraction or repulsion.

Understanding the effects on the iron surrounding the coils of wire should enable you to comprehend effects of the motor's stator and rotor. The magnetic fields produced by coils of wire are surrounded by the iron of the motor. This iron is highly permeable; thus it allows the lines of force of the magnetic field to move easily through it. The result is that the iron in the motor is also temporarily magnetized by the coil's electromagnetic field. When current is flowing in the coil, the surrounding iron becomes a magnet with the center, or the strongest part, at the center of the coil. When current flow through the coil stops, electromagnetic effects also stop. The surrounding iron does not retain the magnetic effects, because of low retentivity, and again returns to nonmagnetized iron. In an induction motor, this process of magnetizing the iron in one direction and then remagnetizing in the opposite direction occurs many times per second.

ROTATING MAGNETIC FIELD

Consider a two-pole stator. This means there are two main poles: one north and one south magnetic polarity. (See Figure 1-10.) Figure 1-10 indicates top and bottom starting windings and left and right running windings. If two voltages are applied to these coils separately and the voltages are split by 90 degrees electrically (see Figure 1-11), an ideal rotating magnetic field results as follows:

Position 1—Peak current flows in top and bottom coils and maximum magnetic poles will result. The top coil is north (using the

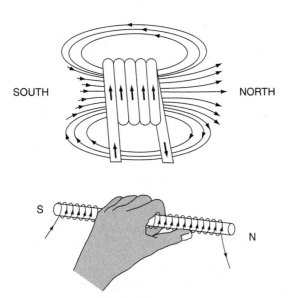

FIGURE 1-9 Left-hand rule for coils indicates magnetic polarities. Thumb points north.

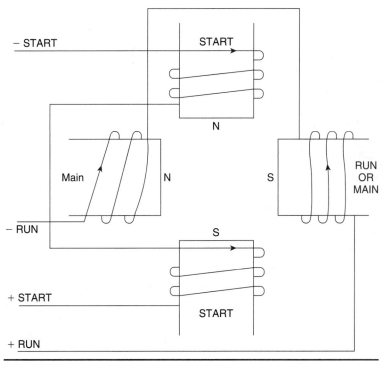

FIGURE 1-10 Counterclockwise rotating magnetic field.

left-hand rule for coils in electron flow theory) and the bottom coil is south. No side poles result.

Position 2—Current is reduced in top and bottom poles, and the magnetic field is also weakened. Current has increased in the side poles to 70 percent of peak. Now the left side is north and the right side is south. This actually magnetizes the stator iron with two main poles with the center located between the windings of magnet wire coils.

Position 3—Top poles are at zero current flow and side poles have reached peak. The left winding is now the center of the north magnetic pole in the stator. The magnetic pole has shifted 90 mechanical degrees.

Position 4—The top and bottom poles become energized again, but notice the voltage waveform is now reversed from the original position. The north magnetic pole is halfway between the left and bottom windings.

Position 5—The side poles are again at zero voltage and the top and bottom poles are at maximum strength. The bottom is now the main north pole and the top is the main south pole. The magnetic field has rotated 180 electrical and mechanical degrees.

The foregoing discussion explains the rotating field with an ideal 90-degree phase split. However, even a 30-degree split in the phase energization of the coils will produce a rotating magnetic field. With two poles (one north and one south) the field will rotate at 3600 RPM if 60 Hz is applied.

Synchronous Speed

The **synchronous speed** of the rotating field will make one complete revolution for each complete sine wave applied to the main poles of a two-pole motor. For a two-pole motor, each 360 electrical degrees of the sine wave will cause

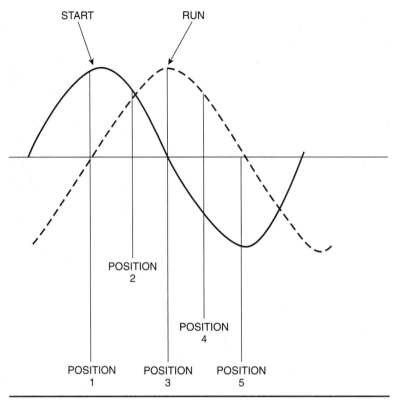

FIGURE 1-11 Ninety-degree phase-shifted coil currents applied to motor coils.

one complete revolution of the magnetic field. Therefore, 60 complete sine waves per second (60 Hz) will yield 60 revolutions of the magnetic fields in one second. Because motor speed is measured in revolutions per minute (RPM), the field will rotate 3600 times in one minute.

The mechanical degrees of rotation of the magnetic field for a two-pole motor are equal to the same number of electrical degrees through which the sine wave travels. The formula in Figure 1-12 shows the relationship between electrical degrees of the applied waveform and the mechanical degrees of rotation of the magnetic field in the stator.

To calculate the synchronous speed of the magnetic field for a four-pole motor, use the same formula. Count the number of main poles, or running windings, and insert that number into the formula for the number of poles. Using the formula, you will notice that 360 electrical degrees will rotate the magnetic field only 180 degrees around the stator. This

$$\left(\frac{2}{\text{Number of poles}}\right) \text{Electrical degrees} = \text{Mechanical degrees}$$

FIGURE 1-12 Formula to convert electrical degrees to mechanical degrees.

$$\text{Synchronous RPM} = \frac{120 \text{ (frequency)}}{\text{Number of poles}}$$

FIGURE 1-13 Formula to find synchronous speed using applied frequency and number of poles.

means that the rotating magnetic field is only rotating half as fast as it did for the two-pole motor, or 1800 RPM. Using the formula again for a six-pole motor, the rotating magnetic field will rotate only 120 mechanical degrees for each 360 degrees of sine wave applied. This means that the field will rotate only one-third as fast as a two-pole motor, or 1200 RPM.

Another formula used to calculate the speed of the rotating magnetic field uses the same basic formula but provides some conversion factors as constants. The formula in Figure 1-13 has a constant factor of 120. This constant is derived by using the number of poles factor of 2, times the 60 needed to convert revolutions per second to revolutions per minute. This is the speed of the rotating magnetic field only, not the speed of the rotor.

Reversing the Rotation

To reverse the direction of rotation, one must change the way the starting and running coils become energized. By reversing the lead connections to the starting winding, the instantaneous magnetic fields will be reversed in relation to the run windings and the rotation will be reversed. Refer to Figure 1-10.

After the rotating magnetic field has established the direction of rotation and started the rotor moving, the starting winding can be removed from the circuit and the rotor will continue to spin even though there is no longer a rotating field, but only a pulsating main field. This is a result of the rotor's momentum.

The starting winding is removed by a switch mechanism. Typically, these switches are mechanical switches inside the motor. The weights on the switch mechanism are thrown outward away from the shaft as the rotor begins to spin. The centrifugal force on the weights causes them to move and pull open a switch that disconnects the starting winding after the motor is approximately 75 percent of running speed. The starting winding is not designed to be left in the motor and will be destroyed if the switch does not open. Conversely, if the switch does not close mechanically, or if the contacts are destroyed, the starting winding will never be energized and no rotating field will be established. This means that the motor will not turn. (More details on starting switches are presented later in this chapter.)

SQUIRREL CAGE ROTOR PRINCIPLES

As already mentioned, the squirrel cage rotor is made up of bars of conductors that are short-circuited at each end. (See Figure 1-14.) Figure 1-15 shows a pictorial view of a squirrel cage rotor. The ends are connected by rings at both ends to create a squirrel cage winding that creates a path for the rotor current to flow. As can be seen in position 2 in Figure 1-15, voltage is induced into some rotor conductors. The induced voltage causes a current to flow down the length of the rotor, around the short-circuited end, and back the length of the rotor. This current flow produces its own magnetic poles in the rotor. The + signs on the end of the rotor bars indicate electrons flowing away from you (like the tail feathers of an arrow). The · on the bar means current is flowing toward you (like the head of an arrow). Use the left-hand rule for a coil (electron flow theory) and your thumb will indicate the position of the north pole on the rotor. Note the location of the rotor pole in relation to the stator pole. The rotor poles are perpendicular, or 90 degrees away from the stator pole.

FIGURE 1-14 Cutaway view of squirrel cage rotor showing rotor bars.

Position 1—With the top and bottom poles energized first (the starting windings), there is no moving magnetic field, so there is no voltage induced into the rotor conductors and no current flows in the rotor bars.

Position 2—As the stator magnetic field begins to rotate counterclockwise (as discussed earlier), the magnetic lines of force travel through the iron of the rotor from north stator pole to south stator pole and cut the rotor conducting bars. This produces a current flow as shown in position 2. The left-hand rule for conductors will establish the direction of current flow in the rotor conductors. This current flows in the rotor under the magnetic poles of the stator. Using the left-hand rule for a coil, with fingers in the direction of current flow, the north pole of the rotor (which will be 90° from the stator) can be determined. Notice that it is ahead of the stator north pole.

Position 3—The top and bottom poles are deenergized, but the magnetic field is still moving and voltage is still induced into the rotor bars, resulting in current flow that produces a rotor magnetic field ahead of the stator magnetic field.

Position 4—Again, the auxiliary windings are energized but are reversed from the original polarity because of the AC sine wave of current that is applied. The rotor continues to be pushed and pulled by the rotating magnetic field.

Position 5—Now the side poles are at zero magnetic polarity in this two-pole motor. The rotor poles are established midway between the stator poles.

Position 6—As the side poles are reenergized, the magnetic polarity has been reversed from the original polarity in position 2. This allows the rotating field to continue at its synchronous speed.

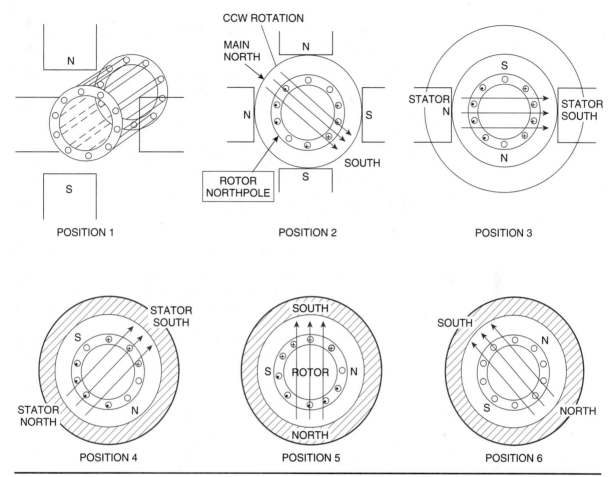

FIGURE 1-15 Counterclockwise rotation of stator field and induction of rotor magnetic field.

SPLIT-PHASE MOTORS AND SLIP

The speed of the rotating field of the rotor is 3600 RPM in this two-pole (main poles) motor. However, the rotor does not travel as fast. The rotor must **slip** behind the synchronous speed in order to have voltage induced into it. In other words, the rotor bars must be cut by the magnetic field of the stator to induce voltage. Induced voltage creates rotor current that establishes the rotor magnetic poles that are pushed and pulled by the stator poles. If the rotor traveled the same speed as the stator field, no current would be induced and no magnetic field would be set up to produce the twisting effort called torque.

The amount of slip depends on many factors: the design of the rotor, type of rotor conductors, air gap between the rotor and the stator iron, and mechanical load on the shaft. Slip is often expressed as a percentage and is calculated by the formula in Figure 1-16. The percent of slip indicates the percent of

$$\% \text{ Slip} = \frac{\text{Synchronous speed} - \text{Actual rotor speed}}{\text{Synchronous speed}} \times 100$$

FIGURE 1-16 Formula for percent slip of induction motors.

$$\% \text{ Speed regulation} = \frac{\text{No-load speed} - \text{Full-load speed}}{\text{Full-load speed}} \times 100$$

FIGURE 1-17 Formula for percent speed regulation for motors.

synchronous speed that the rotor is falling behind. For most split-phase motors, this ranges from 4 percent to 6 percent.

Speed Regulation

As the mechanical load on the motor shaft increases, the rotor slows down. This allows the rotor conductors to be cut at a higher rate of speed (the difference between synchronous speed and actual rotor speed). This effect of relative speed change induces more voltage in the rotor, creating more rotor current and increasing the strength of the rotor magnetic field. This enhances the magnetic pull and increases torque to compensate for the increased mechanical load.

The effect of changing load on the operating speed of the squirrel cage rotor is reflected in the formula for percent speed regulation. This figure represents how well a motor will maintain its design speed over a wide range of load, from no load to full load. (See Figure 1-17.) A perfect motor would have 0 percent speed regulation. This means there would be no speed variation between no-load and full-load speed.

This is not possible with an induction motor. There must be some change in speed to produce a change in current, which produces a change in torque to maintain a new, slower speed for a heavier load.

Connections

Split-phase motors also come in dual voltage ratings. The motor connections are simple and are interchangeable from one voltage to another. If this is the case, the two sets of running windings are designed to operate in parallel with each other at 115 Voltage Alternating Current (VAC), or in series at 230 VAC. The leads are marked with numbers and are sometimes marked with colors according to the standards set by NEMA (National Electrical Manufacturers Association). (See Figure 1-18.)

The schematic diagram for a split-phase motor helps explain the motor connections lead markings and the method of reversing.

If the motor is a single-voltage motor, 115 volts is applied to the main winding T_1 and T_2 and also to the starting winding. If the motor is a dual-voltage-rated motor, there are two sets of running windings. These running windings are designed to create the desired magnetic fields when they are connected in parallel to each other. For the low-voltage connection, all the windings are connected in parallel to each other. This will provide 115 volts to both sets of running coils and 115 volts to the starting winding through the centrifugal switch.

Most dual-voltage motors only use one starting winding. By placing the running windings in series with each other and applying 230 volts, 115 volts will be present across

Single-Phase Motors

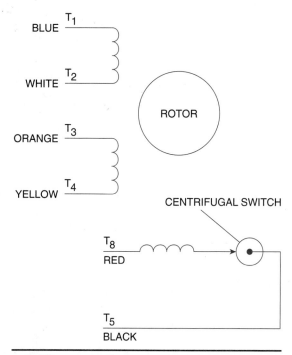

FIGURE 1-18 Dual-voltage, single-phase motor with lead colors. Running winding leads: T_1 = Blue, T_2 = White, T_3 = Orange, T_4 = Yellow. Starting winding leads: T_5 = Black, T_8 = Red.

each half of the winding. Connect the starting winding across the 115 V present across one coil of the running windings. In other words, you will connect the starting winding from one line lead to the center of the two running winding coil connections.

Some dual-voltage motors may also have two starting windings. The connection patterns are shown in Figure 1-19.

Reversing

To reverse the direction of rotation of the rotor, the direction of the rotating magnetic field at time of starting must be reversed. See Figure 1-20 for lead connection patterns.

Reversing the direction of a dual-voltage motor running at the higher voltage, but with only one starting winding, is more complicated. (See Figure 1-21.)

Remember, the starting switch must open and disconnect the starting winding from the circuit or the starting winding will be damaged. Also, the centrifugal switch must be closed during *locked rotor* condition (in other words, rotor is not moving). If the starting switch is open, the starting winding will not be energized and no rotating magnetic field will be produced. The rotor will not spin and the motor may

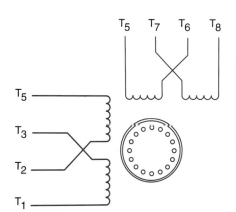

VOLTAGE RATING	L_1	L_2	TIE TOGETHER
115 VOLTS	$T_1, T_3,$ T_5, T_7	$T_2, T_4,$ T_6, T_8	
230 VOLTS	T_1, T_5	T_4, T_8	T_2 AND T_3, T_6 AND T_7

FIGURE 1-19 Connection diagram for dual-voltage motors with two starting and two running windings.

ELECTRIC MOTORS AND MOTOR CONTROLS

Forward	Reverse
115V	
L_1 to: T_1, T_3, T_5, T_7	L_1 to: T_1, T_3, T_6, T_8
L_2 to: T_2, T_4, T_6, T_8	L_2 to: T_2, T_4, T_5, T_7
230V	
L_1 to: T_1, T_5	L_1 to: T_1, T_8
(connect T_2 to T_3)	(connect T_2 to T_3)
L_2 to: T_4, T_8	L_2 to: T_4, T_5
(connect T_6 to T_7)	(connect T_6 to T_7)

FIGURE 1-20 Forward and reverse connection pattern for dual-voltage start and run windings.

Forward	Reverse
115V	
L_1 to: T_1, T_3, T_5	L_1 to: T_1, T_3, T_8
L_2 to: T_2, T_4, T_8	L_2 to: T_2, T_4, T_5
230V	
L_1 to: T_1	L_1 to: T_1
(tie T_2, T_3, T_5 together)	(tie T_2, T_3, T_8 together)
L_2 to: T_4, T_8	L_2 to: T_4, T_5

FIGURE 1-21 Forward and reverse connections for dual-voltage motor with single start winding.

burn out. You can check this by hand spinning the rotor in either direction with power applied (be careful of mechanical parts). If the rotor begins to run only after you spin it, the starting winding is not being energized.

STARTING WINDING SWITCHES

Starting switches and relays for single-phase motors are used to disconnect the starting windings of a motor as the motor approaches running speed. The most common type is the centrifugal switch, which is enclosed in the motor housing and reacts to the physical speed of the rotor to centrifugally throw weights away from the shaft, causing a switch to open. See Figure 1–18 for the electrical location of the switch.

Other methods can be used where the switch is mounted externally to the motor. This may facilitate easier maintenance, or the arcing of the switch may be objectionable if placed too near the motor. In this case, a voltage (or current-operated) mechanical switch, or an electronic switch sensitive to motor current, may be used.

Voltage-operated switches are connected across the starting winding. (See Figure 1-22.) Although these switches are used only on capacitor start motors, they will be explained here with other start switches.

When the motor is energized, the current is allowed to flow through the start winding, which is a very low impedance circuit, but the value of voltage dropped across the winding is too low to energize the relay or pick up the relay contacts. This is because the capacitive reactance of the capacitor is in series with the impedance of the starting winding. The large inrush current creates a large voltage drop on the capacitor and a lesser voltage drop on the starting winding coil. As the rotor speed increases, the voltage across the start coil increases (more voltage is induced back into the stator by the spinning rotor in the form of counter electromotive force, which causes a higher volt drop), and the coil picks up or energizes. This opens the circuit to the start winding. The spinning rotor with its magnetic field induces a large enough voltage into the coil to hold open the contacts.

Another style of starting switch is the current-operated coil. The current-operated coil is connected in series with the run winding and the contacts are in series with the start winding. (See Figure 1-23.) As power is applied, the large inrush current to the run winding causes the relay to pick up. This closes the

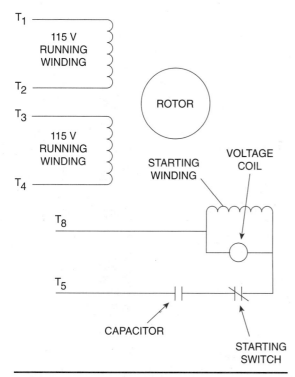

FIGURE 1-22 Single-phase motor using a voltage-operated coil in starting circuit.

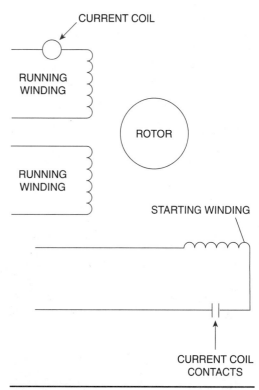

FIGURE 1-23 Single-phase motor using a current-operated coil in starting circuit.

contacts to the start winding. With these contacts closed, the motor begins to spin. As the speed increases, the inrush current decreases (as described later) and the current coil releases the contacts to the open position and removes the start coil from the circuit.

Starting Currents and Torque

To understand the operation of the motor, one must be familiar with the inrush or locked rotor current. When a motor is first connected to the power source, there is a large inrush of current. This current is created by the design of the rotor itself (code letter) and the fact that the rotor has not yet begun to spin. A large current is drawn by the rotor. This is comparable to a transformer primary (the motor stator) inducing voltage, and therefore current, into a shorted secondary winding (the short-circuited bars of the rotor). This large rotor current and large stator current produce strong magnetic poles that produce the torque necessary to start the rotor spinning. Starting torque for a split-phase motor is approximately 150 percent to 200 percent of running torque. The starting current is typically 600 percent of the normal full load running current.

As the rotor comes up to speed, the rotor current decreases because of less induced voltage (less slip) and the decrease in stator current and line current. Also, a major effect of the spinning rotor is the counter electromotive force (CEMF) or "back EMF" produced by spinning a magnetic field (the rotor field) inside a coil of wire (stator windings). The generator effects of the rotor field spinning produce a counter voltage that opposes current from the line. The faster it spins, the more CEMF is produced, and

INDUCTION MOTOR WITH CODE LETTER A

THIS TYPE OF MOTOR HAS A HIGH-RESISTANCE ROTOR WITH SMALL ROTOR BARS NEAR THE ROTOR SURFACE. THIS MOTOR HAS A HIGH STARTING TORQUE AND LOW STARTING CURRENT.

APPLICATIONS:

METAL SHEARS, PUNCH PRESSES, AND METAL DRAWING MACHINERY.

INDUCTION MOTOR WITH CODE LETTERS B-E

THIS TYPE OF MOTOR HAS A HIGH-REACTANCE AND LOW-RESISTANCE ROTOR. THIS MOTOR HAS A RELATIVELY LOW STARTING CURRENT AND ONLY FAIR STARTING TORQUE. IT HAS LARGER CONDUCTORS DEEP IN THE ROTOR IRON.

APPLICATIONS:

MOTOR-GENERATOR SETS, FANS, BLOWERS, CENTRIFUGAL PUMPS, OR ANY APPLICATION WHERE A HIGH STARTING TORQUE IS NOT REQUIRED.

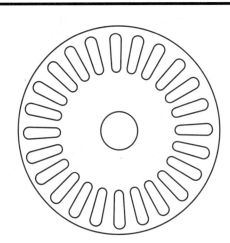

INDUCTION MOTOR WITH CODE LETTERS F-V

THIS TYPE OF MOTOR HAS A RELATIVELY LOW-RESISTANCE AND LOW-INDUCTIVE REACTANCE ROTOR. THIS MOTOR HAS A HIGH STARTING CURRENT AND ONLY FAIR STARTING TORQUE. IT HAS LARGE CONDUCTORS NEAR THE ROTOR SURFACE.

APPLICATIONS:

MOTOR-GENERATOR SETS, FANS, BLOWERS, CENTRIFUGAL PUMPS, OR ANY APPLICATION WHERE A HIGH STARTING TORQUE IS NOT REQUIRED.

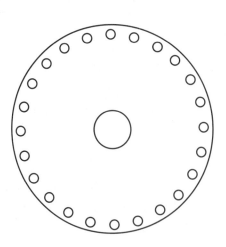

FIGURE 1-24 Rotor types that correspond to the motor code letters.

less current is drawn from the line. This is a *no-load* or *light-load* condition. A heavier load slows the rotor, less CEMF is produced, and more current is drawn from the line. The current is self-regulating.

As a motor's load becomes heavier, its speed will slow and torque will increase until the point is reached where there is not enough twisting effort left to maintain the magnetic effect between rotor and stator. At this point there is an abrupt drop in speed and the motor fails as it reaches **breakdown torque**. Breakdown torque is typically 200 percent of running torque for the split-phase motor.

Code Letter and Starting Current

A quick explanation of the code letter and the relationship to the motor design is necessary to understand the importance of selecting a proper replacement motor. With reference to NEC® Table 430-7(B) you will note that a code letter A motor draws a maximum of 3.14 KVA/HP whereas a code L motor draws a maximum of 9.99 KVA/HP. To illustrate the effects of proper selection, let's calculate actual current draw for 1 HP at 230 VAC:

To find the locked rotor current for a motor at start:

Locked rotor KVA per horsepower × HP = Total KVA

Total KVA × 1000 = Total voltamps (VAs)

VA/Volts = Locked rotor amps

For example;

Code A 3.14 KVA = 3140 VA for each HP

$$1 \text{ HP} = \frac{3140}{230 \text{ V}} = 13.65 \text{ A}$$

Code L 9.99 KVA = 9990 VA for each HP

$$1 \text{ HP} = \frac{9990}{230 \text{ V}} = 43.43 \text{ A}$$

All other parameters are equal except the code letter. The code letter A motor draws only 13.65 A at the start while the code L motor draws 43.43 A. It is obvious that the two motors do not perform the same under starting conditions although they are both 1-HP 230-V motors. This difference is produced by changing the design of the motor, especially the construction of the squirrel cage rotor. (See Figure 1-24 for rotor designs.)

Note that code letter A motors are more expensive than code L motors. Code letters will affect how the motor is protected during starting (refer to NEC® 430.52 with exceptions).

CAPACITOR-START, INDUCTION-RUN MOTORS

Capacitor-start motors are used when single-phase power is available, but there is a need for more starting torque than a split-phase motor can deliver. The capacitor in series with the starting winding allows this motor to have superior starting torque. It has the maximum starting torque per ampere of all types of single-phase motors. By using a capacitor in series with the starting winding, the relative phase angle between the starting and running winding is larger than in the split-phase motor.

A capacitor causes current to lead voltage, so by placing it in series with the start winding, the current in the start winding will lead the line voltage by approximately 35 degrees electrically. The run winding is similar to the split-phase motor. It has a current that lags the line voltage approximately 45 degrees. This corresponds to a phase split of approximately 80 degrees. This is closer to the ideal 90-degree split. (See Figure 1-25.) In Figure 1-25 the current and voltage waveforms show relationships after a few cycles and the capacitor has charged. The leading current (ahead of the line voltage) in the capacitor circuit for the starting winding is explained in basic AC theory.

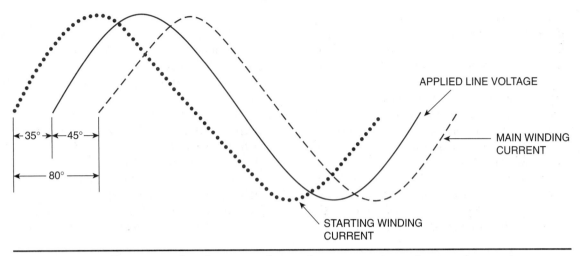

FIGURE 1-25 Representative sine wave of capacitor-start motor voltage and current characteristics.

This phase splitting accounts for the better starting torque of approximately 250 percent to 300 percent of full-load torque and requires slightly less line current to produce the increased torque. Line current drawn is approximately two-thirds that of the same rating of a split-phase motor.

As is the case with the split-phase motor, the capacitor and starting winding are switched out of the circuit after starting. The capacitors are AC electrolytic capacitors and are not intended to be left in the circuit. They do not affect the running characteristics of the motor. Typical values of capacitors vary with HP of the motor. See Table 1-1.

Capacitors are normally mounted externally but may be mounted internally. See Figure 1-26 for an example of an externally mounted capacitor.

Capacitors can fail and create an open or shorted condition. If the capacitor fails while open, the effect is the same as if the centrifugal switch is open. In this case the motor would fail to start. Do not allow the motor to stay on the line in this condition (not spinning) or the running windings will burn out. If properly

TABLE 1-1 Typical capacitor values for motor starting

HP	Microfarad (μF)
1/8	75–85
1/6	85–105
1/4	100–150
1/3	160–180
1/2	215–250
3/4	375–425
1	400–475
1.5	450–600
3	625–780

protected, the motor should be disconnected automatically from the line because of the sustained starting current draw.

If the capacitor fails while shorted, the motor may fail to start because there is no longer a large phase split. It might start but will take a longer than normal amount of time to obtain running speed. A shorted capacitor

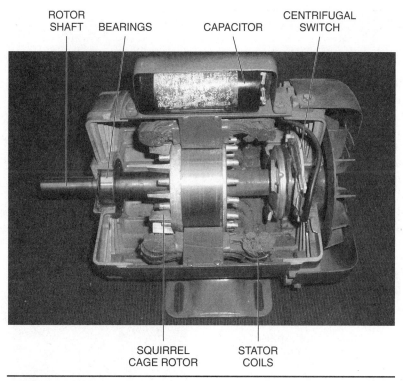

FIGURE 1-26 Cutaway view of capacitor-start, induction-run motor.

should result in higher than normal starting current and (if properly protected) the motor would be disconnected automatically from the line before damage occurs.

Starting capacitors are designed for intermittent duty. If they are started more than 20 times at three seconds each start per hour, they heat up, dry out, and fail. Therefore, capacitor-start motors are not recommended for applications that require frequent starting. Electrolytic starting capacitors are designed for 125-V operation. Too small a microfarad rating capacitor increases starting time and causes premature failure of the capacitor. Too large a capacitor reduces starting torque by splitting the phase too far. The proper capacitor should have approximately 125 percent of the line voltage developed across it during the starting cycle.

Chapter 2 explains how to check a capacitor for rating and proper function. (Details are also given on page 000.)

The starting switch should be a snap-action type that positively disconnects the capacitor from the circuit without any flutter.

Capacitor-start motor connections are similar to the split phase. The lead markings and the connection patterns are identical. Remember that the capacitor and the start winding are connected in parallel with one-half of the run windings. This is similar to the split-phase dual-voltage motor.

Reversing the motor is accomplished the same way as the split phase. Reversing the connections to the start winding will only reverse the rotating magnetic field direction and cause the rotor to reverse too.

OTHER CAPACITOR MOTORS

Two-Value Capacitor Motors

Two-value capacitor motors employ the principle of splitting the single-phase line into two phases by using a capacitor in series with one winding. The effect is to start the motor using an electrolytic capacitor to maintain the good starting torque associated with capacitor-start induction-run motors. In fact, there is a second oil-filled running capacitor in parallel with the electrolytic. As the electrolytic capacitor is switched out of the circuit by a starting switch, the oil-filled capacitor stays in the circuit to increase the running power factor of the motor. The second effect of the running capacitor is to increase breakdown torque as the result of having a continuously rotating magnetic field in the rotor.

(See Figure 1-27 for a schematic of a two-value capacitor motor. See Figures 1-28A

FIGURE 1-28A Electrolytic motor starting capacitor.

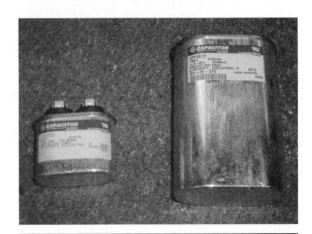

FIGURE 1-28B Oil-filled motor capacitor used in running coil circuit of motor.

and B for sample electrolytic and oil-filled capacitors.)

Permanent Split Capacitor Motors

Motors may be constructed with a single capacitor that stays in series with one of the windings. Because this capacitor stays in the running circuit, the capacitor is an oil-filled type with a lower microfarad capacity. This lower capacity rating gives the motor less starting torque than the other two capacitor motors. It has the

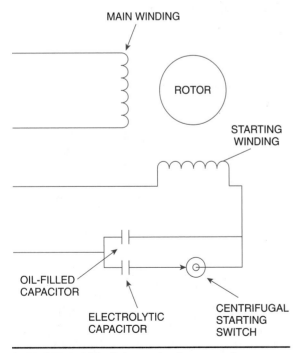

FIGURE 1-27 Schematic diagram of two-value capacitor motor.

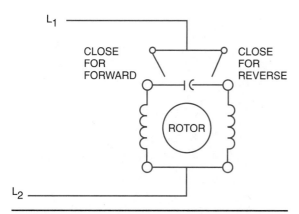

FIGURE 1-29 Permanent split capacitor motor with forward and reverse switches shown.

advantages of good running torque (similar to the two-value capacitor motor) and no starting switch that could burn out (open or shorted).

To reverse the permanent split capacitor motor, simply change the connection to place the capacitor in series with the other running winding. (See Figure 1-29.)

SHADED POLE MOTORS

Shaded pole motors are one of the simplest and cheapest motors to construct. The principle of operation uses the effects of induction not only into the squirrel cage rotor, but also into parts of the stator, which creates a rotating magnetic field from a single phase of input voltage. These motors are typically fractional HP ratings and are used in applications that do not require a great deal of starting torque.

In a simple unidirectional motor there is a ring (shading ring) of solid conductor short-circuited and embedded in one side of a stator winding. (See Figures 1-30A and B.)

As the voltage is applied to the top and bottom coil, the shading coil has voltage induced into it. **Lenz's Law** states that the effect of induction always opposes the cause of induction. Therefore, the magnetic field developed by the shading ring as current flows through its shorted winding opposes the main flux. This causes the main magnetic field to be shifted away from the shading ring.

As the applied voltage waveform begins to decrease from its peak value, the magnetic lines of force also decrease. The effect on the shading ring is the opposite. As main current flow decreases, the magnetic effect of the shading ring tends to keep the same polarity as the main pole but increases in strength. This causes the stator pole to move from the main (unshaded) pole toward the shaded pole. (See Figure 1-31.)

Position 1—Shading coil produces south pole, main pole north. (Use Figure 1-30A as reference.)

Position 2—Zero shading coil current, main pole north.

Position 3—Shading coil current and main pole both north.

Position 4—Main coil has zero flux, shading remains north.

This produces a clockwise rotation of north pole flux.

To reverse the direction of rotation on shaded pole motors, the relationship between the shading ring and the pole face must be reversed as the field always moves from the unshaded to the shaded portion of the stator pole. Another set of shading rings on the opposite side may be shorted.

Controlling the speed of shaded pole motors is easy. Either alter the voltage that is applied to the stator winding to produce less volts per turn on the winding or change the number of turns by using tapped windings and maintaining the same applied voltage. Either method changes the volts per turn on the motor. Less volts per turn means less flux, more slip, and a slower speed under load. See Chapter 2.

Shaded pole motors are typically used for fans where the blades are directly mounted to the rotor shaft and the air passes over the motor. These fans require little starting torque and the air over the motor helps keep it cool. If these motors run hot, or are in a high-vibration

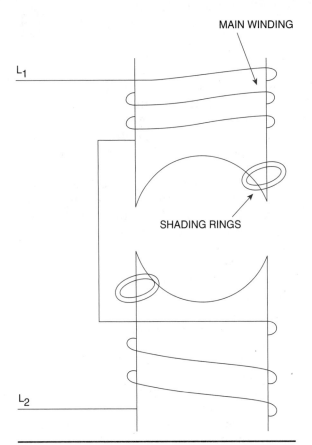

FIGURE 1-30A Schematic of shaded pole motor.

FIGURE 1-30B Photo of typical shaded pole motor.

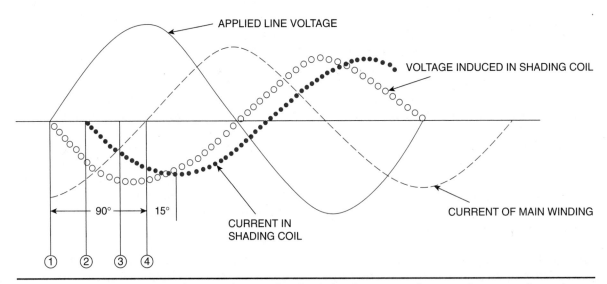

FIGURE 1-31 Representative waveforms of a shaded pole motor voltage and current: line voltage, main coil current, induced shading coil current relationships in an operating motor.

area, the shading rings (which are soldered rings of conductors) can open up and cause the motor to fail.

UNIVERSAL MOTORS

Universal motors are conduction motors. In other words, the current is connected to the rotating rotor through conductors rather than relying on induction to deliver power to the rotor. The name "universal" indicates that the motor will operate on AC or DC or many frequencies of AC. The operating characteristics are not identical for all waveforms. These motors are sometimes referred to as series motors.

The principle of operation for universal motors is similar to a DC series motor. The rotor is wound with coils (not a squirrel cage), but this is not a wound rotor motor. (The name "wound rotor motor" is actually a special three-phase motor to be discussed later.) Because current is conducted through the rotor windings via a commutator and brushes, the poles can be controlled easily. (See Figure 1-32.)

A universal motor is designed with laminated iron material to reduce the effects of eddy currents. The iron is also formulated to reduce the effects of hysteresis when AC is applied to the coils. Some AC motors use a compensating winding placed 90 degrees electrically from the main winding coils. The compensating winding creates a field that fluctuates with load current to reduce the voltage drop in the armature due to reactance. This helps keep the rotor voltage more constant under varying loads and also reduces the sparking.

To reverse the direction of rotation for the series or universal motor, the relative poles of the rotor and stator must be reversed. To do this, the connections to the brushes at the commutator must be reversed to allow current to flow the opposite way through the rotor while current flows in the same direction in the stator coils. Shift the brushes, after reversing connections to readjust the neutral electrical plane, to reduce sparking. The neutral plane is the position where the armature voltages balance out and create a location where there is the smallest difference in potential and, therefore, the least amount of brush sparking.

Series motors are varying speed motors. This means they have drastic speed changes with load. As the motor is running at no load (no mechanical load) it has very high RPM. The RPM may be so high that the centrifugal force on the rotor begins to pull the commutator and windings off the rotor. Thus, the motor is usually connected to some mechanical load at all times. The load may consist of gear boxes or other devices permanently connected so there is load at all times (for example, drills, saws, vacuum cleaners, and mixers). At full load the speed is very low, but the torque is high. (See Figure 1-33.) At high load, there is heavy current draw from the line. The breakdown torque of a series motor is approximately 175 percent. The starting torque is extremely high at 300 percent to 450 percent of running torque. The series motor also has a good operating power factor.

Speed control is common in universal motors. It is accomplished by controlling the voltage applied to the motor. Placing resistors in the circuit or electronic speed control systems that change the applied voltage value. Less applied voltage means less current and less torque developed by the weaker magnetic fields; thus reduced rotational force slows the motor.

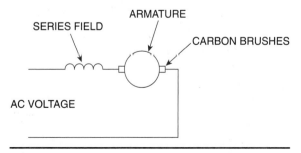

FIGURE 1-32 Series (universal) motor schematic.

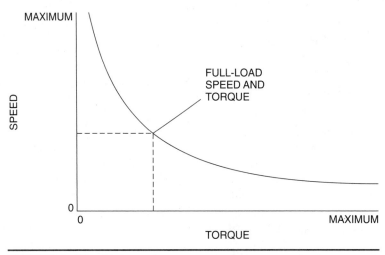

FIGURE 1-33 Universal motor speed versus torque curve.

SUMMARY

This chapter emphasized that the information found on the motor nameplate is critical to the proper selection and application of a motor for a particular purpose. There are many styles of single-phase motors and each style fits a definite purpose for application, utilization, or cost.

This chapter explained the process that makes a single-phase motor turn, and how to make the proper connections for available voltage and desired direction of rotation. Through this chapter's guidance, you should have begun using the National Electrical Code® to correlate information found on the motor to practical applications for installation and troubleshooting.

Summary of single-phase motor types

Type of Motor	Phase Shift Between Run and Start Winding	Starting Torque
Shaded pole	15°	Low
Split phase	30°	Medium
Capacitor start	80°	High
Universal motor		Very high

QUESTIONS

1. What is the formula for synchronous speed of an AC induction motor?
2. Explain what the code letter of a motor indicates.
3. What does the frame number on a motor nameplate tell you about the motor?
4. Split-phase motors use what technique to split a single sine wave?

5. Give two examples of induction motors.

6. How would you reverse the direction of rotation of a capacitor-start, induction-run motor?

7. How does a squirrel cage motor receive a magnetic field for the rotor?

8. Explain what is meant by CEMF in a motor.

9. Running current for a motor at 230 V is (one-half or twice) the current when operating at 115 V.

10. What kind of capacitors are used for motor starting duty?

11. What is the function of a centrifugal switch?

12. Give an application for shaded pole motors.

13. Explain how to reverse the direction of rotation of a universal motor.

14. If a split-phase motor fails to start unless you begin to spin the shaft, what is a probable cause?

15. What would happen if a starting capacitor were replaced by one that was too small?

16. Why is Lenz's Law important to induction motors?

17. How does SF affect the operation of a single-phase motor?

18. Explain the left-hand rule for a coil.

19. What are the synchronous speed and the slip of a four-pole induction motor operating at 60 Hz?

20. Find the locked rotor amps of a 1.5-HP, single-phase, Code K, capacitor-start motor operating at 230 V.

CHAPTER 2

THREE-PHASE MOTORS AND GENERATORS

OBJECTIVES

After completing this chapter and the chapter questions, you should be able to

- Explain the principles of three-phase generation
- Calculate generator frequency and determine phase rotation
- Explain the proper procedure to parallel alternators using monitoring devices
- Describe the principle of operation of a brushless generator
- Determine the three-phase power ratings of motors and generators
- Explain how a three-phase motor creates a rotating magnetic field
- Determine the proper connections for multispeed motors

KEY TERMS IN THIS CHAPTER

Delta: A method of connecting phase coils of a three-phase system. The vector diagram of the phase coils forms the Greek letter delta (Δ).

Multispeed Motor: A motor designed to operate at multiple specific speeds. This is not a variable-speed motor.

Phase Sequence: The sequence of generated sine waves, representing which of three waveforms is developed first, second, and third in a three-phase system.

Power Factor: A factor or a ratio that expresses the relationship between true power, measured in watts, and apparent power, measured in voltamps.

Prime Mover: The mechanical power source that is needed to supply power to a generator.

Synchroscope: A device used to monitor the difference between sine wave peaks of an on-line generator and one to be paralleled to it.

Torque: The twisting or turning effort exerted on a rotating shaft.

Wye: A method of connecting phase coils of a three-phase system. The vector diagram of the phase coils creates a Y pattern.

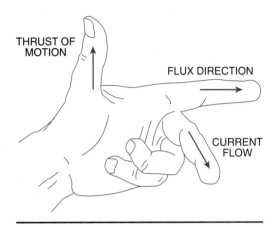

FIGURE 2-1 Left-hand rule for generators using the electron flow theory.

INTRODUCTION

The generation of electricity is based on a simple principle. If you force a conductor through a magnetic field, cutting magnetic lines of force called flux, a voltage will be induced in the moving conductor. Connect an electrical load to the conductor ends and a current will flow. The electron flow theory of electricity indicates that electrons will flow from a negative terminal to the positive terminal of a power supply.

This is the premise of all generators and alternators (AC generators) used to convert energy. Generators are energy converters, changing mechanical energy to electrical energy, just as motors convert electrical energy to mechanical energy.

THREE-PHASE POWER GENERATION

To have electromechanical generation, three factors are required. First, there must be a conductor that will be connected to the load (eventually). The second factor is a magnetic field. The field can be produced either by a permanent magnet or (more likely) an electromagnet. The third factor is relative motion. The conductor must move past the magnetic field or the magnetic field must move past the conductor.

If you use the electron flow theory, the direction of motion and resultant current flow is established by the left-hand rule for generators. (See Figure 2-1.) The thumb represents the thrust or direction the conductor moves through the magnetic field. The first finger represents the direction of the magnetic flux (north to south) through the air. The center finger represents the direction of the current in the moving conductor. This will indicate the positive and negative instantaneous value of the voltage at the generator terminals.

As the conductor moves within a generator's magnetic field, it is going to move in a circular rotation, thereby moving directly under strong magnetic centers of poles and weaker sides of poles. Sometimes it also will travel parallel to the lines of magnetic force and produce no voltage. The simplest generator to visualize is the single-phase AC generator with a revolving armature. The armature is the part of the generator where voltage is induced. It is not always the rotating member. In fact, in most three-phase generators the armature does not move. However, it is easier to visualize how the sine wave is produced by moving the

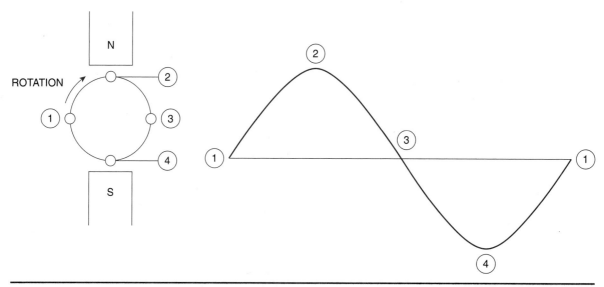

FIGURE 2-2 Representative single-phase sine wave generation using a moving conductor and a stationary field.

conductor. (See Figure 2-2.) The numbered positions on the pictorial generator diagram correspond to the numbered positions on the associated sine wave diagram. As the rotor conductor completes one revolution, it will provide one complete sine wave.

Many small single-phase generators are used for portable power, but the large majority of generators used for commercial use are three-phase generators. Typically, the three-phase generator is a rotating-field-type machine, and the three-phase coils used for the induced voltage are permanently mounted in the stator. The field is a DC energized field that produces a steady magnetic field of constant polarity. As the DC magnetic pole passes a stationary coil (conductor), voltage is induced. All three requirements for generators have been met.

1. A conductor—in the form of a coil
2. A magnetic field—the DC electromagnetic field on the rotor
3. Motion—the field is moving past the conductor, or the conductor can move past the field

As mentioned, this rotating field/stationary armature is the preferred method for larger generators. The DC field can be placed on the rotor and rotated easier than spinning a large armature coil and the associated iron. Also the armature windings on the stator are larger and have heavier currents and higher voltages associated with them. Making these coils stationary helps make permanent, solid connections to them. If the coils were on the rotor, there would need to be sliding contacts to make connections. This would not be practical for large currents and voltages. Also, instead of just a single coil used to collect the induced voltage, there are three separate coils wound into the stator that produce three different sine waves.

As the field rotates past the stationary armature coils, it will produce a sine wave in each coil. The coils are separated so that the 360-degree periphery of the generator is divided in thirds. Each sine wave peak is 120 degrees behind or ahead of the adjacent phase. (See Figure 2-3.)

As you notice, each positive peak is 120 electrical degrees apart as are the zero

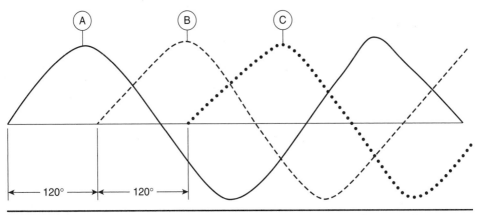

FIGURE 2-3 Representative three-phase sine wave generation waveforms.

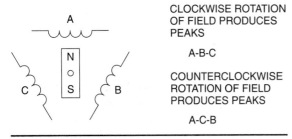

CLOCKWISE ROTATION OF FIELD PRODUCES PEAKS

A-B-C

COUNTERCLOCKWISE ROTATION OF FIELD PRODUCES PEAKS

A-C-B

FIGURE 2-4 Rotating magnetic field produces phase sequences dependent on the direction of rotation.

crossing points and the negative peaks. The phases or sine waves are generated in a specific order. In other words, A phase peaks, then B phase peaks, then C phase peaks, creating an A-B-C rotation of phases. This is determined by the direction of rotation and the coil designations of the generator. Refer to Figure 2-4, and you will notice that if you reverse the direction of rotation of the generator, an A-C-B rotation will result. This means the sine wave will peak A, then C, then B. This phase rotation is important because phase rotation determines the direction of a three-phase motor.

FREQUENCY

As the rotor spins past the armature conductors, it creates positive peaks and negative peaks at a specific number of times per second. The frequency of the peaks is measured in complete sine waves or cycles per second. These cycles per second are measured and referred to as Hertz after the physicist Heinrich Hertz. The abbreviation Hz refers to the number of complete cycles of a wave form per second.

The number of waveforms is directly related to the speed of the field rotating past and the number of poles on the field. The formula is shown in Figure 2-5. This formula is used to determine the frequency output of a generator.

In a two-pole generator operating at 3600 RPM, the frequency output is

$$2 \times 3600 \text{ RPM} \div 120 = 60 \text{ Hz}.$$

The speed needed to maintain 60 Hz is critical. This speed is called the synchronous speed of the alternator (AC generator) and it must be maintained at all current draws. The frequency is closely monitored and must be maintained within strict limits in order to keep equipment, motors, clocks, and other devices that rely on the frequency of generated power operating properly.

Three-Phase Motors and Generators

$$F = \frac{(P \times S)}{120}$$

F = Frequency measured in Hz
P = Number of poles on the rotor field
S = Speed of the rotor in RPM
120 = Constant to convert the number of poles to pairs of poles to create a positive and negative peak and RPM to revolutions per second

FIGURE 2-5 Formula for generated frequency, when speed and number of poles are known.

If the generator slows down, the frequency will drop; if the speed increases, the frequency will increase. A version of this same speed formula will be used to determine the speed of the motor.

OUTPUT VOLTAGE AND CURRENT CONTROL

The voltage generated in the coil of a generator depends on the strength of the magnetic field and the rates at which the magnetic lines of force are cut, or rate of movement. The amount of voltage is determined by the induced voltage calculation, where one volt is induced when a conductor cuts 100 million lines of magnetic flux per second. To increase the voltage, more conductors can be cut. This is the principle used when wires are coiled to get multiple conductors into the lines of flux. A second method increases the strength of the magnetic field to get more lines of magnetic flux. The third method is to increase the speed of the rotating field. Because changing the speed cannot be done without also changing the frequency, speed change is not an option for an alternator connected to operating equipment. The first option of changing the number of conductors is not a viable option either because in most cases the generator is already built with a certain number of coils and changing them is not feasible. This leaves the option of changing the strength of the magnetic field as the method used to control output voltage.

Increasing the DC voltage, and therefore the current to the rotor winding, will increase the DC field strength and increase the number of lines of magnetic force per square inch. As the rotor spins, the lines of force cut the coil winding in the armature and increase the output voltage. The opposite also is true. That is, decreasing field strength will decrease output voltage.

The limitation of this control is that the magnetic material (iron) in the rotor will eventually magnetically saturate so that increasing field current will not create any increase in the magnetic flux. This is the output voltage limitation of the alternator.

Similarly, the field current can be reduced so much that the output voltage of the generator is unusable. A side effect of changing the field current is that the power factor of the alternator can be changed. (See "Parallel Operation of Alternators.") Output voltage is normally regulated so that changes in output current will not affect the output voltage.

As in most electrical circuits, there is resistance in the line wires and impedance in the generator's coils. As current on the loaded generator increases, the voltage available at the terminals tends to decrease because of internal voltage drop in the generator coils. This internal volt drop is greater when the load on the alternator is an inductive load with a low power factor. The low power factor affects the impedance of the coils and causes a large internal drop. If the load is capacitive and has a leading power factor, the terminal voltage tends to rise.

To control fluctuating voltage under various loads, most alternators will have an automatic voltage regulator. The output voltage is

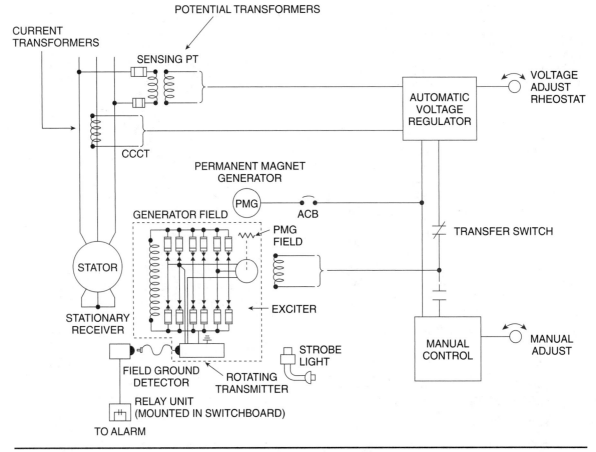

FIGURE 2-6 Automatic voltage regulator connected to the generator output. Also, current transformers are needed to monitor output current to the load. *(Courtesy of Dresser & Rand Electric Machinery)*

monitored at the generator terminals. (See Figure 2-6.) For large alternators with high output voltages, potential transformers are used to measure the accurate output voltage. As the output voltage drops, the automatic regulator circuit changes the DC field excitation current to the DC field of the alternator, thus increasing the field flux to bring the output voltage back to a preset value. Likewise, an increase in voltage out (decreased load) would reduce the field strength and decrease the output voltage to the preset value.

GENERATOR OUTPUT POWER

True power in an AC system is not just the current multiplied by the voltage. **Power factor** (PF) is another element. Because inductive or capacitive loads affect the power factor, the VA output capacity of an alternator will be different than the watt output capacity. Most large system alternators are designed to be operated around 90 percent PF.

Changing the voltamp (VA) output of an alternator may not change the watt output of an alternator because of power factor changes. By changing the exciter current (current to the DC field) the output voltage could change. However, the objective is to keep the line voltage constant for all the other electrical loads on the system. Changing the field current should not affect the system line voltage, but only the amount of current allowed to flow to the load, therefore changing the VA output power. The result is that current also must be monitored to allow more current to flow to inductive loads while holding the output voltage constant. (See Figure 2-6.)

Output KW is changed in another way. If there is an increase in watt load, more true power must be converted from mechanical energy to electrical energy. When more electrical load is placed on an alternator, several things happen. The generator must do more work. The effect is that the generator tends to slow down. This is called the motor effect of the generator. As the conductors of the generator deliver more current to the load, the current actually produces a counter **torque** within the generator. (See Figure 2-7.) The right-hand rule for motors illustrates that an increased current flow in the conductor will produce a torque opposite to the left-hand rule for generators. More input mechanical energy is required to keep the alternator moving at the synchronous speed required to hold the frequency constant. If you listen to diesel-driven generators, you can hear them increase input power to match demands on output power.

Output voltage also tends to drop with an increased load on the generator's output. The automatic voltage regulator (previously discussed) compensates for internal line drop.

To summarize: The increase in output power must be matched by an increase in input power with no change in the input speed that will change frequency. An analogous situation is a vehicle using cruise control. Set the speed for 60 mph (60 Hz) and cruise along on flat ground (steady load). As you begin to climb a hill (increased load) the speed stays steady, but the input power must increase to maintain the same speed. As you descend the hill (decreased load) the input power is reduced in order to maintain a constant speed. The alternators must have speed controls on them to hold the speed of the alternator to a preset synchronous speed.

EXCITERS

Many large commercial alternators have separate DC generators connected to the same shaft as the rotating field. The DC generator called an exciter is used to produce the DC needed to supply the DC field of the alternator. With this type of generation, using separate exciters, the DC is fed to the rotor via slip rings and brushes.

Generators also can produce electrical power using a brushless system. As the name implies, no rotating slip rings or brushes are used to supply the variable DC voltage to the exciter, or DC field. (See Figure 2-8.) There is a permanent magnet mounted on the rotor. As the rotor spins, it induces a three-phase AC voltage into a stator winding, creating a pilot exciter AC voltage. The pilot exciter voltage is

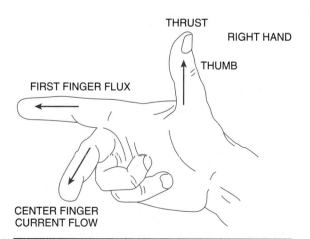

FIGURE 2-7 Right-hand rule for motors illustrates counter torque in generators under load.

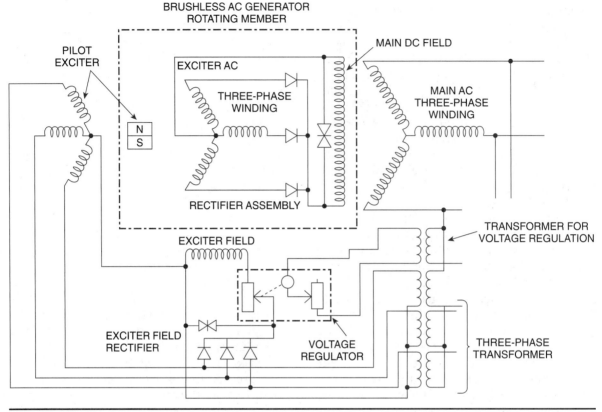

FIGURE 2-8 Simplified block diagram of a brushless AC generator.

rectified using a three-phase rectifier circuit, and then the DC output is controlled to provide the proper DC exciter current. This exciter current is fed to stationary DC field poles located on the stator. The main stationary exciter field induces AC into the moving rotor's three-phase AC coil. The output current from the rotating coils is rectified by rectifiers mounted on the rotor. The three-phase rectified DC is then fed to a field coil on the rotor. The DC field acts as a rotating DC electromagnet so the generator now works the same way as the rotating-field-type alternator previously described.

As the output voltage increases, a portion of the AC output is fed back to the stationary exciter field through a three-phase transformer connection. Also, another transformer is used to monitor output voltage and make adjustments to the exciter field's DC current control.

PARALLEL OPERATION OF ALTERNATORS

It is most efficient to operate alternators at the designed rated loads. If there is only a small load on the power system, small alternators can supply the load. As the power demand grows, more alternators are needed to supply all the demand on the system. This requires you to use parallel alternators. When operating alternators in parallel, output controls must be monitored to prevent an alternator from being overloaded and others not providing enough output. In order to place alternators in parallel operation, the following synchronizing conditions must be met:

1. The phase rotations must be the same. The phases must peak in the same order. The phase rotation must be A-B-C or C-B-A. Either rotation is acceptable, but the two generators must match.
2. The AC RMS voltages of both generators must match. Both generators must emit the same waveform, which is typically a sine wave, so the peaks must match. Actually, the generator that is being added should have slightly higher voltage than the on-line generator. This prevents the incoming generator from becoming an immediate load.
3. The frequencies must match. The incoming generator should be slightly faster than the on-line generator. To verify the frequencies, the use of a **synchroscope** and frequency meters are employed.

Phase rotation meters (Figure 2-9) can be used to check for proper phase rotation. Bring the leads of the three-phase systems to opposite sides of the synchronizing breaker as in Figure 2-10. *Do not close the breaker!* Check the leads with the phase rotation meter

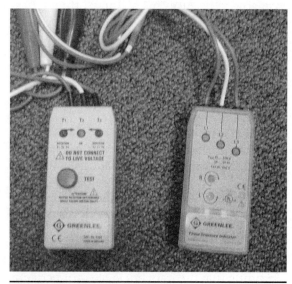

FIGURE 2-9 Motor direction indicator and phase rotation tester.

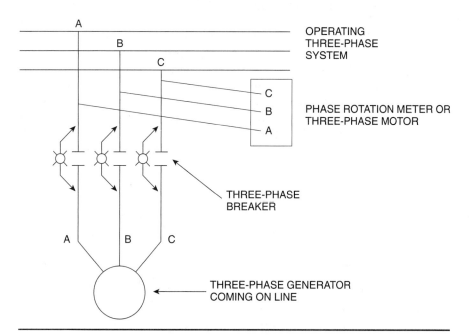

FIGURE 2-10 Phase rotation checks across an open breaker.

discussed in "Motor Rotation Checkers" in this chapter. It will indicate A-B-C or C-B-A. Make sure the leads that will contact, on breaker closure, have the same A- B- or C- designation. If a phase rotation meter is not available, the use of a small three-phase motor serves the same purpose. Again, bring the leads to opposite sides of the breaker. Make sure the three-phase motor leads are marked and connect to the on-line generator. Note the direction of the motor shaft rotation. Then make the identical lead connections to the oncoming generator. The motor rotation must be identical.

Another method is to use sets of lamps. By connecting lamps across the open breaker (Figure 2-10) phase rotation can be checked. If the phases are correctly matched and the phase rotation is the same, the three lights will brighten and darken in unison. If the phase rotation is not correct, the lights become bright and dark at different intervals. (This is assuming that the frequencies are not perfectly matched.)

Caution: Check the voltage ratings of the meters, motors, or lamps. When connecting across the breaker (as with the lights), the test devices must have twice the voltage rating of either of the alternators.

To correct the differences in phase rotation, interchange any of the two leads from the incoming generator. The two alternators should now be in-phase with each other and are ready for the remaining synchronization steps.

The second check on the operating system is to match the voltages. As a standard precaution, a check of all three voltages using a selector switch verifies that all three phases are producing voltage and that it is delivered to the paralleling breaker. The actual voltmeter is a single-phase meter and it is often panelmeter-connected through the use of a potential transformer (PT). The incoming generator should be 2 to 5 volts higher than the on-line one. The speed will be a little faster on the incoming

FIGURE 2-11 Synchroscope face. The indicator should be moving clockwise (fast) for incoming generator.

generator. As it comes on line it should slow down a little. This will bring the voltages together.

The last check and the final visual check before closing the paralleling breaker is to test for proper frequency. This can be done employing lights and the same arrangement as in Figure 2-10. The lights should blink in unison. It is difficult to use this method because it is not obvious which generator has the higher frequency. The incoming generator should be the faster machine or have the higher frequency. A better method is to use the synchroscope, as in Figure 2-11. The synchroscope is a single-phase device that is used to indicate when two identical phases are matched with each other. Remember, the incoming alternator is at a slightly higher frequency so peaks do not match at all times. (See Figure 2-12.) However, the same phase peaks will match once, at each full rotation of the generator. At this instant the synchroscope should be at the top, dead-center position. It should creep in the fast direction, indicating that the incoming generator is running faster. As the pointer approaches the top-center position, close the paralleling breaker. The synchroscope will stop.

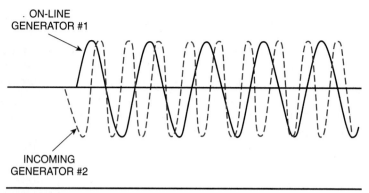

FIGURE 2-12 Frequencies of two alternators are different. The incoming generator is faster (blue).

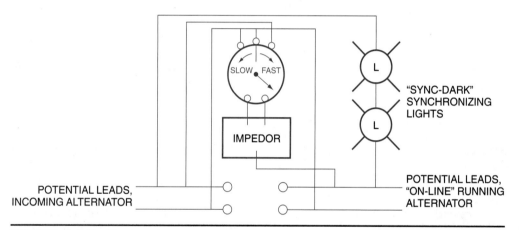

FIGURE 2-13 The synchroscope matches identical phases of two three-phase generators.

The synchroscope is often used in connection with a set of indicating lights. This is a double-check system for the synchroscope. Figure 2-13 shows the connection for the synchroscope and the two clear lamps. The lights are clear so that you can observe the filaments. The typical connection is to use the lamps across the same phases of the alternators and at exact synchronous match, the lamps will be completely dark. You will notice any slight filament glow if the lamps are clear and you can discern exactly when synchronism occurs. This method is called *sync-dark* because the lamps are completely dark.

To synchronize light, the lamps could be connected to alternate phase leads (for example, A to B); the lights would be at their brightest when synchronism occurs. This method is not as accurate because one must determine when the lamp is brightest, which is not a simple task.

Input Energy

The mechanical energy used to drive the generator is provided by the **prime mover.** The prime mover is typically a steam or a water-driven turbine for large commercial generation.

For smaller generation, cogeneration, or standby generation, the prime movers often are engine driven. Diesel or natural gas engines often are used. When paralleling alternators, the speed/load characteristics should be similar. This means that the operating characteristics of the prime movers should match. The same drooping speed characteristics are essential if alternators are to continue to share the load equally or proportionally. This simply means that each alternator should respond to the increased load by slowing a proportional amount. If they do not slow proportionally, one alternator (the one that holds speed better) will assume more and more of the system load.

To change electrical load proportions that parallel alternators' supply to a fixed electrical load, change the prime mover power. To increase load, increase the prime mover power. Remember, you do not want to change speed or frequency will be affected. Simultaneously, you must decrease prime mover power from the one surrendering the electrical load. If you do not adjust prime mover (PM) power, the frequency of the fixed system or the voltage will change.

Changing field excitation of alternators in parallel affects the alternator's power factor, but not the voltage. By adjusting each field excitation to the DC field, the alternators produce higher or lower voltage within the generator coils. However, this voltage is not seen on the parallel system voltage grid. The fields should be adjusted to deliver lowest overall line currents (cumulative value) of all alternator outputs. If you have a constant load, you will see that changing the field currents alters the generator output current to the fixed grid. This current in excess of load current is referred to as the *circulating current* and passes back and forth between alternators. It consumes energy and loads the capacity of the system but does no real work and therefore should be minimized.

LOSS OF ALTERNATOR OUTPUT

There are several conditions that could cause loss of alternator output. Loss of prime mover power results in the loss of output power. If one generator loses mechanical power and remains attached to the power system grid, the other generators try to keep the frequency at the preset value. The other system generators try to drive the faulty generator at synchronous speed even though the failed generator acts as a motor. If the prime mover is free to spin or freewheel and there is still DC on the rotor, the generator acts as a synchronous motor. The failed generator starts to consume rather than supply power. The synchronous motor (failed generator) tends to run in the same direction as the alternator was being driven. If the prime mover is heavy or connected to broken equipment so it cannot freewheel, the synchronous motor will not remain in synchronism. The rotor begins to slip poles as the rotor falls behind the synchronous rotating magnetic field of the stator. The slipping poles induce high voltages into the rotor, which already has DC supplied to it. The high voltage begins to break down the field insulators and also the slipping poles try to resync with the stator field, thus creating heavy torque pulses.

Torque pulses cause severe mechanical damage to the mechanical structures. Also, high current surges circulate between parallel alternators, causing damage to other alternators on the system. This effect of driving an alternator as a motor is called *motorization of the generator*. The generator must be removed from the operating power system to prevent damage to all connected generators.

Another problem that could cause motorization of the generator is loss of field excitation. As the output of the generator decreases, the generator tends to run as a motor. If the prime mover power is still supplying input energy, the generator acts like a motor with no load. The generator still acts as a load to the power system and circulating currents flow between generators. The physical damage is less severe than loss of PM power, but the generator should be removed from the line.

In most generating situations where alternators are connected to supply lines, protective relays will disconnect the generators from the line. Reverse power relays and loss of excitation relays (as well as other protections) are used to remove failed alternators from the operating power grid.

STANDBY AND COGENERATION SYSTEMS

As is the case where facilities are using cogeneration to relieve peak demands on the utilities supply or their own generation to produce power, all these controls must be operational and tested regularly. If companies are using cogeneration, the equipment is often connected or synchronized to the operating system by use of automatic synchronizers, instead of eye and hand coordination.

Some installations *require* a standby system. Legally required standby generators must start and supply specific loads in critical applications such as hospitals and other locations where continued electric supply is essential. These generators would supply loads for operating rooms, smoke removal systems, critical communication systems, and firefighting equipment. (See Figure 2-14.) Refer to NEC Article 700 for Emergency Systems, Article 701 for Legally Required Standby Systems, and Article 702 for Optional Standby Systems.

FIGURE 2-14 A 1.8-million-watt diesel engine emergency power generating system.

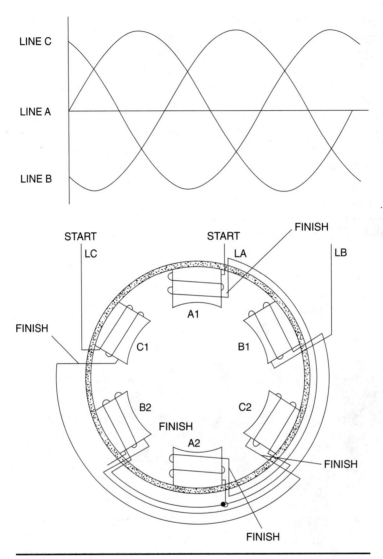

FIGURE 2-15 Three-phase generation produces three sine waves, 120 degrees apart.

Three-Phase Power (Wye)

Whether generators are delivering three-phase electrical power or motors are consuming three-phase electrical power, three-phase power calculations are necessary. As mentioned in the section on generators, the three phases occur along a time line so that each peak is 120 electrical degrees apart. A quick look at these waveforms in Figure 2-15 proves that the three phases balance each other. For example, when phases A and C are one-half value and positive, B phase is full-value negative. The sums of all the instantaneous values will be zero. This reaffirms that there is just as much current being delivered to a load as there is current returning from the load at any instant in time. This is true in any electrical system.

The three-phase system has power delivered more frequently than a single-phase system, by using two phases and returning on the third. This explains why there is more power delivered in a three-phase system than a single-phase system, which uses the formula volts × amps.

First consider a **wye**-connected system. This can be either a generator or a motor or any balanced load for that matter. The three phases of the generated voltage are represented by vectors as illustrated in Figure 2-16. All vectors are shown leaving the star point, 120 degrees apart. This vector representation showing magnitude and direction is not the actual case, as is evident by viewing actual waveforms of the generated three-phase voltages. To better illustrate this, reverse one of the three-phase vectors (B) so that B voltage is opposite the A and C voltage vectors, as verified by the actual waveforms. Now, add vector B to vector C. If you allow each vector to be the same value of generated voltage (for example, 10 volts), the vector addition of B to C is 17.3 volts. Likewise, any of the coil voltage vectors added together is 1.73 times the

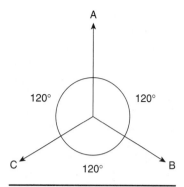

FIGURE 2-16 Vector diagram of three-phase voltage.

original phase-generated voltage. This calculation determines the line-to-line voltage available from wye-connected coils. If one measures line-to-line voltage on a wye system, it will be 1.73 times the phase voltage. An example of this situation is the U.S. standard three-phase voltage as shown in Figure 2-17.

This information illustrates that the line-to-line voltage is higher than the phase-generated voltage. It also is essential to understand that the phase current flow is equal to the line current. Even though the voltage

120 V × 1.73 = 208 VOLTS (NOMINAL)

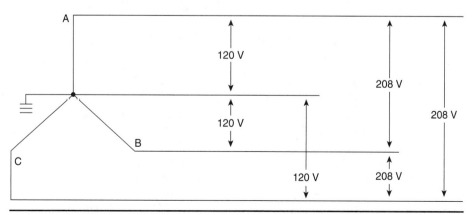

FIGURE 2-17 Three-phase, four-wire wye system.

is 1.73 times the generated voltage, the generated coil current is the same as the line current. This is also true for wye-connected motors. Line current equals the actual motor coil current, but the motor coils have only .58 (the reciprocal of 1.73) times the line voltage applied to them. This leads to the actual calculation of three-phase VA or apparent power.

The actual power delivered or consumed in a three-phase power situation can be calculated by measuring each of the single-phase watt loads and adding them together. If the three phases are balanced (as in motors), calculate each coil voltage and coil current as a watt calculation and multiply by 3 for the three-phase load.

For example, each phase coil voltage × each phase coil current × 3 = three-phase voltamps. Or, because the line voltage is 1.73 × phase coil voltage, the formula becomes the extracted formula as shown in Figure 2-18. Remember that in a wye or star system, coil current is equal to line current, so substitutions make the formula evolve to (line voltage × line current × 1.73) ÷ 1000 = three-phase KVA. This is the formula that is used when dealing with three-phase balanced load systems.

Three-Phase Power (Delta)

Three-phase power for a **delta** (Δ) system uses the same formula. The only difference is in the line and coil substitutions. In a Δ system the line voltage is equal to the individual coil voltage (see Figure 2-19). The line current splits at the delta corner and the same 1.73 factor applies to the vector addition of the current. For a delta system, the line current is 1.73 × coil current. The formula then derived is as shown in Figure 2-20. The end formula is the same for wye and delta, but the individual coil current and voltages are different.

Three-Phase Watts

If the motor or generator is balanced (all phases have equal current and voltage), the three-phase VA formula can quickly reveal the three-phase watts with a known power factor. The three-phase true power or watt formula is shown in Figure 2-21. This formula is used for calculating how much three-phase true power is entering the motor.

If you do not know the power factor, you will need to measure the power factor with a power factor meter, or the voltamps of

---

COIL VOLTAGE × 1.73 = LINE-TO-LINE VOLTAGE
    OR
COIL VOLTAGE = LINE-TO-LINE VOLTAGE × .58
    AND
COIL CURRENT = LINE CURRENT

Using substitution:
(coil voltage) × (coil current) × 3 = three-phase VA
(line voltage × .58) × (line current) × 3 = three-phase VA
(line voltage) × (line current) × (.58) × (3) = three-phase VA
(line voltage) × (line current) × 1.73 = three-phase VA

FIGURE 2-18 Derivation of three-phase power formula for a wye system.

Three-Phase Motors and Generators

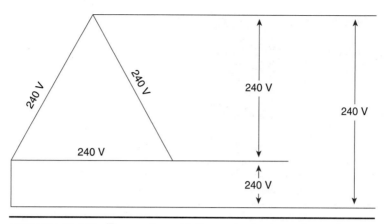

FIGURE 2-19 Three-phase delta system where line voltage equals phase voltage.

COIL VOLTAGE = LINE VOLTAGE
COIL CURRENT × 1.73 = LINE CURRENT
OR
COIL CURRENT = LINE CURRENT × .58

Using substitution in the three-phase power formula as follows:
(coil voltage) × (coil current) × (3) = three-phase voltamps (VA)
(line voltage) × (line current × .58) × (3) = three-phase VA
(line voltage) × (line current) × (.58) × (3) = three-phase VA
(line voltage) × (line current) × 1.73 = three-phase VA

FIGURE 2-20 Derivation of three-phase power formula for a delta system.

WYE OR DELTA
THREE-PHASE WATTS = LINE-TO-LINE VOLTAGE × (LINE CURRENT) × 1.73 × $\left(\dfrac{\text{PERCENT POWER FACTOR}}{100}\right)$

FIGURE 2-21 Formula for calculating three-phase watts.

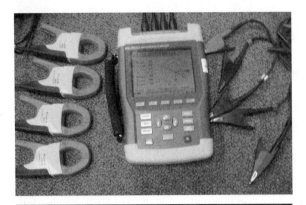

FIGURE 2-22 Clamp-on meters for measuring watts, VA, VARS.

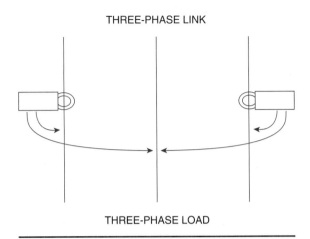

FIGURE 2-23 Measuring three-phase power by a two-wattmeter method.

apparent power and then use a wattmeter to measure the true power. Use a clamp-on wattmeter as illustrated in Figure 2-22. The power factor consists of the watts of true power divided by the voltamps of apparent power. This ratio × 100 will be expressed as the percent power factor.

When using a single-phase, clamp-on wattmeter you will need to make two readings to determine the three-phase watts. The method used takes advantage of Blondell's theorem. The theorem states that in a system of N conductors, N − 1 meter elements will measure the power, but the voltage reference coils must have a common tie point to the wire that has no current coil measurement.

Essentially this method states that the two wattmeter elements may be used to measure three-phase power. This is the method used in three-phase watt hour meters at the service entrance. This method can be used to measure three-phase watts using two measurements. See Figure 2-23 for connections.

When using this method, you need an approximate idea of the motor power factor. If the power factor is between 50 percent to 100 percent, simply add the two wattmeter readings to get the three-phase watts. If the percentage of PF is exactly 50 percent, one wattmeter will read zero. If the PF is below 50 percent, as can be the case in unloaded motors, one wattmeter will try to read downscale. Reverse this wattmeter and subtract the upscale reading from the other wattmeter reading. To test for low PF, try loading the motor slightly. Both meters should move upscale if the percentage of PF is over 50 percent.

To calculate PF, measure the three-phase watts, the three-phase line voltage and line current, and apply the three-phase VA formula. Use the power triangle as in Figure 2-24 to calculate percentage of PF. If you have a PF meter, you may simply calculate the watts by

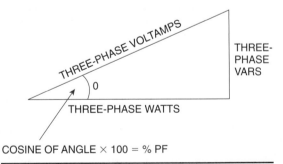

COSINE OF ANGLE × 100 = % PF

FIGURE 2-24 Power triangle shows relationship between true power (watts), apparent power (VA), and reactive power (VARS).

determining the three-phase voltamp using the formula and then apply the percentage of PF to find the three-phase watts.

Summary of Power System Differences

The difference between wye and delta systems is determined by the needs of the customer. The wye system is versatile in that 277/480 volts are easily obtained to supply 277-V lighting systems and 480-V motors. The wye system can also deliver 120/208 V for a 120-V single-phase receptacle load and 120-V lighting, as well as 208-V three-phase for light motor loads. The advantage of a delta system is that a higher voltage (240 V) is delivered compared to 208-V three-phase voltage, for example. This will drive a motor load with slightly higher voltage and therefore less current. Less current means smaller size wire, switchgear, conduits, and so on. With the addition of a center tap on one phase of the delta, a 120/240-V single-phase voltage is available. This 120-V supply may be used to supply control voltage for a 240-V three-phase motor starter. Each system has advantages depending on the customer needs.

THREE-PHASE MOTOR THEORY

Motors used in commercial and industrial applications usually take advantage of the three-phase power that is delivered to the facility. Three-phase motors are physically smaller, lighter, and have fewer mechanical malfunctions than similar horsepower single-phase motors, which are used extensively in residential applications. Because of the operating principle, three-phase motors also tend to be more efficient in energy conversion.

Three-phase motors are the reverse process of the three-phase generator. As you know from discussion of three-phase generators, there are three separate coil windings (or phases) wound on the stator core to collect the induced voltage and deliver it to the load. Similarly, the three-phase motor has three coil or phase windings on the stator to take the three-phase delivered voltage and induce voltage into the motor's rotor.

To see how a rotating magnetic field is produced in a three-phase wye motor, refer to Figure 2-25. Because all the coils are wound identically and then inserted in the slots of the stator, the end connections have been designated start and finish, or S and F. By connecting like ends (finish to finish), opposite magnetic polarities will be established in the coils. This creates the two magnetic poles needed to produce the magnetic flux required in the stator's center air space. Remember that

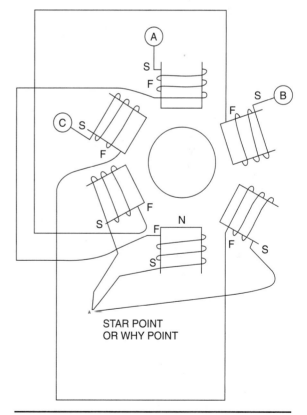

FIGURE 2-25 Three-phase, two-pole motor coil representation.

flux lines are said to travel from north pole to south pole through the air's magnetic path.

This representation shows a two-pole motor. Even though there are six magnetic poles being produced, each sine wave produces only two poles and the synchronous speed of this motor is based on the number of poles per phase. The same formula is used as that for single-phase motors. Additionally, the same formula is derived to determine necessary alternator speeds. Use the formula

$$Ns = \frac{120 \cdot \text{freq}}{\text{\# poles}}$$

120 × the applied frequency ÷ number of poles (per phase) = synchronous speed (Ns).

Ns is the synchronous speed of the rotating magnetic field.

120 is a constant conversion factor.

F is the frequency of the applied voltage.

P is the number of magnetic poles developed per phase.

Three-Phase Rotating Field Description

Three-phase induction motors use principles that are similar to those used in single-phase motors. The stator is wound with magnet wire to produce magnetic fields that create a rotating magnetic field. The major difference is that there is an equal number of poles for each of the three phases. These electromagnetic poles are always in the circuit and are therefore creating a continuously rotating magnetic field. There are two general types of three-phase motors—the wye connection and the delta pattern.

Y-connected, three-phase, two-pole motor explanation assumptions follow. (See Figure 2-26.)

- If the wave form is above the zero reference line, then current is flowing from the line into the motor.
- If the AC waveform is below the zero reference line, then current is returning from the motor and flowing back toward the generator.

The following description uses Figures 2-25 and 2-26.

Position 1—Phase A (zero current flow), phase B (current out), phase C (current in). Note that phases B and C are equal and opposite when phase A is zero. Using the left-hand rule for coils, grasp the coil with left hand, fingers in the direction of the electron flow and thumb pointing to the north pole of the coil. The left coils B and C both produce a north magnetic pole in the stator opening. The opposite side of stators B and C produces south magnetic poles. These poles are not full strength as the maximum point of the sine wave voltage is not applied at this point.

Position 2—Phase A (current in), phase B (current out), phase C (current in). Phase B has the peak value of voltage supplied and is the strongest magnetic pole. Phase A has current flow now that produces a N pole at the bottom, but it is a weak pole. Phase C retains the same polarities as position 1, but the magnetic strength is weaker. B is the center of the large magnetic pole, but the magnetic influence is felt from the left C phase to the bottom A phase creating one large N pole. The opposite side of the stator poles are all south polarities of the same strength, so two magnetic poles are produced with the centers, slightly counterclockwise from position 1.

Position 3—Phase A (current in), phase B (current out), phase C (zero current). Note that phase A is equal and opposite to phase B. As is the case in three-phase power, all the phases create a balanced condition. In other words, two of the phases combined will be equal and opposite to the third. If one is zero, then the other two are equal and opposite. In this position, phase A creates a stronger north at the bottom than position 2, and phase B has a weaker north pole on the left than in position 2. Now the center of the north pole has moved counterclockwise (CCW) to the iron of the stator

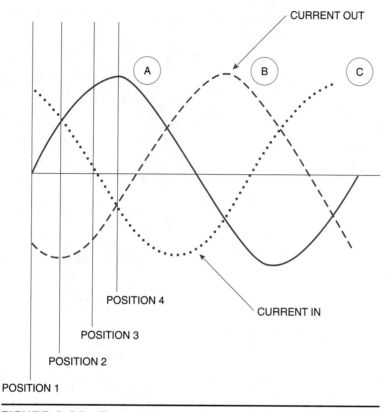

FIGURE 2-26 Three-phase waveforms produce a rotating magnetic field in a three-phase motor.

between the phase B and A coil windings. The corresponding south pole has also moved CCW.

Position 4—Phase A (current in), phase B (current out), phase C (current out). At this point phase C has reversed the current through the phase C coils. This condition produces an opposite polarity as it did in the previous positions. Therefore, the right hand C coil now has a north polarity, the A coil is at maximum strength, and the B coil has weakened moving from position 3. This moves the center of the magnetic north pole directly over the center of the bottom A coil, but the north pole encompasses the B and C coil as well. The two poles have again shifted in the counterclockwise (CCW) direction.

This explanation has moved 90 degrees on the A phase sine wave. If the applied frequency is 60 Hz, then the time elapsed is 1/4 of 1/60 of a second, or 1/240 of a second. To get back to the original starting point, phase A crossing zero going positive, takes 1/60 of a second and the field will rotate counterclockwise (CCW) one revolution. In one second it will rotate 60 times and in one minute it will rotate 3600 times. The synchronous speed (of the rotating magnetic field) of a two-pole, three-phase, 60-Hz motor is 3600 RPM.

To reverse the three-phase motor, interchange the phase line connections to the motor connections. For instance, connect the L3 lead (phase sine wave C) to coil A connection, connect L1 lead (phase sine wave A) to coil C connection. By doing this, the sequence in which the coils became energized is reversed.

- Position 1: Coil winding A and B, bottom and left have north poles.
- Position 2: B is again strongest, but left-hand C now has a north pole.
- Position 3: The coil A connected to phase supply C is now at zero magnetic flux and the north pole has rotated clockwise with the center between B and C on the left side.

Any two reversed phase connections to the lines would result in the reversal of rotation of a three-phase motor.

If the motor is apart so that there is no rotor installed, you can check the direction of rotation with a simple experiment. Place a steel ball (large enough to not get stuck in the stator slots) into the hollow stator. Energize the motor leads for a very short period of time. The steel ball should follow the rotating magnetic field. You may need to give the steel ball a slight push to start the movement. Use a nonmagnetic material to reach in and push the ball. Magnetic materials will become magnetized and will stick to the stator.

If the motor is assembled, it is best to check the direction of rotation before connecting it to a load. Many loads are unidirectional and damage could result by trying to drive them the wrong way. A simple check with line power applied, noting shaft rotation and line and motor connections, is the most common method of checking.

MOTOR ROTATION CHECKERS

When a motor is installed, it is best to know the direction that it will rotate as the three-phase line connections are made and before the power is applied. With the use of a motor rotation tester, the accurate phase rotation can be determined by connecting the tester to the

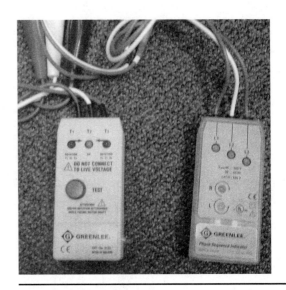

FIGURE 2-27 Meter used to check phase rotation A-B-C or C-B-A.

motor first. See Figure 2-27. By spinning the motor in the desired direction of rotation, a small voltage is produced that indicates an A-B-C or a C-A-B rotation based on the lead markings connected to the motor and the desired rotation of the rotor. With the phase rotation tester, as in Figure 2-27, the proper phase rotation of the line can be established to give you the proper direction of rotation for the motor as power is applied.

THREE-PHASE MOTOR CONNECTIONS

There are two main AC motor patterns. These are the wye- and delta-connected three-phase induction motors. More specialized wound rotor and synchronous motors are discussed in Chapter 8.

Refer to Chapter 7 for wye- or star-connected motor lead markings and connections for high or low voltage on dual-voltage motors. Delta-connected motors are also checked for lead marking and connection patterns for high and low voltage.

There are other connections that need to be considered. If motors need to be speed-changing motors, the method of speed control must first be considered. If motors are to have a continually changing speed or need a variable speed from very slow to full load, a variable-frequency drive is the method most used. To change the speed with this method, the frequency of the supplied AC is varied and the speed of the rotating magnetic field is also altered.

$$Ns = 120 \times F / \text{\# poles}$$

This is the formula version of the method.

This type of control is becoming increasingly popular, especially in variable air volume (VAV) ventilation systems and process control manufacturing systems. (See Chapter 5.)

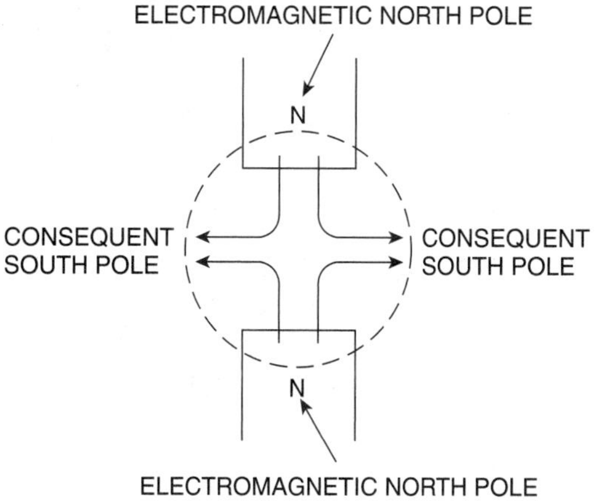

FIGURE 2-28 Consequent pole technique to create four magnetic poles from two coils.

MULTISPEED MOTORS

If a variable-frequency drive is not needed, and the motor is just changed from high to low speed or specific speeds, then the motor is reconnected to provide for a different number of magnetic poles. If this is the case, special **multispeed motors** are used. Often these motors are single-voltage motors with more leads brought out for connection to the line. The two basic connections refer to the method that is used to produce the various speeds. If one winding is used to produce the two speeds, the ends of the coils must be brought out for reconnection. The actual internal coils of each phase are reconnected so that opposite coils produce the same magnetic pole. (See Figure 2-28.) If this method (one winding) is used, the motor has consequent poles. As a consequence of making opposing coils the same magnetic polarity, two south poles are established midway between the two actual poles. This method will double the number of original poles (which are called alternate poles) to make consequent poles. The motor can be changed from a two pole to a four pole, or four pole to eight pole. The speeds will always be approximately one-half of the second speed. For example, consequent pole motors change from 3600 to 1800, or 1800 to 900 RPM synchronous speed.

The other method of producing various speeds is the separate winding style of the motor. This method of creating poles is called the alternate pole method. These motors actually have two separate windings installed, and each is wound for a different number of poles. These motors change the internal windings. For example, a four-pole winding is disconnected and a six-pole winding is reconnected. This allows the user to create closer speed variations, such as 1800 to 1200 RPM, rather than such a large difference in synchronous speed as is provided by consequent pole motors, where speeds are always one-half of the other speed. Of course, you may want to have more than two definite speeds. If this is the case, both methods can be combined to produce three- or four-speed motors. Using two windings and the consequent pole idea will give you many options.

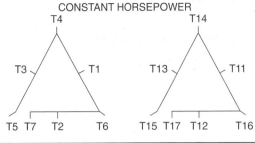

SPEED	L1	L2	L3	OPEN	TOGETHER
LOW	T1	T2	T3	ALL OTHERS	T4, T5, T6, T7
2ND	T6	T4	T5, T7	ALL OTHERS	
3RD	T11	T12	T13	ALL OTHERS	T14, T15, T16, T17
HIGH	T16	T14	T15, T17	ALL OTHERS	

(A)

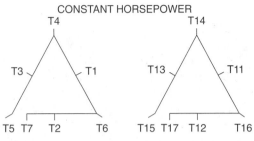

SPEED	L1	L2	L3	OPEN	TOGETHER
LOW	T1	T2	T3	ALL OTHERS	T4, T5, T6, T7
2ND	T11	T12	T13	ALL OTHERS	T14, T15, T16, T17
3RD	T6	T4	T5, T7	ALL OTHERS	
HIGH	T16	T14	T15, T17	ALL OTHERS	

(B)

CONSTANT TORQUE

SPEED	L1	L2	L3	OPEN	TOGETHER
LOW	T1	T2	T3, T17	ALL OTHERS	
2ND	T6	T4	T5	ALL OTHERS	T1, T2, T3, T7
3RD	T11	T12	T13, T17	ALL OTHERS	
HIGH	T16	T14	T15	ALL OTHERS	T11, T12, T13, T17

(C)

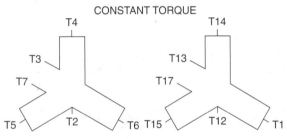

SPEED	L1	L2	L3	OPEN	TOGETHER
LOW	T1	T2	T3, T17	ALL OTHERS	
2ND	T11	T12	T13, T17	ALL OTHERS	
3RD	T6	T4	T5	ALL OTHERS	T1, T2, T3, T7
HIGH	T16	T14	T15	ALL OTHERS	T11, T12, T13, T17

(D)

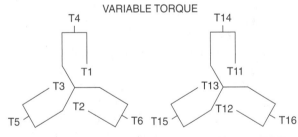

SPEED	L1	L2	L3	OPEN	TOGETHER
LOW	T1	T2	T3	ALL OTHERS	
2ND	T6	T4	T5	ALL OTHERS	T1, T2, T3
3RD	T11	T12	T13	ALL OTHERS	
HIGH	T16	T14	T15	ALL OTHERS	T11, T12, T13

(E)

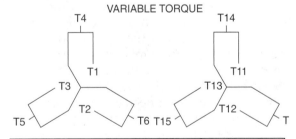

SPEED	L1	L2	L3	OPEN	TOGETHER
LOW	T1	T2	T3	ALL OTHERS	
2ND	T11	T12	T13	ALL OTHERS	
3RD	T6	T4	T5	ALL OTHERS	T1, T2, T3
HIGH	T16	T14	T15	ALL OTHERS	T11, T12, T13

(F)

FIGURE 2-29 Constant torque, constant horsepower, and multispeed motor connections. *(Courtesy of Square D/Schneider Electric)*

Horsepower, Torque

As shown in Figure 2-29, there are various combinations of wiring motors to give the desired results. By changing the number of poles either by reconnection or by consequent poles, the operating characteristics of the motor change. If you want the same horsepower at all speeds, more leads are required to produce the added torque required at lower speed to maintain horsepower. If constant torque is a requirement, then horsepower can change to produce the same torque at different speeds. Variable torque allows the horsepower and torque to change with speed.

When making these connections be sure to note that some leads are left open during the other speed operations. Often these leads are brought to a motor starter and will have induced voltage at the leads. *Be careful!* Do not assume there is no voltage on the motor leads just because the contactor is not pulled in.

Rotors

Three-phase induction motors work the same way as single-phase induction motors in that voltage is induced into the squirrel cage rotor. The squirrel cage rotor will produce a current flow through the shorted bars just as in single-phase motors in Chapter 1. There will be the same number of poles produced in the rotor as there are poles per phase in the stator. Remember there must be some slip; therefore the rotor cannot follow the stator field at synchronous speed in an induction motor. The torque produced by the rotor is a result of how much current flows in the rotor, which is dependent on the amount of induced voltage. At standstill (or locked rotor) the rotor has not yet begun to move and there is large voltage and current produced. This gives a relatively large torque (starting torque). As the rotor begins to spin, the induced voltage and the rotor current decrease and torque goes back down to a value of running torque. The torque is determined by the formula:

$T = Kt \times \phi s \times Ir \times \cos \phi R$

T = torque measured in foot-pounds of twisting effort

Kt = a design constant based on the physical design of the motor

ϕs = the strength of the stator flux

Ir = rotor current

$\cos \phi R$ = the rotor power factor, based on the rotor design.

Torque is the twisting effort needed to drive the load. The combination of torque and RPM is used in the formula to determine horsepower. The horsepower formula is the same as for single-phase motors as presented in Chapter 2.

$HP = FRN/5252$

HP = output horsepower

F = force in pounds on scale

R = radius of pulley in feet

N = operating RPM of motor 5252 is a constant in the formula

Torque is normally measured in foot-pounds or the product of $F \times R$. For small motors the measurement may be in ounce-inches. If this is the case, convert ounce-inches to foot-pounds by dividing by (16 × 12) or 192; for example, 192 oz-in/192 = 1 ft-lb.

Torque also is produced in a motor based on the factors in the torque formula. Torque will increase as speed slows, until a point is reached where the motor can no longer turn under the heavy load and the motor abruptly drops in speed. The following sequence occurs as a motor is loaded. Developed torque, T_d, is the torque developed by the motor. Retarding torque, Tr, is the drag placed on the rotor by the mechanical load.

If T_d is less than Tr, then motor slows down.

If T_d is greater than Tr, then motor speeds up.

If $T_d = Tr$, motor speed is constant.

By adding more mechanical load, T_r is increased; therefore the speed of the shaft begins to decrease. As the speed decreases (increased slip), the rotor bars begin to cut magnetic lines of flux at a greater rate. The induced electromagnetic frequency (EMF) in the rotor increases and the rotor current increases. Also, the stator has less counter electromotive force impressed on it from the slower speed rotor, so the line current and the stator flux increases.

By reviewing the formula for torque, you can see that several factors change. The strength of the stator flux—ϕs—increases. The rotor current also increases, producing a stronger rotor flux. The rotor power factor will decrease slightly based on the fact that the induced EMF of the rotor is at a slightly higher frequency. However, the magnetic attraction of the stronger stator field and the stronger rotor field will increase the developed torque, T_d. The torque will increase to match the T_r at the new, lower speed. Remember, the T_d increased because there was a decrease in speed. This is the response to a changing load that is present in an induction motor. How much it changes or how well the motor holds the design speed is based on the design of the motor. Again, this sequence will occur until the motor magnetic fields no longer have enough attraction to carry the load. At this point there is an abrupt drop in speed and the breakdown torque of the motor has been exceeded. (See Figure 2-30.)

The way to compare different motors on the ability to hold full-load speed is to compare speed regulation. As explained in Chapter 1, speed regulation compares a motor speed at no load and at full load, and expresses it as a percent.

$$\% \text{ speed regulation} = \left[\frac{(NL:RPM - FL:RPM)}{FL:RPM} \right] \times 100$$

If a motor speed varies greatly from NL to FL, then the speed regulation is poor. No change in speed would be the ideal, but as you have just seen, induction motors cannot have 0 percent regulation.

MOTOR DESIGN LETTERS

Figure 2-30 shows the speed torque curves for typical AC induction motors. Design letters are not the same as code letters. Design letters are intended to indicate operating characteristics whereas code letters are used to size equipment based on the locked rotor KVA per horsepower.

Design A motors have relatively constant speed, or very low percent speed regulation. They have high starting torque and rapid acceleration, but higher starting current. The breakdown torque also is one of the highest at approximately 275 percent.

Design B motors are the general-purpose-design motors. They have lower starting torque than design A but also have lower starting current. Breakdown torque is somewhat lower.

Design C motors use a double squirrel cage to produce higher starting torque, but lower starting current. The motor uses the high impedance of one squirrel cage winding located deep in the slots to keep current low during starting. Because the rotor frequency is high at starting, the deep conductors that have high impedance also have high inductive reactance. The outer bars are small conductors with high resistance, but low reactance. During starting, most of the current flow will be in the outer high resistance bars. The phase relation between rotor and stator is nearly in phase because of the resistance of the bars. This allows the magnetic fields to occur at the same time and produce a high torque without a high current. As the rotor speed increases, the XL of the inner bars decreases as rotor frequency decreases. The inner windings are lower resistance and now also lower reactance, so the combined impedance is lower than the outer

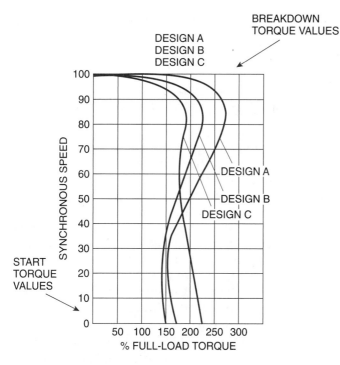 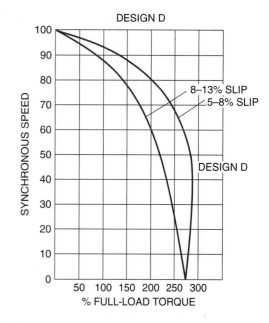

Design A–Design B–Design C
The typical speed-torque graph above illustrates the characteristics of the designs. Design A is very similar to Design B, except breakdown torque and starting current are higher on Design A motors. Design B is the standard, general-purpose design. Design C motors have high starting torque and low starting current. All have normal (0 percent to 5 percent) slip.

Design D
Design D motors are characterized by their high-starting torques combined with high slip. Starting torque for 4-, 6- and 8-pole design motors is 275 percent or more of full-load torque. Applications using stored energy flywheels are best matched with Design D motors, reducing shock to both driven load and power system in two standard values: 5 percent to 8 percent or 8 percent to 13 percent.

FIGURE 2-30 Design letters determine operating characteristics of AC motors. *(Courtesy of Wesco)*

bars. Now rotor current flows in the inner bars. This allows the motor to react quickly to load changes so it has good running torque and high breakdown torque. Of course, this motor costs more to purchase.

Design D motors use a high-resistance rotor to produce high starting torque but also have a high slip. These motors are typically suited to loads that drive high inertia loads, such as flywheel loads. Starting current is high and these motors should not be used for equipment requiring frequent starting.

POWER FACTOR OF INDUCTION MOTORS

Power factor is an indicator of the amount of apparent power that is used by the motor in the conversion of electrical to mechanical energy. At no load (not doing any work at output) there is still energy consumed in the motor windings, friction, windage, and core losses. At no load most of the energy is lost due to the inductive component of the motor or energy in the form of VARS: Volt · Amps

Reactive. This is the energy needed to create the rotating magnetic field. As most of the energy is not consumed, the power factor is very low. Typically, 10 percent PF is normal at no-load conditions. As the motor is loaded, more energy is consumed in the form of watts compared to the VARS of the motor circuit. The percentage of the power consumed increases, compared to the magnetizing power and the percent power further increases to approximately 85 percent to 90 percent. It is better to run the motor at full load for higher PF and better efficiency.

$$\frac{W}{VA} \times 100 = \% \text{ Power Factor}$$

MOTOR EFFICIENCY

After you have determined the true power input (watts) into the motor, you need to know the actual mechanical work done at the output shaft in order to determine efficiency. You may use all the same formulas described in horse-power testing in Chapter 2.

The output horsepower is a mechanical measurement of work done. Convert this to electrical measurements to compare to electrical energy input. Multiply the horsepower output by 746 watts to get horsepower watts output. The formula to compute efficiency is:

output watts/input watts × 100 = % efficiency.

Remember, motor efficiency is the percent of true power output compared to the true power input. Do not confuse power factor with efficiency. Power factor is the percent of true power input compared to the amount of apparent power input. Efficiency compares input watts to output watts. Motors are most efficient at full load. This is the design point where the highest amount of input watts are used for output watts (horsepower) with the internal losses of the motor at the smallest percentage. Typical values of standard motors are 75 percent to 90 percent efficient and new high-efficiency motors run at 90 percent to 95 percent efficiency.

TROUBLESHOOTING

The three-phase induction motor is rugged and reliable. Because there are no internal switches (as there are on many single-phase motors) and there are no brushes or commutators (as on DC motors), few things can go wrong. If the bearings are breaking down, the motor may be working too hard to overcome bearing friction and carry a mechanical load. Overheating occurs, and winding damage can occur. Megohmmeter checks can be made as described in Chapter 4 to detect gradual breakdown. Overheating also can occur if the motor shell is left full of grease or dirt. The shell or frame is designed to dissipate the heat of the motor windings. Winding failure can occur if the heat is not allowed to escape. Another common failure occurs when the motor does not receive all three phases delivered to it. This can occur through a lost phase from the service, a poor connection at the starter, or loose motor terminal connections. If two of the three phases are delivered to the motor, the condition is called *single phasing*. The motor will typically not start by itself, but if you spin the shaft of an unloaded motor it may run. This condition will not carry a full load and damage will occur on either the start or running of the motor. Many critical motors may have phase monitors to protect them from the chance of single phasing. Figure 2-31 shows a phase loss protection relay. This device is connected to the motor circuit. It protects against low-voltage conditions, high-voltage conditions, phase loss, phase reversal, and voltage unbalance. The output contacts are connected to the motor control circuit and interrupt the starter control circuit if the power supply is not correct.

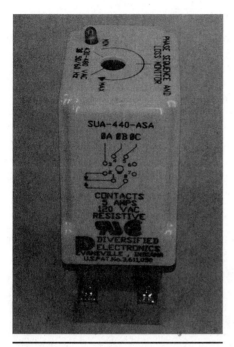

FIGURE 2-31 Phase protective relay monitors phases for imbalance, failure, or reversal.

SUMMARY

Three-phase power generation was explained to help you understand where the three separate phase voltages are developed. The coils, whether connected wye or delta, produce waveforms that are 120 electrical degrees apart. The generation and control of AC power at a specific frequency were discussed, and the methods used to parallel alternators were presented to allow you to place standby or cogeneration generators on-line.

Three-phase power calculations were explained using the wye and delta patterns to find the proper three-phase voltamps, watts, power factor, and reactive power.

General information on three-phase motor rotation and synchronous speed was explained. Multispeed motors use different methods to create the multiple speeds. Torque and horsepower were explained in relation to the multispeed motors. Information on squirrel cage rotors was presented to enhance your knowledge of why motors have different operating characteristics. Power factor and efficiency were also presented with general motor operations.

QUESTIONS

1. Name the three factors needed to induce a voltage.
2. What is meant by the term "stationary armature"?
3. Write the formula to calculate frequency of a four-pole generator.
4. What is the speed called that will deliver a desired set frequency?
5. How does changing the DC field affect the AC output of an alternator?
6. Explain how true power output is changed in an alternator.
7. What is an exciter generator used for?
8. List the conditions that must be met before alternators may be paralleled.
9. Explain one method of determining three-phase phase rotation.
10. What can you do if the two generators have different phase rotations?
11. What is the function of a synchroscope?
12. Name two problems that will create motorization of a generator.
13. How are three-phase motors reversed?

14. What is the difference between consequent pole motors and alternate pole motors?

15. When using two-speed two-winding motors you must be careful around the deenergized starter. Why?

16. What is the difference between starting torque and breakdown torque?

17. How is a design letter different from a code letter on a motor?

18. How are phase voltage and line voltage related in a wye generator or motor?

19. Write the formula for three-phase KVA of a system.

20. How is motor efficiency computed for three-phase motors?

CHAPTER 3

BASIC MOTOR CONTROL

OBJECTIVES

After completing this chapter and the chapter questions, you should be able to

- Identify and use manual controllers for single-phase motors
- Identify and size overload protective systems for motors
- Use the NEC® to help determine equipment requirements
- Determine manual and automatic reset applications
- Begin constructing control circuits, using two- and three-wire control
- Properly size fuses and fuse holders
- Make a complete single-phase motor installation in compliance with the NEC®

KEY TERMS IN THIS CHAPTER

Contactor: The device that is used to make or break the circuit to an electrical load through the use of power contacts.

Controller: A device that governs the delivery of power to an electrical apparatus in some predetermined manner.

Jogging Control: Sometimes called inching, this is the process where the contactor is closed for very short durations to allow the motor to move in small increments. This control does not allow the motor to remain running. Because the high current for starting is frequently applied during jogging, the process of jogging is hard on the motor because cooling and electrical insulation of the windings are stressed. The mechanical torque forces from starting frequently also create wear on the motor shaft and bearings.

Low- or No-Voltage Protection: Sometimes called undervoltage protection, this is the method that is used to release the contacts of a controller if the supply voltage falls but will not reclose the contacts when proper voltage returns. This method is used with three-wire control.

Low- or No-Voltage Release: Sometimes called undervoltage release, this method is used to release the contacts of a controller when the supply voltage falls too low to properly supply the load. The contacts may reclose upon restoration of normal voltage. This method is used with two-wire control.

Maintained Contact: A contact that is moved mechanically, or push-button operated, from NO to close or NC to open and stays that way after the mechanical force is removed until it is moved back again.

Momentary Contact: A contact that changes position only when actuated. After the actuating event, the contact returns to its normally open or closed state.

Motor Starter: A controller designed to start the motor and bring it up to normal speed. The motor starter also contains the overload protective devices.

Pilot Device: A device that directs the operation of another device. A pilot device is typically a low-power device that creates a circuit to control a larger power-handling device.

Three-Wire Control: A means of providing control signals to the motor starter. This method uses three wires from the start/stop stations to the starter.

Two-Wire Control: A means of providing control signals to the motor starter. This method uses two wires from the pilot device to the starter and is used for nonsupervised operation.

INTRODUCTION

Motors provide the means of changing electrical energy to rotational mechanical energy. Motors are therefore limited as to the functions they can perform without various controls. Control of motors is a very important function, and the design and proper implementations of the controls is an ongoing and ever-changing field. Speed control and precise motion control are essential to today's manufacturing.

MANUAL CONTROLS—CONTROLLER DESIGN

The simplest form of control is the manual motor control. This consists of a basic on–off switch mechanism that controls current to the motor. This type of control uses a mechanical action to close the switch and apply full-line voltage to the motor. The mechanical energy needed to move the switch open or closed is provided by your hand; thus the name "manual control."

Although manual controllers are used for small-horsepower AC or DC motors, they must comply with the NEC® rating for motor controllers as specified in NEC® Article 430.83. These ratings determine the basic electrical characteristics of the controller. They shall have an HP rating at the application voltage not lower than the motor's HP rating. In other words, the controller must be able to start and stop the motor at the voltage that is supplied.

The exception to the rule is that stationary motors (not portable) rated at 2 HP or less and 300 volts or less can utilize a general-use snap switch if the switch ampere rating is at least twice the full-load current of the motor. (See Figure 3-1.) Multiply the full-load current of the motor by two. The rated switch current must be greater than or equal to the product.

AC circuits may use a switch for the same motor conditions, if the switch is rated "AC only" and the motor is not more than

Basic Motor Control

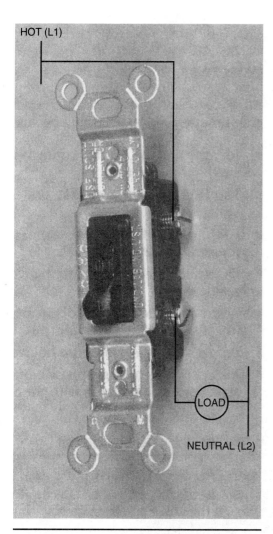

FIGURE 3-1 General-use snap switch with correct voltage and current rating may be used as a controller.

80 percent of the switch rating. For this application, multiply the rated switch current by .8. The motor full-load current must not exceed this product.

Switches can be used for controllers in applications for bathroom fans, garbage disposals, and so forth, where the aforementioned conditions apply. When snap switches are used as controllers, the motor and branch circuit overload protection required in NEC® Article 430.31 must be considered. Manual controls are used when the operator is present and the connected load may be on or off for long periods. They are used when precise control or varying control is not needed.

OVERLOAD PROTECTION

Overloads are considered as excessive load on the motor while running. These are caused by equipment jams or too heavy a mechanical load on the motor. These overloads cause the motor to draw too much current and the motor will begin to overheat.

NEC® Article 430.32a(2) or 430.32b(2) allows the use of a thermal protector integral with the motor (part of motor assembly) to be used to protect the motor and branch circuits from overloads and failure to start. These thermal protectors are mounted in the motor to directly sense motor heat. Many of these sensors are designed to open the motor circuit in the motor when the motor has been overloaded. However, if the actual circuit-opening device (switch) is separate from the motor, the thermal device shall open a control circuit and interrupt current to the motor. (See Figure 3-2.)

Manual **controllers** also can provide running overcurrent (overload) protection. Some manual starters, as in Figure 3-3, use a thermal actuating element that is sized specifically to the motor nameplate full-load current. The thermal element heats as the motor load current passes through it. If an overload of current persists, it physically moves, thereby releasing a switch and breaking the circuit to the motor. (See section on "Overload Selection.") Usually, after an overload occurs, the switch is left in a tripped position. Wait a few minutes to allow the thermal element to cool, then push the lever to off (or reset), and then turn on again.

Caution: If you try to reset too soon on some starters, you may cause the eutectic alloy or solder pot to slip and will not be able to reset the switch. An eutectic alloy is a combination

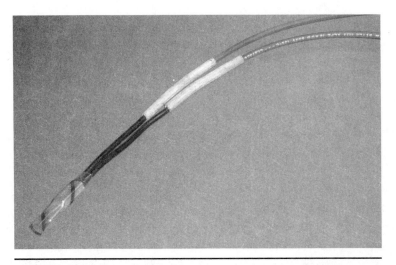

FIGURE 3-2 Thermal protector used to sense when motor windings become too hot.

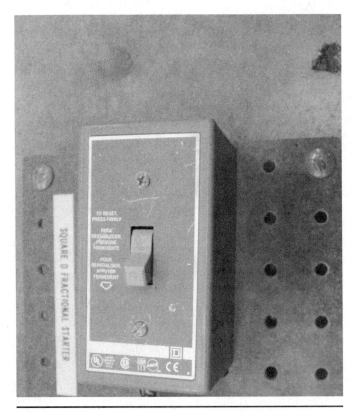

FIGURE 3-3 Single-phase manual motor starter.

Basic Motor Control

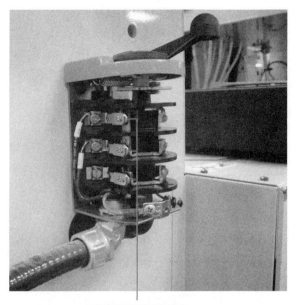
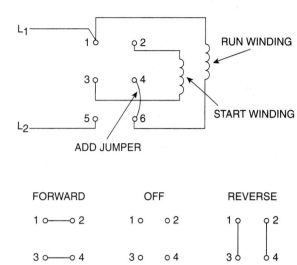

FIGURE 3-4 Manual drum-switch controller with associated control diagram.

of metals that has a low melting point, such as solder. The alloy changes from a solid to a molten metal at a specific temperature. The thermal sensing element will melt the alloy and allow the overload relay to open the power circuit to the motor. The alloy will reharden after the heat is removed. If the alloy cools but does not hold the overload contact closed, you will have to remove the alloy and reheat it to reset the latching mechanism.

Manual starters can look like toggle-switch-style controllers. (See Figure 3-3.) Manual controllers can also be a drum-switch style as shown in Figure 3-4. Drum controllers make and break contacts by sliding a set of movable contacts to connect sets of stationary contacts. The stationary contacts are points connected to the motor or to the source of power. As indicated in the Figure 3-4 connection diagram, different sets of contacts are connected in the forward and the reverse direction. The connection diagram in Figure 3-4 shows how to use the drum controller to reverse a single-phase motor.

Manual controllers can look like a magnetic starter even though the same principle (mechanical energy is used to open or close the contacts under manual operations) is used. One style of manual controller uses a magnetic coil to hold the contacts in a closed position. (See Figure 3-5.) The magnetic coil is used only after the contacts are manually closed. The coil holds the contacts shut as long as normal line voltage is present at the controller. This particular style of controller will provide low-voltage release even though it is a manual controller. Low-voltage release means that if the voltage of the line goes too low, the starter will release the contacts and prevent damage to the motor caused by a low-voltage condition. Most manual starters will not provide low-voltage release.

Manual controls have many applications. They are used where an operator is present to start and stop the motor. They are usually employed where motors are to be left running for long periods and where precise control is not

ELECTRIC MOTORS AND MOTOR CONTROLS

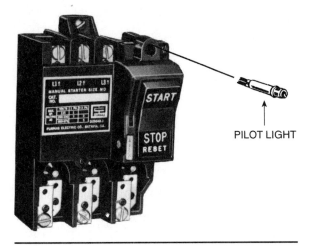

FIGURE 3-5 Manual motor starter with low-voltage protection.

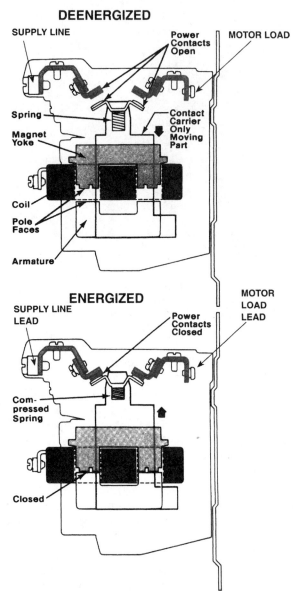

FIGURE 3-6 Power circuit contacts in a magnetic starter shown deenergized (open) and energized (closed). *(Courtesy of Rockwell Automation)*

required. Manual controls are cheaper than other styles of controllers.

Automatic Control

As the name of this style of control implies, the control is more automatic or not manual. To accomplish this, the use of a magnetic contactor is used. A magnetic contactor not only contacts or connects the power to the motor through the power circuit (Figure 3-6), but also has a second electrical circuit that provides the current to a magnetic coil for control (Figure 3-7). The electromagnet is energized by closing a control circuit and the magnet circuit now draws the contactor's armature to a closed position. This armature operates a set of movable contacts that closes the power circuit, delivering power to the motor. The word **contactor** is used at this point rather than "starter" because a motor starter is made up of a contactor and overload sensing system that will open the contactor on a motor overload (OL).

There are several styles of magnetic contactors. All provide the same basic function, which is to close contacts, but they differ in operational mechanisms. Controller design varies according to the manufacturer. However, the

Basic Motor Control

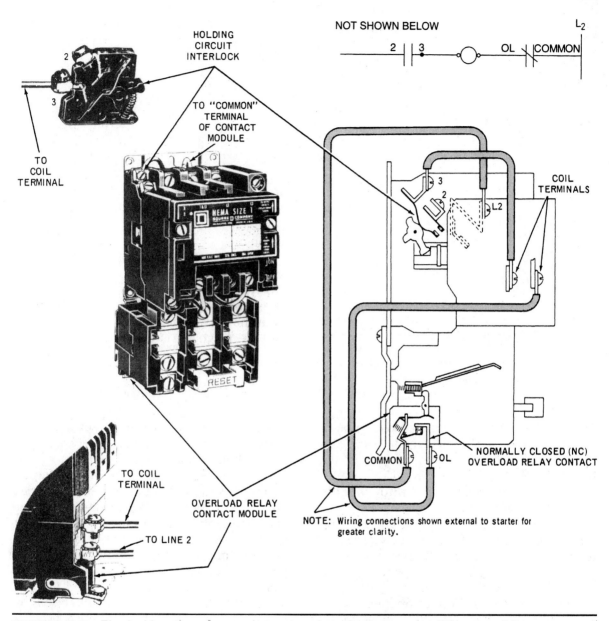

FIGURE 3-7 Physical location of control circuit and overload contacts. *(Courtesy of Square D/Schneider Electric)*

proper controller for the job is based on the application and requirements of the NEC®. Some contactors must be oriented or mounted in the correct position in order to provide proper operation. (See Chapter 4 for more on contactor types.)

TWO-WIRE AUTOMATIC CONTROL

Contactors and starters utilize control circuits to energize the magnetic coil to close and hold the power contacts closed, or to deenergize the

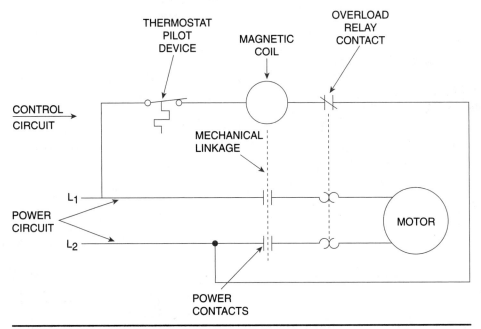

FIGURE 3-8 Two-wire control using a thermostatic switch and a magnetic starter. This type of control has no- or low-voltage release but *not* no or low voltage protection.

magnetic coil and release the power contacts. One method of automatic control is called a **two-wire control**, so named because of the number of wires needed from the starter enclosure to the operating control or pilot device. (See Figure 3-8.)

Two-wire control is an on/off-type control. A **pilot device** closes a set of contacts (as shown in Figure 3-8) to complete the electrical circuit to the magnetic starter. See Appendix A for a chart of pilot devices and symbols.

The pilot device is the component of the electrical unit that initiates the change to the controller circuit. For example, as the temperature falls, the thermostat (pilot device) contacts close; the magnetic coil is energized and allows the motor to start blowing heat into a room. As the temperature rises, the thermostatic switch opens the control contacts, breaks the circuit to the electromagnet, and allows the power contacts to open, removing power from the motor.

Any number of devices can be used as an actuating element or pilot device. (See Chapter 4.) See symbols in Appendixes A and C.

This type of control is automatic because no one has to be there to operate the starter. This type of control also provides low-voltage release to protect the motor. Again, if the voltage at the starter (therefore at the motor) is too low, the magnetic circuit will not hold the power contacts shut and the motor will be protected. If the voltage is restored to adequate levels, and the pilot device is still closed, the motor will restart. The motor also has overload protection provided by the starter.

Caution: This two-wire control example does not provide **"low"** or **"no" voltage protection.** Definition: "Low-voltage protection" means that the contactor will release the motor from the power circuit when too low a value of voltage is provided; however, the contactor will not reenergize when the power is restored.

THREE-WIRE AUTOMATIC CONTROL

Three-wire control is used to provide motor control and where low-voltage protection is needed. A typical three-wire control (as shown in Figure 3-9) is so named because three wires are used to connect the pilot (start/stop) station to the controller.

As you can see, the use of an additional contact is necessary to provide another additional circuit path in the control circuit. The additional contact is on the motor contactor and is identified as the auxiliary contact. This

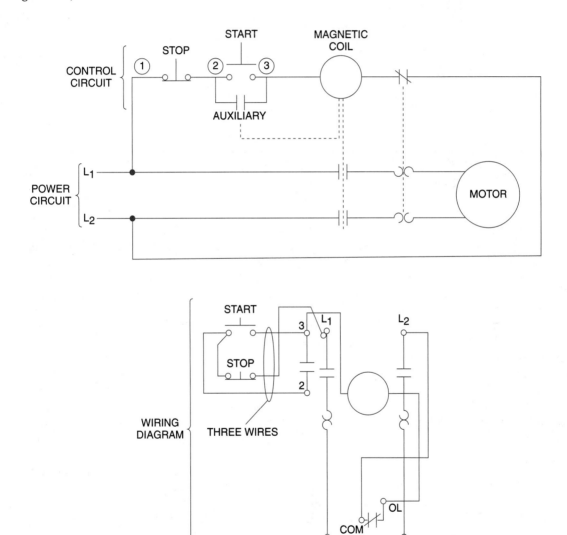

FIGURE 3-9 Schematic and wiring diagram for three-wire control using a momentary contact, start/stop switch. This type of control has both no- or low-voltage release and no- or low-voltage protection.

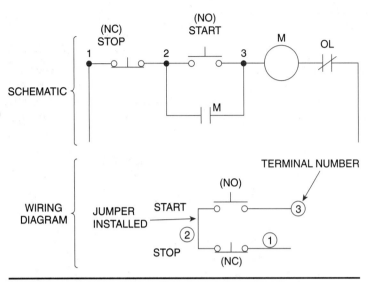

FIGURE 3-10 Momentary start/stop push-button switch shown in schematic diagram and again in wiring diagram.

type of control circuit uses a **momentary contact** pilot device (*not* maintained contact). This means that switches change position from open to closed or closed to open only when you apply pressure momentarily. Otherwise they remain as shown in Figure 3-10. A typical control is a start/stop station. The stop switch is a normally closed (NC) momentary contact and the start switch is a normally open (NO) momentary contact switch. The connection between the two is at point number "2." This connection is usually in place when the start/stop station is purchased.

Schematic Diagrams

Wiring diagrams and schematic diagrams differ in appearance. The schematic diagram illustrates how the electrical path is laid out in a logical sequence. We call the standard motor control schematic diagram a ladder diagram because the outside runners of the diagram provide the rails (voltage source) for the rungs of the ladder (the control sequence lines). (See Figure 3-11.)

When 120 V is applied to the **motor starter** (motor starter includes an overload mechanism) 120 volts is available at the stop station. (See Figure 3-12.) As long as the stop station contact is closed, a circuit is complete to the common point of the start station and the starter's auxiliary contacts. By momentarily closing the start station, a circuit is completed to the magnetic coil and through the OL contacts to energize the magnetic coil. Before the start button is released to open again, the magnetic contactor closes all the contacts including the auxiliary contacts. The auxiliary contacts close to provide a holding circuit around the start station contacts. There is a constant circuit for current to flow, through the stop station, auxiliary contacts, magnetic coil, and overload contacts, and the start push button can be released. The stop station always must be master. This means that if the stop station button is depressed, the magnetic coil must be released, no matter what other conditions are present in the electrical circuit. Standard practice is to always place the stop station or other master controls first in line as in Figure 3-12. This will break the

Basic Motor Control

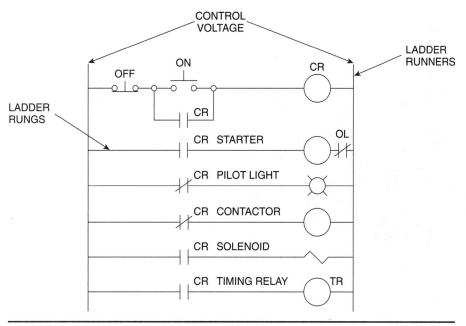

FIGURE 3-11 Ladder diagram shows rails, or runners, and ladder rungs in a control circuit. All CR contacts change condition when the CR relay is energized.

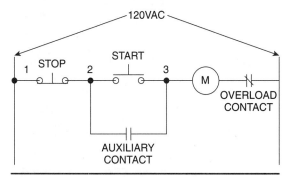

FIGURE 3-12 Stop PB is normally closed and is placed first in the electrical rung. Pressing the stop PB must have "master control" over the rest of the circuit.

circuit at the earliest point and prevent any other controls from operating the contactor, if you are trying to stop the motor.

Caution: Some old contactor and control designs did not appear this way. When providing repair or maintenance, be careful!

Caution: Never push the contactor closed to see if the motor will spin. If the control circuit does not operate, find out why! There may be a critical reason the controls are not functioning.

It is evident that there are separate circuits provided for control and power through the controller. The magnetic coil only uses several voltamps (milliamps at 120 V), but the power contacts of the starter may deliver thousands of voltamps of power to the motor. The actual contacts for power and auxiliary control also are different. The power contacts have large surface areas and possible arc control features. (See Figure 3-13.) The control contacts are typically small button contacts with no arc control. Do not interchange them and try to use the control contacts to deliver motor power.

Three-wire control is also used for multiple start/stop locations. When you need to control motors from more than one location, a modification to the standard start/stop

FIGURE 3-13 Control contacts have relatively small contact surface area compared to area for power contacts.

station is needed. Remove the jumper between the start and stop (control point No. 2) from all the stations except the one farthest from the starter.

Remember stop stations must be master. To accomplish this, place all stop stations in series so that if any of the stations is momentarily depressed, the circuit is broken and will not restart until one of the start buttons is pressed. By placing the push button with the jumper at the most distant location, wire is saved by running three wires to the most distant location rather than four wires (as is required if the jumper were in the other stations closer to the motor controller). (See Figure 3-14.)

Motor control circuits are covered in Part F of Article 430 of the 1993 NEC®. A definition of

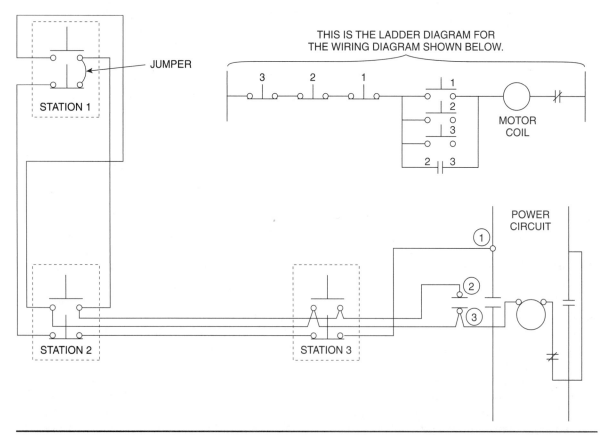

FIGURE 3-14 Jumper is removed from all multiple start/stop stations except the most distant push button.

a motor control circuit is contained in Article 430.71. Control circuit conductors must be protected according to Table 430.72b. Several conditions exist that determine when to use column A, B, or C. Remember, the control circuit may be using much smaller wire than the power circuit and must be protected using the rules of Article 430.72. Two- and three-wire controls are used extensively when dealing with magnetic starters. Manual controllers or snap switches are most used in residential situations.

JOG AND REVERSING CONTROLS

Jog stations or **"inch" stations** are used to provide momentary movement of the motor rotation. Jog stations provide a connection for the magnetic starter's energizing circuit but prevent the holding circuit from being completed. The jog push button is made up of a two-part switch. As stop stations always are placed in series to maintain the master electrical function, start stations always are placed in parallel in a multiple station control. This allows any start station to complete the energizing circuit to the motor's magnetic starter. The holding contact on the starter maintains the circuit.

Figure 3-15 shows how a jog station would be wired into a three-wire control circuit. The jog station's bottom part NO contact acts like a start switch that closes to apply power to the magnetic coil. However, the top part of the switch opens at the same time and prevents a holding circuit from being established through the auxiliary contacts. Jog stations normally are placed at only one location, within sight of the motor. Remember, stop stations must remain master and starts or jogs must be prevented if the stop station is depressed.

Reversing single-phase motors becomes more complicated because the power connections to the motor must be changed. The basic operation of this motor must be understood to determine control schemes (see Chapter 1).

The typical reversal for a single-phase motor is to reconnect the motor windings so that both forward and reverse connections provide the same connection to the main or run windings but reverse the connections to the start windings. Reversing controls must

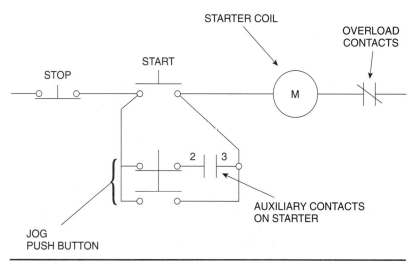

FIGURE 3-15 Schematic diagram of start/stop station with a "jog" push button.

disconnect live power between the forward and reverse modes of operation.

Reversing power to the motor leads does not have to reverse the start windings and leave the run windings the same. In some motors, it may be more convenient to reverse the running winding and leave the connection to the starting winding the same. A manual contactor with maintained contacts as in Figure 3-16 can be used as a reversing controller.

Another common type of manual controller used with motors is the drum controller. In order to connect a drum controller to the motor, you must be able to read switch diagrams such as the one presented in Figure 3-4.

Remember that the controller must keep the same line wires connected to the run winding in forward and reverse and change the orientation or instantaneous current flow to the

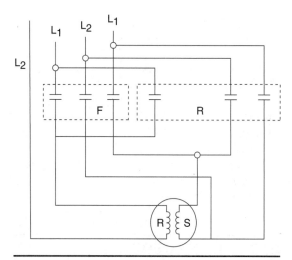

FIGURE 3-16 Diagram for a manual reversing starter for a single-phase motor.

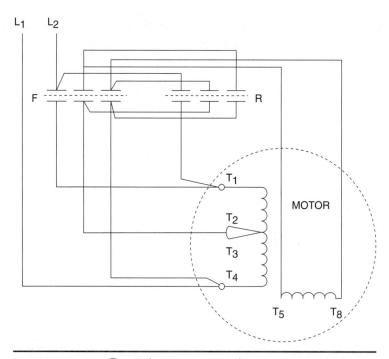

FIGURE 3-17 Reversing starter used to reverse a single-phase motor with high-voltage run windings, but a single low-voltage starting coil.

start windings. Many switch patterns are available and you must determine the proper connection for the application you encounter. It is important to remember that the controller need not open all conductors to the motor as stated in NEC® Article 430.84 unless the controller is also serving as the disconnection means.

More complex switching is needed where the controller is used to start a single-phase motor, connected for high voltage but using a low voltage start winding. (See Figure 3-17.)

In Figure 3-17, the running winding coils are connected in series to provide for the higher line voltage. If there is only one starting winding, it is to be connected across one-half of the running winding. This will provide the proper voltage to the starting winding. If 230 volts is applied to the running winding, there will be 115 volts available to the starting winding. The controller must reverse the current flow to the starting winding but maintain the proper level of voltage.

OVERLOAD SELECTION

As you know, motors need to be protected from overheating, which could cause a fire or destroy the motor. Overloads occur when the motor is forced to do too much mechanical work, or the load is too heavy and the motor fails to start. In either case, the motor shaft is not turning at rated speed. This causes the CEMF provided by the rotating armature to be too low, thereby allowing too much current to flow in the motor windings. As too much current flows, the motor will heat. See NEC® Article 430.31 for general information on overloads.

(*Note:* Overload protection is not required for fire pump motors. Allow them to run as long as possible even if they are overloaded by pumping water to firefighting equipment.)

Overload protection comes in various styles. The main principle for all the overload protective devices is the ability to monitor the current that is drawn by the motor. As mentioned earlier, the slower the motor turns (compared to the design speed), the more current it will draw from the line. Therefore, the various devices used to protect the motor from overheating are calibrated according to motor current.

One common type of overload sensor uses the principle of a heater to heat a melting alloy (eutectic alloy) designed to become fluid at a certain temperature. Often this type of overload protection (OL) is called a solder pot because the alloy resembles solder, which becomes fluid when heated and rehardens when cooled. One manufacturer uses a dual cylinder shaft with the inner shaft attached to a ratchet wheel. (See Figure 3-18.)

When you size the OL protection to protect a specific motor, you are actually selecting the heater that covers the outer shaft. The heater is designed to produce a specific heat with a specific amount of current flowing through it (electric heat).

As the heater gets hot because of too much motor current, it melts the alloy and allows the inner shaft and ratchet wheel to turn. This allows a set of contacts to open, which opens the control circuit interrupting the power circuit to the motor. When the current to the motor is interrupted, the motor and the solder pot begin to cool. As the solder rehardens, the inner shaft is again held in place and the contacts can be pushed closed again. A manual reset is required where you physically have to push the contacts back to the closed position. Remember, resetting this type of overload too soon may cause the shaft to spin and not reset the solder pot.

Another common type of OL also uses an electrically heated element to sense motor current. However, rather than using a melting alloy, it employs the principle of bimetal expansion. It is called a bimetal strip overload relay. The heater is placed in the overload part of the starter. The heat directly affects a strip of metal that is actually two dissimilar metals adhered together. The two metals have different

ONE-PIECE THERMAL UNIT

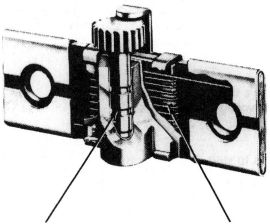

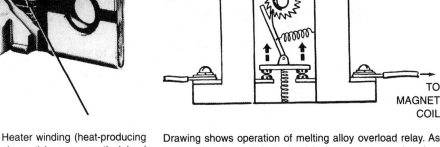

THERMAL RELAY UNIT
TO MOTOR
TO MAGNET COIL

Solder pot (heat-sensitive element) is an integral part of the thermal unit. It provides accurate response to overload current, yet prevents nuisance tripping.

Heater winding (heat-producing element) is permanently joined to the solder pot, so proper heat transfer is always insured. No chance of misalignment in the field.

Drawing shows operation of melting alloy overload relay. As heat melts alloy, ratchet wheel is free to turn—spring then pushes contacts open.

FIGURE 3-18 Diagram of one style of "solder pot" melting alloy overload sensing mechanism. *(Courtesy of Square D/Schneider Electric)*

coefficients of expansion when heat is applied. That is, they expand at different rates when heated. This causes the strip to bend as one metal begins to expand faster than the other (similar to a home thermostat). As the metal bends, it reaches a point where it will open a set of contacts to open the control circuit to the motor. (See Figure 3-19.)

This particular style of OL has the advantage of having automatic reset available. As the bimetal strip cools, it could allow the contacts to remake and allow the motor to restart. Why would this be an advantage? In heating applications or refrigeration applications, it may be an advantage to allow the device to automatically reset after the power to the motor has been interrupted and the motor has been allowed to cool. The motor is still protected, but the system is allowed to operate and keep the environment controlled, although not in normal operation. These bimetal strips can also be set for manual reset mode, so that you will know that there are problems with the motor that needs attention.

Basic Motor Control

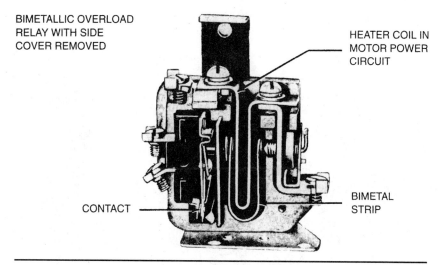

FIGURE 3-19 One style of "bimetal strip" type of overload. As the motor current heats the bimetal strip, it bends, allowing the overload contact to open. *(Courtesy of Square D/Schneider Electric)*

Another style of thermal protection looks like a disk. The disk is warmed by a circular heater. The disk is a bimetal-style device that changes its shape from concave to convex when heated the proper amount. The change causes a set of contacts to open, as do the other styles of thermal elements. (See Figure 3-20.)

A newer style of OL is the electronic overload. The same basic principle is used to protect the motor from too much current. An electronic current sensor is placed in the power circuit to the motor. As the current increases above acceptable levels, the electronic relay will open the control circuit to the motor controller.

As you will note, all of these overload monitoring devices are based on current to the motor flowing through the OL sensor. Typically, this is shown by a heater symbol as in Figure 3-21.

The associated contact that this heater causes to open is usually an NC contact shown alongside the heater symbol. The actual style of the overload protection is not usually shown on the electrical diagram, but you need

to know which style you are using to order replacements or make adjustments. The contacts are in the control circuit of the motor starter. The OL contacts do not carry the motor current and care must be taken when making connections at the starter to be sure motor current does not pass through the OL contacts.

To size the OL device, refer to NEC® Article 430, Part C. For example, Article 430.32(a) specifies that continuous duty motors, more than 1 HP, shall have devices (OL contacts) selected or rated to trip at specific percentages of motor nameplate full-load current. It is important to note that the code is specific when it states motor nameplate ratings. Use the actual motor information, not a generic code book value. (See NEC® Article 430.6(a).)

Motors that have a service factor of 1.15 or more must have protection rated to open motor power when 125 percent of nameplate current is sensed going to the motor. The same is true for motors that have a marked temperature rise 40 °C or less. When using the heater selection chart that accompanies most starters, simply use the motor nameplate current for

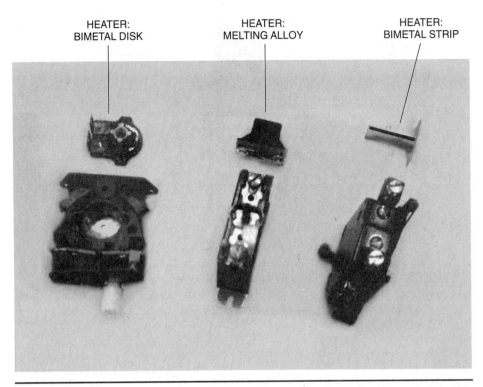

FIGURE 3-20 Thermal overload relays. Shown are the older bimetal disk, the melting alloy, and the bimetal strip.

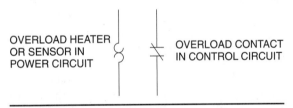

FIGURE 3-21 Heater symbol and associated overload contacts. These parts of the overload are connected to different circuits.

the proper voltage and select the heater from the table. (See Figure 3-22.)

Manufacturers already calculate the heater to trip at 125 percent of the chart value. If your motor does not fit these conditions, use 115 percent as the trip value. This is accomplished by dropping one size on the heater selection chart. When using electronic overloads and some adjustable range thermal overloads, the overload sensor is purchased to create protection within a certain range. The range adjustment is located on the overload relay section and is adjusted after installation.

Thermal protectors integral with the motor also may be used (see Chapter 2 for description). These thermal protectors are sized according to Article 430.32(a-2). Make sure you use the proper conditions for selecting the OL device according to Part C of Article 430 of NEC®.

Note that Article 430.34 allows you to increase the size of the heater by one size to 140 percent, or 130 percent if the OL is not sufficient to start the motor or carry the load. Be certain there is not another problem with the motor. Measure the motor current and the motor voltage. Are they within nameplate values?

TABLE 110

Heater Type Number	Full-Load Amperes	
	Size 0	Size 1
W10	0.18	0.18
W11	0.20	0.20
W12	0.22	0.22
W13	0.24	0.24
W14	0.27	0.27
W15	0.30	0.30
W16	0.33	0.33
W17	0.36	0.36
W18	0.40	0.40
W19	0.44	0.44
W20	0.48	0.48
W21	0.53	0.53
W22	0.59	0.59
W23	0.65	0.65
W24	0.71	0.71
W25	0.78	0.78
W26	0.86	0.86
W27	0.95	0.95
W28	1.05	1.05
W29	1.16	1.16
W30	1.27	1.27
W31	1.41	1.41
W32	1.55	1.55
W33	1.71	1.71
W34	1.89	1.89
W35	2.08	2.08
W36	2.30	2.30
W37	2.53	2.53
W38	2.79	2.79
W39	3.07	3.07
W40	3.38	3.38
W41	3.73	3.73
W42	4.11	4.11
W43	4.51	4.51
W44	4.96	4.96
W45	5.44	5.44
W46	5.98	5.98
W47	6.57	6.57
W48	7.21	7.21
W49	7.92	7.92
W50	8.70	8.70
W51	9.57	9.57
W52	10.5	10.5
W53	11.6	11.6
W54	12.7	12.7
W55	14.0	14.0
W56	15.4	15.4
W57	16.8	16.8
W58	18.3	18.3
W59	....	19.9
W60	....	21.7
W61	....	23.6
W62	....	25.7
W63	....	28.0

FIGURE 3-22 Sample heater selection chart to choose overload heaters for overload protection. *(Courtesy of Rockwell Automation)*

Fuses can be used to protect the motor from overloads. (See "Fuse Protection" section.) NEC® Article 430.37 specifies the minimum number of overload units that are required for motor installations. NEC® Article 430.38 requires that the OL device causes a sufficient number of ungrounded conductors to open and interrupt the current to the motor.

In the case of OL relays, where current causes a mechanical device to open and prevent further current flow to the motor, there is a time lag between heat sensing and OL trip. This time lag is desirable. It allows the initial starting inrush current to the motor to flow without tripping the overload device. Without this delay, every time the motor attempts to start the OL would trip. However, the delay is too long to protect the motor against short circuits and ground faults. Another type of protection is used for short-circuit protection, such as fuses.

FUSE PROTECTION

In motor installations, fuses are generally used to protect the motor branch circuit conductors, the motor controller, and the motor against high currents due to short circuits or ground faults (NEC® Article 430, Part D).

Fuses have many different ratings and applications. All 20-amp fuses do not provide the same protection. The first rating to consider is the ampere rating. Fuses are constructed so that when normal current is passed through them, no excessive heating occurs in the conducting element. (See Figure 3-23.) As current begins to rise, the fusible (melting metal) element will melt at some predetermined temperature. Sometimes the element is surrounded by a sand filler that is both a heat absorber and an arc quencher. Small current overloads that heat with short duration can be absorbed by the filler and prevent the element from burning. Short-circuit conditions that have very high currents will cause the meltable element to burn open, and the arc resulting from breaking the circuit is quenched by the filler.

Fuses have an inverse time to heat ratio. This means the higher the heat produced, the shorter the time to burn open. This is important when determining all the different ratings.

Fuses not only have ampere ratings that indicate the normal melting point of the element but also have voltage ratings. The second rating to consider is the voltage rating. Usual low-voltage fuse voltages ratings are 250 V for circuits up to 250 V between conductors and 600-V ratings for circuits 600 V or less between conductors. The lower-voltage fuses are less expensive and fit less-expensive fuse holders, so these are used where practical. The voltage rating affects the physical size of the fuse. Higher voltage ratings have to clear or suppress larger arcs, so are generally longer than the same amperage 250-V fuses.

Interrupting capacity is the third major requirement of which to be aware. This rating is the highest current at rated voltage that can be safely interrupted. Using a fuse with too low an interrupting rating may not protect downstream equipment from severe damage if a fault or short circuit occurs. It would be a serious mistake to replace a fuse rated at 50,000-amp interruption rating with one rated 10,000-amp interruption rating. A short-circuit calculation is necessary to determine the proper short-circuit interrupting rating. Too low a rating may not clear a full short circuit.

Fuses are built with different dimensions and different styles to help prevent misapplications. A major design characteristic between fuse dimensions is the contact point between the fuses and the fuse holder. Cartridge-type fuses as shown in Figure 3-24 show that the ends of the fuses are designed differently. A ferrule end looks like a cap on the fuse. Ferrule end fuses are used for 60 A or less ampere rating. Fuses that are 250 V with 1/10–30-A current ratings are two inches long and the ferrule is 9/16-inch diameter. Fuses rated 30 to 60 amps are 3 inches long and the ferrule

Dual-element fuse for short-circuit and overload protection

Figure shows two functional elements: (1) Short-circuit protective elements for high-current, short-time protection; (2) overload element for lower current, longer time protection.

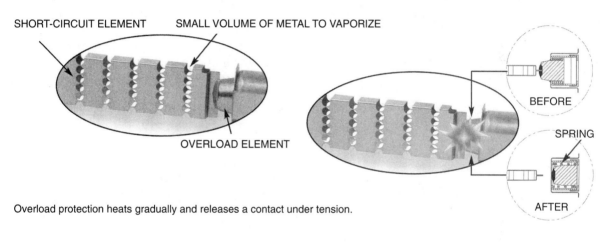

Overload protection heats gradually and releases a contact under tension.

Short-circuit protection has many small current paths that vaporize to create many small arcs which are quenched by the filler.

FIGURE 3-23 Cutaway view of fuse shows short-circuit and overload protection. *(Courtesy of Cooper-Bussmann)*

changes to 13/16-inch diameter. Compare these ferrule ends to 600-V rated fuses: 1/10 to 30 A are 5 inches long and 13/16-inch ferrule and the next category 31 to 60 amps is 5½ inches long with 1-1/16-inch ferrule.

Therefore, the fuses are not interchangeable because the fuse holders also are different for the various fuses. Fuses over 60 amp up to 600 A have knifeblade ends. (See Figure 3-24.) The sizes of the knifeblades and the dimensions

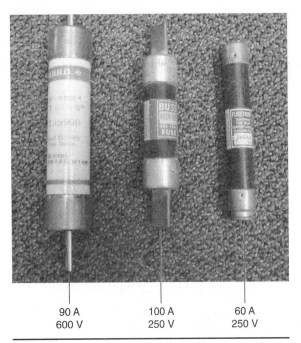

FIGURE 3-24 Fuses with knifeblade ends on left and ferrule ends on right.

also change in increments. NEC® Article 240.60 describes requirements for fuses and fuse holders of the cartridge type.

Fuses also are distinguished by class. Time-delay fuses are designed to carry a specific overload current for a minimum time without opening. However, it will open short-circuit currents quickly. These fuses (dual element—time delay) are referred to as acceptable protection for motors as required in NEC® Article 430, Part D. Article 430.52(a) refers to Table 430.152. Time-delay fuses are used in the second column to derive correct protection percentages. Notice the non-time-delay fuses, or single element, have much higher percentage ratings. This is because the fuses must carry the starting current of the motor. Without a high rating, the non-time-delay fuse will burn open each time you start the motor. The time-delay fuse can survive this large inrush current but still protect the motor from short circuits. By using the time-delay fuse column, often a smaller ampere rat-

ing fuse can be used. This will also allow you to provide a smaller size fuse block in the disconnect and therefore keep costs down.

Fuse class is the last variable in fuse selection. The class of the fuse refers to the operating characteristics under load. Class CC, T, and J fuses have the lowest melting-point current and the lowest peak let-through current. These fuses would provide maximum protection, especially for IEC starters. Class K (standard fuse) and RK1 fuses with a rejection slot in the contact block often are used for NEMA-rated starters.

When selecting fuse protection for your motor installation, be aware of the following factors: ampere rating (use appropriate column from NEC® Table 430.52); voltage rating (rating between conductors); interrupting rating (based on short-circuit conditions); class of fuse (application of protection); size of fuse and associated fuse holders; and cost of fuses and fuse holders.

NEC® AND SINGLE-PHASE MOTOR APPLICATIONS

Single-phase motors are used extensively in residential and commercial operations. Motors need basic protection to prevent overheating and fire. At the beginning of Article 430 in the NEC® is a chart (Figure 430.1) that lists the basic requirements and how they relate to each other. When connected to a branch circuit, the motor will first need a disconnecting means. The requirements are found in Part IX of Article 430. In general, you must provide a way to disconnect the motor and its controls from the source of power. If the motor uses a separate controller, the disconnect must be within sight of the controller.

Note: "Within sight" means you must be able to see it *and* it must be within 50 feet (see NEC® definition: In Sight From).

Article 430 PART IX determines the type of disconnect. It should be a horsepower-rated switch, a circuit breaker, or a molded case switch, and shall be a listed device. NEC® Article 430.109 specifies requirements. Often small stationary motors can meet requirements by using exception 1 or 2. Many residential applications make use of Article 430.111.

Part IV of Article 430 relates to motor short-circuit protection. Generally, Table 430.52 is used to select protection of motors. Many times, small single-phase motors are placed on a circuit with other loads. If this is the case, Article 430.53 can be used to allow the motors to be placed on a 15-A or 20-A branch circuit if they are under 1 HP, the controller ratings are not exceeded, and the motor has individual overload protection.

Motor circuit conductors for single motors are specified in Article 430.22 in 430, Part II. For motors and other loads on the same branch circuit, Article 430.24 is used. Add all the full-load current ratings (values found in Article 430.248 for single-phase motors) plus 25 percent of the highest rated motor current; then add the other load currents. Use the wire charts in Article 310 to determine the correct wire to carry the needed current (ampacity).

A controller used to start and stop the motor is defined in Article 430, Part VII. The controller shall have a horsepower rating not lower than the motor horsepower. Article 430.83(c) allows us to use snap switches as a controller if all the provisions are met. Make sure the voltage rating is adequate for the application of the controller. The controller does not need to open all conductors to the motor, but if you do open the grounded conductor to the motor you also must open all ungrounded conductors. Article 430.91 provides a chart to guide the electrician in choosing the proper NEMA enclosure type for other than hazardous locations.

Overload protection is covered in Part III of Article 430. Generally, the specific motor must be protected based on its nameplate data. Consider the following when making the proper selection: horsepower rating of the motor, if it has a marked service factor or temperature rise, or if it has integral protection. All of these factors dictate the amount and style of overload protection required. If a fuse is used for short-circuit and overload protection, the fuse must be installed according to Article 430.36.

Note: Often dual-element fuses can be used for both protections. Use Table 430.37 to determine number and location of overload units required.

Article 430.42 determines the requirements for motors on general-purpose circuits usually in single-phase, residential applications. The motors themselves are installed as required in Article 430, Section I.

Of particular importance to all motor installations is Article 430.6. Basically, this directs us to the wire ampacity table, but, most important, Article 430.6A tells us to use Table 430.247 to 430.250 including notes to size all *but* the motor overload protection. Separate overload protection should be based on the actual motor nameplate. *This particular part of Article 430 allows an electrician to size all parts of the job based on standard or nominal values, except the specific motor protection. This practice allows replacement of the motor with another brand motor without having to resize and reinstall the entire installation. Only the motor and the motor overcurrent protection would need to be changed.*

Another important point to consider is the actual motor location. Article 430.14 refers to ventilation and maintenance. Remember to provide adequate ventilation to prevent overheating. Also try to plan enough room to provide motor inspection and maintenance. Proper cleaning of the ventilation slots and proper maintenance of the bearings will extend the life of the motor and provide more trouble-free operation of the controls.

SUMMARY

This chapter explained the basics of motor control, protection, and single-phase motor installation. You should know the difference between manual and magnetic starters. You should know where each of these devices might be applied. Methods of identifying and sizing the overload protection for motors and properly protecting motors during running and starting were presented, including fuses.

This chapter was designed to acquaint you with the NEC® requirements in order to safely install and maintain a motor. Two- and three-wire control was presented to enable you to begin diagnosing problems in a basic control circuit.

QUESTIONS

1. Explain what is meant by "motor controller."

2. To what does "manual controller" refer?

3. Does two-wire control provide low-voltage release or low-voltage protection? Explain.

4. Describe how a momentary contact switch is different from a maintained contact switch.

5. Describe the current path in a "holding" circuit found in a three-wire circuit.

6. Describe how a motor OL circuit operates. Include elements in the power circuit, and elements in the control circuit.

7. Explain what is meant by making the stop station master.

8. Which type of overload sensor is referred to as the "solder pot"? Why?

9. A motor has a nameplate current of 7.4 amps at 230-V single phase. It has a service factor of 1.15 and 40-degree rise. What is the actual current at which the overload should be designed to trip?

10. Use the NEC® to determine how many overload trip units are required for a 230-V, single-phase motor.

11. At which current ratings do fuses change from ferrule to knifeblade ends?

12. How does a "jog station" prevent the holding circuit from closing?

13. What does "in sight from" mean when locating a disconnect?

14. Give the general rule for sizing a controller to a motor. Where is it specified in the NEC®?

15. When sizing conductors for a motor installation, what motor current values do you use: code book or nameplate?

CHAPTER 4

CONTROLLERS, RELAYS, AND TIMERS

OBJECTIVES

After completing this chapter and the chapter questions, you should be able to

- Select a proper NEMA starter for a motor application
- Explain how a motor starter functions and identify the components
- Identify and use IEC-rated contactors
- Identify standard pilot devices and symbols
- Determine the proper photoelectric control to use
- Identify relay types and begin developing relay control schematics
- Inspect and determine need for contact repair and replacement

KEY TERMS IN THIS CHAPTER

AC Chatter: A chattering noise that can occur with AC magnetic devices when the magnetic effect is lost as the sine wave passes through zero. This effect usually is prevented through the use of a shading ring.

Dropout Voltage: The value of the applied control voltage that will no longer supply enough magnetic current to keep the armature seated. This value is approximately 40 percent of the rated coil voltage.

IEC: International Electrotechnical Commission is an international body that sets performance ratings for its members' products.

Inrush Current: The current that is present as a magnetic coil is first energized. This current is typically six to ten times normal current.

Ladder Diagrams: A method of showing the electrical relationships of multiple controls and relays in an organized fashion. The schematic diagram resembles a ladder with rungs.

NEMA: National Electrical Manufacturers Association is a body that sets performance standards for its members' products.

Pickup Voltage: The minimum control voltage needed to cause the armature on an electromagnetic relay or starter to move. This voltage is typically 80 percent to 85 percent of the rated coil voltage.

Proximity Sensors: Sensors that are designed to determine the presence of a material. These are usually thought of as capacitive- or inductive-type sensors.

Relays: A device that is used to relate one electrical control signal to another electrical circuit or to multiple circuits.

Seal-In Voltage: The minimum control voltage required to cause the magnetic pole faces to seal against the moving armature without chatter or poor contact. This voltage is approximately 75 percent of coil-rated voltage.

Shading Ring: A conducting ring that is inserted in the magnetic face of an AC electromagnet. It is used to maintain magnetism while the AC waveform passes through zero.

INTRODUCTION

Electric motor control may have many different requirements. A simple on–off switch or a complex control scheme may be needed to provide the desired control of the electric motor. Motors are controlled by supplying voltage to the motor terminals at a desired frequency. By controlling the voltage on and off or by adjusting the frequency to an AC motor, the motor operation can conform to almost any control requirement. This is what makes the electric motor so versatile and convenient to use.

For years the standard controller or starter used in the United States has been manufactured under guidelines created by the National Electrical Manufacturers Association (**NEMA**). These standards provide a means for electricians to pick any manufacturer's product for a specific application with the assurance that it will perform satisfactorily. NEMA manufacturers provide a chart that can be used to select the desired NEMA-size starter (from 00 to 9). One must take into account the duty of the motor: jogging or plugging duty; the voltage rating of the circuit; the horsepower of the motor; and whether the motor is single-phase or polyphase. (See Figure 4-1.)

Electromechanical contactors are made by many manufacturers and in many styles. The most common types of contactors produced in the United States are by the NEMA group of manufacturers. Motor controllers are also made to other manufacturing standards that are organized by the International Electrotechnical Commission (IEC). This group is a European-based manufacturing organization. Both NEMA and IEC manufacturers make electromechanical and solid-state contactors and starters.

NEMA CONTACTORS AND STARTERS

First are the NEMA electromechanical contactors. Magnetic contactors and motor starters work on the same principle. A movable iron armature is drawn to the stationary iron by a magnetic coil to complete the magnetic path. The movable iron is attached to a movable set

Controllers, Relays, and Timers

NEMA Size	Continuous Amp. Rating	600 VOLTS MAXIMUM				
		\multicolumn{3}{c}{Maximum Horsepower Rating [2] — Full-load current must not exceed the "Continuous Ampere Rating"}			Maximum Horsepower Rating For Plugging Service [1]	
		Volts	Single Phase	3 or 2 Phase	Single Phase	Three Phase
00	9	120	1/3	3/4	—	—
		208	—	1½	—	—
		240	1	1½	—	—
		480	—	2	—	—
		600	—	2	—	—
0	18	120	1	2	½	1
		208	—	3	—	1½
		240	2	3	1	1½
		480	—	5	—	2
		600	—	5	—	2
1	27	120	2	3	1	2
		208	—	7½	—	3
		240	3	7½	2	3
		480	—	10	—	5
		600	—	10	—	5
2	45	120	3	—	2	—
		208	—	15	—	10
		240	7½	15	5	10
		480	—	25	—	15
		600	—	25	—	15
3	90	120	—	—	—	—
		208	—	30	—	20
		240	—	30	—	20
		480	—	50	—	30
		600	—	50	—	30
4	135	120	—	—	—	—
		208	—	50	—	30
		240	—	50	—	30
		480	—	100	—	60
		600	—	100	—	60
5	270	120	—	—	—	—
		208	—	100	—	75
		240	—	100	—	75
		480	—	200	—	150
		600	—	200	—	150
6	540	208	—	200	—	150
		240	—	200	—	150
		480	—	400	—	300
		600	—	400	—	300
7	810	208	—	300	—	—
		240	—	300	—	—
		480	—	600	—	—
		600	—	600	—	—
8	1215	208	—	450	—	—
		240	—	450	—	—
		480	—	900	—	—
		600	—	900	—	—
9	2250	208	—	800	—	—
		240	—	800	—	—
		480	—	1600	—	—
		600	—	1600	—	—

1. An example is plug-stop or jogging (inching duty), which requires continuous operation with more than five openings per minute.
2. **Nonmotor Loads**—When contactors are required to switch nonmotor loads such as lighting circuits, ovens, transformer primaries, etc., use the Bulletin 702L contactor.

Interrupting Capacity—The double-break contacts will carry rated loads continuously, are designed for full-voltage starting, and are capable of interrupting six times the current corresponding to the horsepower rating.

Coil Performance—Continuous operation at 110% of the rated voltage; and pickup at 85% of rated voltage.

FIGURE 4-1A Electrical ratings for AC magnetic contactors and starters size 00–9. *(Courtesy of Rockwell Automation)*

IEC MOTOR STARTERS (60 Hz)

SIZE	MAX AMPS	MOTOR VOLTAGE	MAX. HORSEPOWER 1Ø	MAX. HORSEPOWER 3Ø
A	7	115 200 230 460 575	1/4 1/2	 1 1/2 1 1/2 3 5
B	10	115 200 230 460 575	1/2 1	 2 2 5 7 1/2
C	12	115 200 230 460 575	1/2 2	 3 3 7 1/2 10
D	18	115 200 230 460 575	1 3	 5 5 10 15
E	25	115 200 230 460 575	2 3	 5 7 1/2 15 20
F	32	115 200 230 460 575	2 5	 7 1/2 10 20 25
G	37	115 200 230 460 575	3 5	 7 1/2 10 25 30
H	44	115 200 230 460 575	3 7 1/2	 10 15 30 40
J	60	115 200 230 460 575	5 10	 15 20 40 40
K	73	115 200 230 460 575	5 10	 20 25 50 50
L	85	115 200 230 460 575	7 1/2 10	 25 30 60 75
M	105	115 200 230 460 575	10 10	 30 40 75 100
N	140	115 200 230 460 575	10 10	 40 50 100 125
P	170	115 200 230 460 575		 50 60 125 125
R	200	115 200 230 460 575		 60 75 150 150
S	300	115 200 230 460 575		 75 100 200 200
T	420	115 200 230 460 575		 125 125 250 250
U	520	115 200 230 460 575		 150 150 350 250
V	550	115 200 230 460 575		 150 200 400 400
W	700	115 200 230 460 575		 200 250 500 500
X	810	115 200 230 460 575		 250 300 600 600
Z	1215	115 200 230 460 575		 450 450 900 900

FIGURE 4-1B IEC motor starter sizes based on amperage, voltage, and horsepower.

Controllers, Relays, and Timers

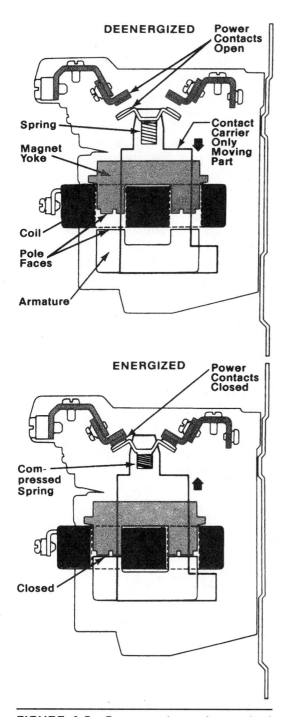

FIGURE 4-2 Contacts shown deenergized and again energized. *(Courtesy of Rockwell Automation)*

of contacts, which in turn meet a stationary set of contacts to complete the electrical circuit. (See Figure 4-2.)

The electromagnetic principle is used in all electromagnetic starters. (See Chapter 1 for a review of magnetic principles.) A coil of wire, usually encapsulated in epoxy, is mounted on a laminated steel core. The coils' ratings are designed to operate with a variety of voltages and possibly different frequencies. The energized coil provides the magnetic field that will draw the armature into position as the magnetic lines of force try to shorten themselves. The armature should seat against the face of the magnet assembly without obstruction. There should be very little AC hum when the armature and the air gap are closed correctly. AC hum is a low-pitch resonance that occurs in laminated magnetic material when AC is applied to the coil.

Because the standard coil operates on AC and the AC sine wave passes through 0 volts 120 times a second on 60 Hz, the electromagnet will lose magnetism 120 times a second. To prevent the armature from falling away from the magnetic core, an additional piece of equipment is installed. A **shading ring** or shading coil is installed at the ends of the magnet assembly. (See Figure 4-3.) As the magnet coil current fluctuates from zero to maximum and back (AC), it induces a voltage into the shading ring. The action is very similar to transformer action where the magnetic field is carried through the iron and induces a voltage into a secondary winding. In this case, the secondary winding is a shorted single conductor. The current produced in the shading ring is 90 degrees out of phase with the main coil current. The magnetic field produced by the shading rings holds the armature assembly in place during the main coil zero magnetism points. (See Figure 4-4.)

Another consideration in the design of the magnetic assembly is the requirement that the armature release from the main core. If main current is interrupted, the magnetic assembly

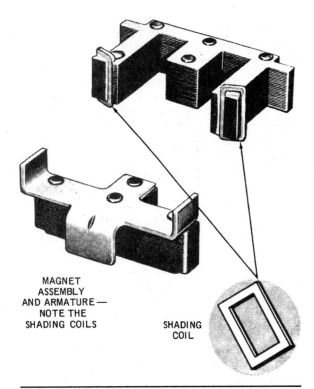

FIGURE 4-3 Shading coil or shading ring installed in the face of a magnetic assembly. *(Courtesy of Square D/Schneider Electric)*

must release the armature. To ensure that this takes place, the armature is usually pulling against spring pressure to close. As the magnetic field is discontinued, the spring pressure will pull or push the armature away. Also, the magnetic assembly has an air gap between the center leg of the "E" iron and the movable armature. This helps break the magnetic circuit so the iron will not stay magnetized. (See Figure 4-5.)

Starter Styles

Different styles of contactors are used by different manufacturers. (See Figure 4-6.)

Clapper-type magnet frames use a pivot-type control. As the armature is pulled in by magnetic action, the contacts close at the top. This style uses a small armature movement to provide larger contact movement. This reduces the amount of magnetic pull required but still provides large separation for the contact surfaces.

Vertical-action type contactors pull the contacts up into position. The added weight of the movable armature helps pull the contacts open on this design. Mounting requirements are critical to contactors of different design. Do not invert this one!

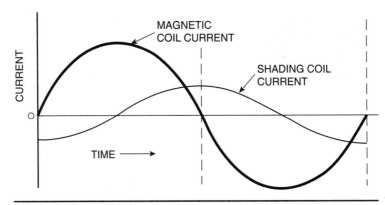

FIGURE 4-4 Waveform produced in the shading coil, out of phase with main coil current. *(Courtesy of Square D/Schneider Electric)*

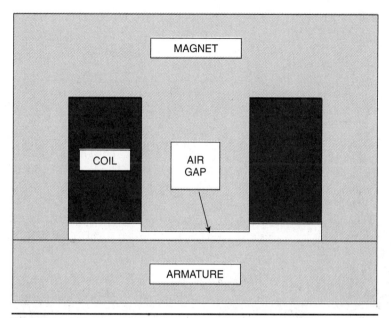

FIGURE 4-5 Magnetic assembly has built-in air gap to allow easier release when deenergized. *(Courtesy of Square D/Schneider Electric)*

Bell crank assemblies are used where the starter may be opened or closed frequently. The armature movement is vertical, but the contacts are closed through a pivot arm. This pivot arm absorbs the shock of the armature pickup and reduces contact bounce to increase contact life.

Horizontal action assemblies are most common. As the solenoid pulls into the closed position, the contacts also move to the closed position. These are the simplest form of contactor and are very easy to assemble and disassemble for repair.

There are many things to look for when maintaining or repairing a magnetic starter. A little information on voltage variation and magnetic requirements is helpful in determining the cause of failure.

When the magnetic coil is deenergized, the magnet assembly and armature have an air gap and the contacts are also open. The magnetic circuit has a low permeability at this point and the L (inductance) and the X_L (inductive reactance) is also low. As current is applied to the coil, a fairly high current is allowed to flow because of a comparatively low X_L. This relatively large current is referred to as the **inrush current**. The inrush current is about six to ten times the normal operating current. As the armature moves toward the magnet assembly's stationary iron, the L (inductance) and the X_L of the coil increase due to the permeability of the iron as it reduces the air gap. This causes the current to decrease in the electromagnetic coil. As the magnetic circuit is complete, the shading ring keeps the armature closed during the main coil zero-crossing current time. The current drawn from the line is now reduced to its normal operating current. As the coil heats from being energized, the coil current drops to approximately 80 percent of normal sealed-in current. Remember that inrush current is six to ten times larger than sealed-in current. Control circuits need to be able to supply and withstand the inrush current.

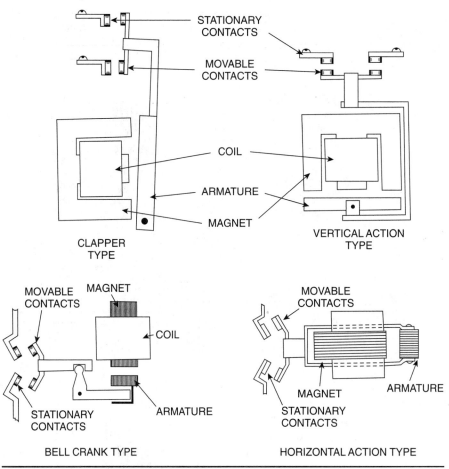

FIGURE 4-6 Four different styles of electromagnetic contactors. *(Courtesy of Square D/Schneider Electric)*

The effects of voltage on the coil are important. Because changing the applied voltage to a coil affects its current, some terms are needed to explain what happens and what is required to cause magnetic circuits to operate.

Pickup voltage is the minimum control voltage needed to cause the armature to move. It is approximately 80 percent to 85 percent of the full rating of the coil. Pickup voltage must be high enough to begin drawing the movable armature toward the stationary iron core.

Seal-in voltage is the minimum required voltage to cause the magnetic pole faces to seat against the armature. Vertical action assemblies require a higher seal-in voltage than pickup voltage to ensure solid contact pressure. Bell crank assemblies have a lower seal-in voltage than pickup voltage. The seal-in voltage is only slightly less than the pickup voltage, approximately 75 percent of normal rated voltage in bell crank contactors. This feature allows the contacts to stay closed under varying conditions that might temporarily reduce voltage to the starter. One such condition is when large motors start. The large starting current causes a line drop in voltage and reduces the voltage available to nearby equipment. This condition may cause contactors to drop out or deenergize.

Dropout voltage is the value of voltage applied to the coil at which the magnetic field is no longer strong enough to hold the armature closed. Depending on design, the dropout voltage is approximately 40 percent of coil-rated voltage. This parameter is used when referring to low voltage release.

Coil-rated voltages are stamped on the coil assemblies. The coils are ordered to meet the control voltages available and at the supply frequency. NEMA standards require the coil to operate satisfactorily from 110 percent of rated voltage and also provide pickup and seal-in at 85 percent of rated voltage. If the coil voltage is over 110 percent of normal, the coil will have too large a current flow. The coil will become hot and burn out. Also, the contactor may slam the armature into place and damage the magnetic faces and cause damage to the contacts.

Coil voltages that are too low reduce the coil current and therefore the magnetic pull. This condition causes too high a current if the magnetic circuit is not completely closed. Because the coil has higher than normal current, it gets hot and may burn out. Also, the contacts do not seat with enough pressure and the contact surfaces will begin to arc and deteriorate. The armature may **chatter** and begin to wear the magnetic faces of the iron and cause further damage to the contacts.

Too high or too low voltage may be a result of line fluctuation in the premise's wiring or the result of improper installation and specification. Make sure you check the actual ratings on the coil and the actual voltage available for the coil. Many starters may be purchased with a dual-rated coil (for example, the coil has connections that allow it to be used on 120 or 240 volts). Be sure you have the jumpers in the proper slots for the control voltage you are using. Also, be aware that 240 coils do not work well on 208-volt systems.

Contactor Maintenance

When maintaining starters and contactors, be sure to remove control voltage and main power voltage. Some things to check: Make sure the armature is free to move, that it does not stick or bind from fully open to fully seated; ensure that all contacts are aligned and move with the armature assembly movement; check the pole faces for wear and misalignment, and be sure they are clean; check to be sure the shading rings are in place and that they have not become an open circuit; check for signs of overheating of the coil; tighten all conductor mountings to reduce heat and poor connection.

IEC STANDARDS

To compete in the world market, U.S. manufacturers need to produce goods for the rest of the world. United States users have also wanted to find cheaper, but adequate, equipment for their electrical systems. The emphasis in Europe has been to produce electrical products that are more specific to need. This has reduced the usage range and the amount of materials needed for the starter. You don't pay for more than you need. However, the product is not considered as rugged under adverse conditions.

The International Electrotechnical Commission (**IEC**) (headquartered in Geneva, Switzerland) is a counterpart to NEMA. IEC standards can be used to select the most suitable controller. The IEC philosophy is that utilization is a critical part of the selection process. The user must choose the utilization category and then select a product in that category that will fit the need. Also, the user must check to determine if the manufacturer's contact life rating is adequate for the purpose. Typical utilization categories are as listed in Figure 4-7. Also, contact life is taken into account. To make the proper selection, use the

Common utilization categories for AC contactors

Utilization category	Typical duty
AC1	Noninductive or slightly inductive loads
AC2	Starting of slip-ring motors
AC3	Starting of squirrel cage motors and switching off only after the motor is up to speed
AC4	Starting of squirrel cage motors with inching and plugging duty

Note: In an AC3 application, the contactor will never interrupt more than the motor's full-load current. If the application requires interruption of current greater than motor FLC, it is an AC4 application.

FIGURE 4-7 Typical utilization categories and duties for IEC starters. *(Courtesy of Rockwell Automation)*

proper application selection graphs as shown in Figure 4-8.

The horizontal axis (Ie) represents the operational current that the power contacts must handle. The vertical axis represents the number of operating cycles that can be expected in the contactor life. Read along the Ie axis to the point of the normal full-load current, then up to the point a particular contactor crosses the current line. Horizontally over to the left supplies the projected number of operations. For example, the AC3 category is for starters that control squirrel cage motors and switch off only after the motor is up to speed. A 5-HP 460-V three-phase motor draws 15.2 amps according to NEC® Table 430.150. Using this data, read 15.2 amps on Ie axis. Read straight up to where the bold line crosses the A18 vertical line. Contactor A18 would provide approximately 1.5 million operations. If that does not seem sufficient, go up to an A24 bold line crossing the 15-amp vertical line. An A24 may provide 3 million operations.

Certain considerations need to be taken into account when selecting NEMA or IEC standards for installations. How much expertise do you have to make proper selection? Do you want to dedicate the increased space usually needed for NEMA controllers? Do you have a good understanding of the application? If you are replacing an existing installation, what other components need to be changed? How much is cost a factor compared to interchangeability? Are overloads adjustable or changeable? These questions along with the difference in price must be considered when determining the proper selection.

PILOT DEVICES

Pilot devices are sensors and/or switches that provide the input to the motor control devices. Without the input devices, the motor could not function. The standard push button will be either a maintained or a momentary switch. The maintained switch stays in the position where it was last actuated. The symbol for the maintained contact start/stop station sample is shown in Figure 4-9. A good use for maintained switches is manual motor control.

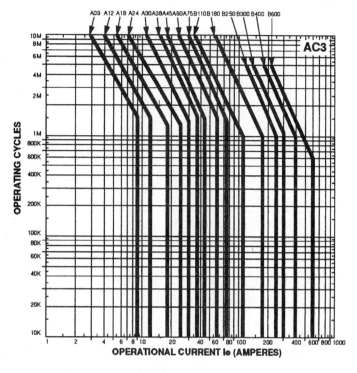
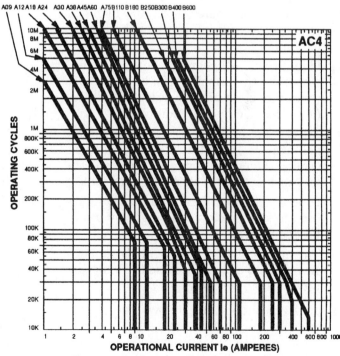

FIGURE 4-8 Use of charts in utilization categories helps in selection of correct starter. *(Courtesy of Rockwell Automation)*

1. Note the AC4 life load curves are based upon the assumption that motors used will have a locked rotor current equal to or less than 600 percent of motor full-load current.

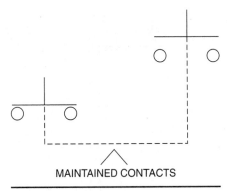

FIGURE 4-9 Symbol for maintained style contacts.

There are other types of pilot devices that are commonly used. Pilot devices include the following:

- Flow switches are mounted as water flow impellers or as sail switches to monitor air flow in ducts.
- Temperature-activated switches, which will either make or break on temperature rise (commonly called thermostats).
- Liquid-level switches or float switches, which use the principle of a buoyant float to open or close a switch as the liquid rises (typical application is a sump pump).
- Air pressure or vacuum switches use a diaphragm to monitor pressure changes. (These switches are commonly used to control air pressures in air compressor tanks.)
- Foot switches are floor-mounted switches used to start or stop equipment by use of foot control.
- The limit switch. (These symbols are a little harder to analyze because the limits are often held in position.)

Figure 4-10A shows the pilot switches in the various positions. Notice that the normally open switch is below the contact line as if gravity holds it open. This switch would have to be mechanically actuated closed. The next switch is a normally open (drawn below the contact line), but it is physically held closed. A mechanical actuation would force this type of switch to the open position. Often these switches are used as safety switches, such as door access switches. If a safety door is removed, the door switch will go to the open position and prevent the equipment from operating. The same description of operation is used for the normally closed limit switch. However, the switch is drawn above the contact point so that in its normal position it appears that gravity is allowing it to stay closed. If a mechanical device travels beyond its normal limit of motion, a mechanical arm will actuate the limit switch, change the switch position, and prevent further travel of the motor-operated device. (See Figure 4-10B.) (See Appendix A for more pilot devices.)

PHOTOELECTRIC CONTROLS

Other than mechanical actuation, other types of sensors also are used. Photoelectric controls are used where physical contact is not needed or desired. Photoelectric control covers a broad spectrum of controls. The light used is not always visible. The light may be anywhere in the light spectrum from ultraviolet at one end to infrared at the other. (See Figure 4-11.)

The methods of gathering or sensing the light may also vary. Photoelectric controls use a variety of devices to detect the presence or absence of light.

Photodiode sensors are light-activated sensors that have a semiconductor diode installed that uses a clear covering for the diode case. Because the diode is operated in reverse bias, it does not allow current to flow in the circuit when there is no light striking the transparent case. However, as light energy increases on the diode, current is allowed to flow in the circuit. This, in turn, controls a transistor to provide the switching needed in the circuit.

Photoresistive cells (or photoconductive cells) are also used. Often the concept is used in photoelectric controls employed to monitor

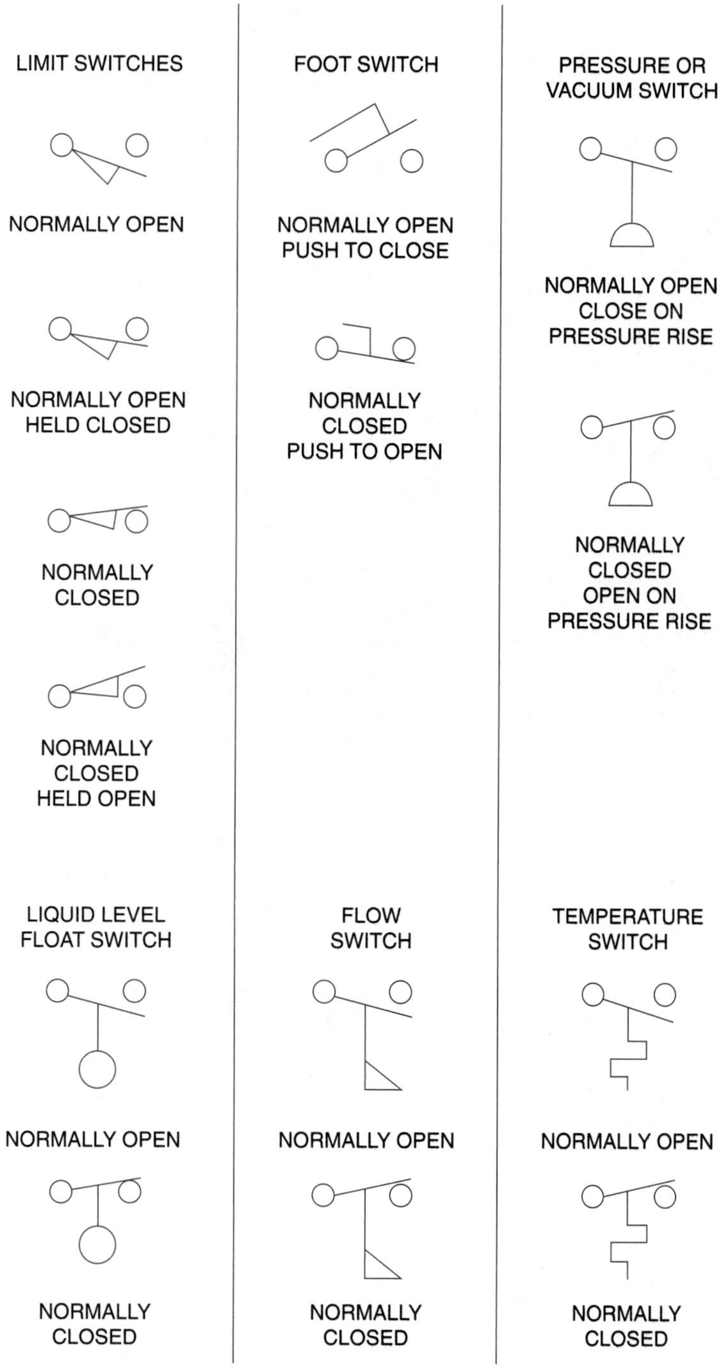

Controllers, Relays, and Timers

LIMIT SWITCHES

NORMALLY OPEN

NORMALLY OPEN HELD CLOSED

NORMALLY CLOSED

NORMALLY CLOSED HELD OPEN

FOOT SWITCH

NORMALLY OPEN PUSH TO CLOSE

NORMALLY CLOSED PUSH TO OPEN

PRESSURE OR VACUUM SWITCH

NORMALLY OPEN CLOSE ON PRESSURE RISE

NORMALLY CLOSED OPEN ON PRESSURE RISE

LIQUID LEVEL FLOAT SWITCH

NORMALLY OPEN

NORMALLY CLOSED

FLOW SWITCH

NORMALLY OPEN

NORMALLY CLOSED

TEMPERATURE SWITCH

NORMALLY OPEN

NORMALLY CLOSED

FIGURE 4-10A Mechanically actuated pilot devices.

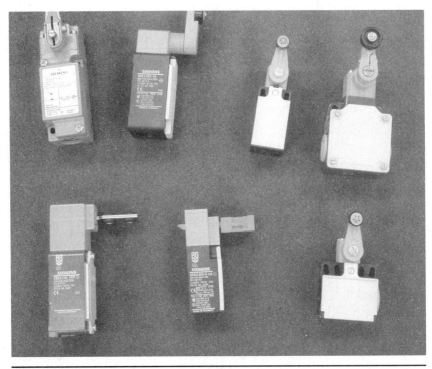

FIGURE 4-10B Mechanically actuated limit switches.

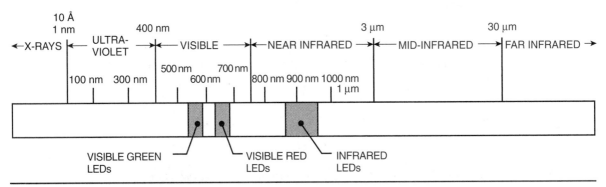

FIGURE 4-11 Light spectrum of photoelectric sensors. *(Courtesy of Banner Engineering Corporation)*

surrounding daylight conditions and control outdoor lighting. When light strikes the photoresistive cell, the resistance of the device drops and allows a larger current to flow. The resultant current flow often picks up a relay (electronic relay) to provide the desired switching result. A cadmium sulfide cell (cad cell) is used because the resistance changes from approximately 50 ohms in direct sunlight to many thousand ohms when dark.

When used in conjunction with more electronic circuitry, the switching point of the relay can be changed to reflect desired light level control. Older types of photo cells often used a metal hood to cover the cad cell. This allowed control of the amount of light on the cell and allowed adjustment of the activation point.

A word of caution: When installing photo cells to control lighting, based on available (ambient) light, make sure the cell faces away from the light source you are controlling. Otherwise, as dusk arrives (and you wish to turn on the outdoor lighting) the cad cell looking at the light source will turn off the lights it just turned on. The result is blinking lights that go on and off all night.

Photovoltaic cells are also used sometimes. As the name implies, the cell actually creates or generates a voltage as the light strikes the cell. Current silicon photovoltaic cells (solar cells) produce approximately .5 V DC at 150 MA for each 1 square inch of cell area. The symbol in Figure 4-12 indicates that the cell actually produces a DC voltage and is useful where an external voltage source may not be convenient.

Sensor style and technique are different based on the type of sensor and the application. Photo sensors that use their own internal light source rather than ambient light use one of several styles of reflection techniques to "see" the light source. The items to be seen, the physical space, and surrounding light will help determine which reflective technique should be used.

Retroreflective scanning uses a light source and receiver mounted in the same enclosure. The light is directed toward a reflective surface such as a common bicycle type reflector. (See Figure 4-13.) The reflected light is detected by the receiver and in turn picks up a **relay**. Often these devices can be arranged to "pick light" or "pick dark." Pick light means the relay is energized with a completed light beam. Pick dark means the relay is picked up when the light beam is obstructed and the light is not reflected.

Retroreflective is a good style to use when you have room for a reflector and the object to be sensed is not reflective itself. Also, the alignment is not critical, so some vibration of the scanner or reflector is tolerated. The typical limit of distance to the target reflector is approximately 40 feet. One advantage of the retroreflective-type sensor is that all wiring is to one device with a nonelectrical reflector placed where convenient. Retroreflective scan also could be used to count objects that have reflective tape applied. As the objects move past the sensor, the reflected light would cause a contact closure in the retroreflective sensor.

Diffuse scan uses the principle of a light source directed at the surface of objects that are

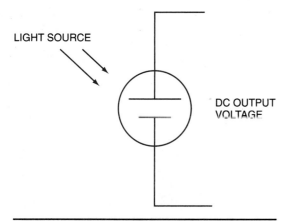

FIGURE 4-12 Photovoltaic cell produces DC output voltage when struck by light.

RETROREFLECTIVE SENSING MODE

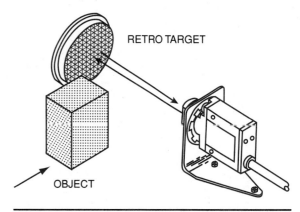

FIGURE 4-13 Photoreflective sensor uses a light source and a reflective target. *(Courtesy of Banner Engineering Corporation)*

not reflective. Most of the light source is absorbed by the object and the rest is diffused in all directions. A receiver is placed at about the same distance from the surface as the light source. The small amount of light that is returned to the receiver indicates presence of the object. Most diffuse scan sensors use a lens to collinate, or make the light rays parallel, so that the receiver can receive more returned light. These sensors are used in close proximity to the object to be detected. (See Figure 4-14.)

A convergent beam is used to detect objects that are near other reflective surfaces, or for scanning small objects. (See Figure 4-15.) Convergent beam uses the light source and sensing device in one head. The light beam is focused or converged at the object to be sensed. This is a fixed distance determined by the sensor and is adjustable by field focus controls. If the object is not at the focal point of the beam, no light is reflected back to the receiver. If an object is at the focal point (not in front of or behind the focal point), a small amount of light is reflected to the receiver. Only one head is used and the surrounding surfaces are ignored. Convergent beam sensors are ideal where the objects

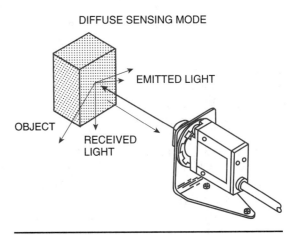

FIGURE 4-14 Diffuse mode sensor with source and reflector in the same housing. *(Courtesy of Banner Engineering Corporation)*

are near other surfaces or where the objects have low reflectivity.

Specular scan uses a light source and a receiver mounted at angles to each other, as in Figure 4-16. The principle is to use a reflected light beam off a shiny object. The focal point is set to reflect light from the light source to the receiver. If an object interferes with that reflection, the receiver "sees" the object. Another

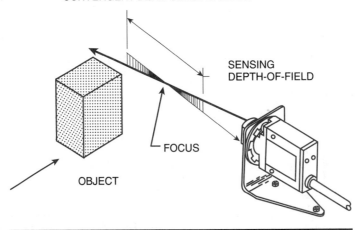

FIGURE 4-15 Convergent beam sensor. Light is focused at a specific point where the object is sensed. *(Courtesy of Banner Engineering Corporation)*

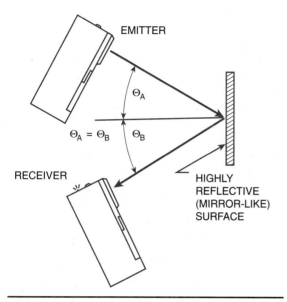

FIGURE 4-16 Specular scan uses an emitter and a receiver to sense objects with shiny surfaces.

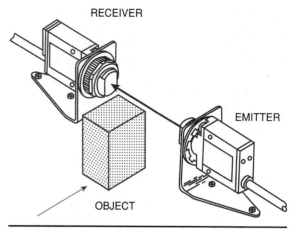

FIGURE 4-17 Direct scan, or opposed sensing, uses an emitter and receiver to detect an object between them. *(Courtesy of Banner Engineering Corporation)*

option is for the object to complete the reflected beam. By adjusting the angle of reflection, a shiny object can cause the beam to be reflected back to the receiver to "see" the object. Mounting angles are very precise and a lot of vibration is not tolerated.

The least complicated sensor is called direct scan. This method uses a transmitter (light source) and a receiver as in Figure 4-17. If an object passes between the transmitter and receiver, the beam is broken and the object is detected. Alignment is critical in these sensors and wiring is required to both the transmitter and receiver. Direct-scan sensors are harder to keep in alignment, but they are used extensively.

Many light sources use visible light, but the options include ultraviolet light and infrared light. When using other than visible light sources, aids are required to help set up, aim, and troubleshoot the devices. To determine if an infrared light source is emitting, a small portable receiver is used to detect light output. If the light receiver is suspected to be defective, a portable light source can be used to transmit a known beam to the receiver. Then troubleshooting regarding focus, improper application, or replacement of the defective device can proceed.

Fiber optics also are used to sense objects. Light is transmitted through a fiber optic cable to the point where an object is to be detected. This allows the sensor to be located out of the way and only the small cables need to be located at the sensing location. Bifurcated fiber optic cable uses the same fiber bundle to send and receive light. (See Figure 4-18.) A "thru"

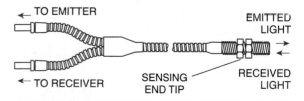

FIGURE 4-18 Bifurcated cable can be used in opposed mode. It uses fiber optics to bring light from the sensor head to the sensed object. *(Courtesy of Banner Engineering Corporation)*

scan has a transmitter and receiver cable as separate cables.

Be sure to check the manufacturer's specifications when determining the proper photoelectric device to use. Specific application is important. Ambient light, temperature, vibration, moisture, and surrounding dust are all important considerations.

PROXIMITY SENSORS

Proximity sensors (controls) are available in two types: inductive sensors and capacitive sensors. Typical inductive sensors are an electronic circuit-type sensor that produces a small magnetic field at a radio frequency (RF). As the magnetic waves are produced, a tuned circuit is set to react to an altered oscillation created by a detected metal object. The tuning of these circuits is adjustable to account for distance of detection and amount of metal that is required before sensing is accomplished. As a metal object approaches the inductive proximity sensor, the magnetic coil in the sensor actually induces eddy currents into the detected target. This causes a change in the RF oscillations, and the amplitude of the RF is reduced or dampened. This dampened signal is detected by the driving electronics and switches the proximity sensor output to an "on" or dampened state. As the object moves away from the sensor, the oscillator output returns to the undampened state, which allows the output to turn off.

Inductive proximity sensors work well in dirty or greasy environments where photo sensors do not work well. They have very fast response and can detect lightweight objects that cannot be detected by mechanical limit switches. Inductive sensors can also be used to detect metal objects behind nonmetallic materials such as glass or plastic. The two drawbacks to an inductive sensor are that the target to be detected must be metal and the distance of detection is limited to approximately 4 inches. The output of these sensors can be a relay contact or other type of output desired for the associated control circuitry. (See Figure 4-19.)

FIGURE 4-19 Proximity sensors.

Capacitive Sensors

Capacitive sensors work because of the effects of different materials on the dielectric constant of a capacitive device. Different nonmetal materials have different dielectric constants. As a material is sensed by a capacitive sensor, the dielectric constant of the circuit changes and starts an oscillation that is detected by the receiver. This altered condition initiates the electronics to cause an output relay to change state (open to closed, or vice versa).

The capacitive proximity sensor has an advantage over the inductive sensor in that it can detect both nonmetallic and metallic objects. It has a larger sensing range and it can be used for detecting liquid targets through nonmetallic barriers. It can be used for looking through glass or plastic tanks to detect liquid or other materials, such as plastic pellets in a feeder hopper.

There are several limitations to the capacitive sensor. It is more affected by moisture and humidity, and if it is embedded in the equipment, the surrounding material must be considered in the calibration. Also, because it has greater sensing range, more space must be maintained between sensors and the target area must be well defined. (See Figure 4-20.)

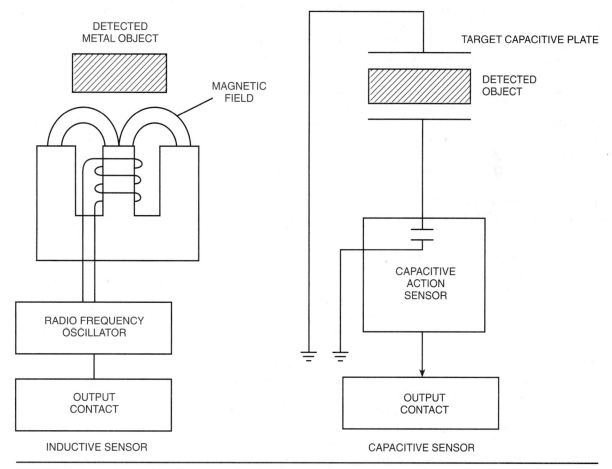

FIGURE 4-20 Simplified diagrams of inductive and capacitive proximity sensors.

There are many types of sensors as well as many types of pilot devices. They are all intended to provide input information to control electrical systems. Up to this point, you have only looked at simple on–off control schemes where the pilot device is the decision point to energize or deenergize a motor starter. If more complex controls are needed with many different inputs controlling many different output functions, another decision maker may be needed, such as a programmable controller. (See Chapter 6.)

ENCLOSURES

Enclosures are designed to keep the electrical mechanisms from direct contact with people and to provide some form of protection from the surrounding atmosphere. NEMA has enclosures for specific purposes listed by a reference number. A partial list and general description of each type and intended uses is provided in NEC® Table 430.91. (See Figure 4-21 for NEMA configurations.) IEC enclosures also have a rating system that incorporates applications. (See Figure 4-22 to cross reference some IP international ratings to NEMA ratings.) Enclosure types do not have to match the controller size. For example, a NEMA size 1 starter may be used in a NEMA 12 enclosure.

HAZARDOUS AREAS AND SAFETY

Hazardous locations are defined in Chapter 5 of the NEC®. Class I locations are areas where flammable gas may be present. The classes are then broken down into groups as described in NEC Article 500.5. Class II locations are atmospheres that contain combustible dusts. This class is also broken down by group as described in Article 500.5. Class III locations are described in Article 500.5. In addition, each class is further described by Division 1 or 2. Division 1 environments are generally atmospheres where the dangerous condition is present as a normal operating condition. Division 2 locations are locations where hazardous materials are handled or stored but generally are confined in closed systems. See Article 500.5.

Enclosures must meet the NEC® requirements for the location where they are installed. The enclosures are constructed to prevent further explosions if the contactor (or switches) contained within does ignite or explode.

Wiring within hazardous locations whether Class I, II, or III and Division 1 or 2 is clearly defined in Chapter 5 of the NEC. Article 501 refers to the required methods used for Class I, Division 1 or 2. For instance, Article 501(A)(3) states that all boxes and fittings in Class I, Division 1 shall be approved for Class I, Division 1 locations. Where flexibility is required, such as for motor connection, flexible explosion-proof flexible conduit is required.

Article 502 refers to the wiring methods employed where there is combustible dust that may explode or ignite. Coal movement and storage or grain milling are generally considered Class II locations. A registered engineer or other qualified authority is required to determine the hazards involved and which class and division are involved.

Article 503 is the location to determine wiring methods and procedures in locations where there are ignitable fibers or flyings, but not in an air mixture that would cause it to be a Class II location. Areas such as carpet mills, or cloth fabric manufacturing could be Class III locations. See Figure 4-23 for hazardous location classification system.

Hazardous areas include class and zone locations, which are hazardous location guides from the IEC. As the NEC can be interpreted as an international code, the adaptation and inclusion of the IEC standards are necessary. Article 505 refers to the classification of systems used in areas that contain flammable gases, vapors, or liquids. Zones 0, 1, and 2 further define the concentrations of gases present.

Controllers, Relays, and Timers   105

**NEMA Type 1
General Purpose
Surface Mounting**

Type 1 enclosures are intended for indoor use primarily to provide a degree of protection against contact with the enclosed equipment in locations where unusual service conditions do not exist. The enclosures are designed to meet the rod entry and rust-resistance design tests. Enclosure is sheet steel, treated to resist corrosion.

**NEMA Type 1
Flush Mounting**

Flush mounted enclosures for installation in machine frames and plaster wall. These enclosures are for similar applications and are designed to meet the same tests as NEMA Type 1 surface mounting.

NEMA Type 3

Type 3 enclosures are intended for outdoor use primarily to provide a degree of protection against windblown dust, rain and sleet; and to be undamaged by the formation of ice on the enclosure. They are designed to meet rain **1**, external icing **2**, dust, and rust-resistance design tests. They are not intended to provide protection against conditions such as internal condensation or internal icing.

NEMA Type 3R

Type 3R enclosures are intended for outdoor use primarily to provide a degree of protection against falling rain, and to be undamaged by the formation of ice on the enclosure. They are designed to meet rod entry, rain **3**, external icing **2**, and rust-resistance design tests. They are not intended to provide protection against conditions such as dust, internal condensation, or internal icing.

NEMA Type 4

Type 4 enclosures are intended for indoor or outdoor use primarily to provide a degree of protection against windblown dust and rain, splashing water, and hose-directed water; and to be undamaged by the formation of ice on the enclosure. They are designed to meet hosedown, dust, external icing **2**, and rust-resistance design tests. They are not intended to provide protection against conditions such as internal condensation or internal icing. Enclosures are made of heavy gauge stainless steel, cast aluminum or heavy gauge sheet steel, depending on the type of unit and size. Cover has a synthetic rubber gasket.

**NEMA Type 3R, 7 & 9
Unilock Enclosure
For Hazardous
Locations**

This enclosure is cast from "copper-free" (less than 0.1%) aluminum and the entire enclosure (including interior and flange areas) is bronze chromated. The exterior surfaces are also primed with a special epoxy primer and finished with an aliphatic urethane paint for extra corrosion resistance. The V-Band permits easy removal of the cover for inspection and for making field modifications. This enclosure meets the same tests as separate NEMA Type 3R, and NEMA Type 7 and 9 enclosures. For NEMA Type 3R application, it is necessary that a drain be added.

1 Evaluation criteria: No water has entered enclosure during specified test.
2 Evaluation criteria: Undamaged after ice which built up during specified test has melted (Note: **Not** required to be operable while ice-laden).
3 Evaluation criteria: No water shall have reached live parts, insulation or mechanisms.

FIGURE 4-21 NEMA descriptions and examples in metal and fiberglass. *(Courtesy of Rockwell Automation)*

**NEMA Type 4X
Non-Metallic,
Corrosion-Resistant
Fiberglass Reinforced
Polyester**

Type 4X enclosures are intended for indoor or outdoor use primarily to provide a degree of protection against corrosion, windblown dust and rain, splashing water, and hose-directed water; and to be undamaged by the formation of ice on the enclosure. They are designed to meet the hosedown, dust, external icing [2], and corrosion-resistance design tests. They are not intended to provide protection against conditions such as internal condensation or internal icing. Enclosure is fiberglass reinforced polyester with a synthetic rubber gasket between cover and base. Ideal for such industries as chemical plants and paper mills.

NEMA Type 6P

Type 6P enclosures are intended for indoor or outdoor use primarily to provide a degree of protection against the entry of water during prolonged submersion at a limited depth; and to be undamaged by the formation of ice on the enclosure. They are designed to meet air pressure, external icing [2], hosedown and corrosion-resistance design tests. They are not intended to provide protection against conditions such as internal condensation or internal icing.

**NEMA Type 7
For Hazardous
Gas Locations
Bolted Enclosure**

Type 7 enclosures are for indoor use in locations classified as Class I, Groups C or D, as defined in the National Electrical Code. Type 7 enclosures are designed to be capable of withstanding the pressures resulting from an internal explosion of specified gases, and contain such an explosion sufficiently that an explosive gas-air mixture existing in the atmosphere surrounding the enclosure will not be ignited. Enclosed heat generating devices are designed not to cause external surfaces to reach temperatures capable of igniting explosive gas-air mixtures in the surrounding atmosphere. Enclosures are designed to meet explosion, hydrostatic, and temperature design tests. Finish is a special corrosion-resistant, gray enamel.

**NEMA Type 9
For Hazardous
Dust Locations**

Type 9 enclosures are intended for indoor use in locations classified as Class II, Groups E, F or G, as defined in the National Electrical Code. Type 9 enclosures are designed to be capable of preventing the entrance of dust. Enclosed heat generating devices are designed not to cause external surfaces to reach temperatures capable of igniting or discoloring dust on the enclosure or igniting dust-air mixtures in the surrounding atmosphere. Enclosures are designed to meet dust penetration and temperature design tests, and aging of gaskets. The outside finish is a special corrosion-resistant gray enamel.

NEMA Type 12

Type 12 enclosures are intended for indoor use primarily to provide a degree of protection against dust, falling dirt, and dripping non-corrosive liquids. They are designed to meet drip [1], dust, and rust-resistance tests. They are not intended to provide protection against conditions such as internal condensation.

NEMA Type 13

Type 13 enclosures are intended for indoor use primarily to provide a degree of protection against dust, spraying of water, oil, and non-corrosive coolant. They are designed to meet oil exclusion and rust-resistance design tests. They are not intended to provide protection against conditions such as internal condensation.

[1] Evaluation criteria: No water has entered enclosure during specified test.
[2] Evaluation criteria: Undamaged after ice which built up during specified test has melted (Note: **Not** required to be operable while ice-laden).

FIGURE 4-21 (cont.)

For instance, Class I, Zone 0 is where gases or vapors are present continuously, or for long periods. Class I are locations where flammable gases are present under normal operating conditions. Class II is an area where flammable gases or vapors are not likely to occur in normal operation, or for only short periods of time.

Article 506 was added to the 2005 NEC to include information on combustible dusts,

Controllers, Relays, and Timers 107

ENCLOSURES
IEC

GENERAL

DEGREE OF PROTECTION

IEC Publication 529 describes standard Degrees of Protection which enclosures of a product are designed to provide when properly installed.

SUMMARY

The publication defines degrees of protection with respect to:
- Persons
- Equipment within the enclosure
- Ingress of water

It does **not** define:
- Protection against risk of explosion
- Environmental protection (e.g. against humidity, corrosive atmospheres or fluids, fungus or the ingress of vermin)

NOTE: The IEC test requirements for Degrees of Protection against liquid ingress refer only to water. Those products in this catalog, which have a high degree of protection against ingress of liquid, in most cases include Nitrile seals. These have good resistance to a wide range of oils, coolants and cutting fluids. However, some of the available lubricants, hydraulic fluids and solvents can cause severe deterioration of Nitrile and other polymers. Some of the products listed are available with seals of Viton or other materials for improved resistance to such liquids. For specific advice on this subject refer to your nearest Allen-Bradley Sales Office.

IEC ENCLOSURE CLASSIFICATION

The degree of protection is indicated by two letters (IP) and two numerals. International Standard IEC 529 contains descriptions and associated test requirements which define the degree of protection each numeral specifies. The following table indicates the *general* degree of protection — refer to Abridged Descriptions of IEC Enclosure Test Requirements below and on Page 193. **For complete test requirements refer to IEC 529.**

FIRST NUMERAL■	SECOND NUMERAL■
Protection of persons against access to hazardous parts and protection against penetration of solid foreign objects.	Protection against ingress of water under test conditions specified in IEC 529.
0 Non-protected	0 Non-protected
1 Back of hand; objects greater than 50mm in diameter	1 Vertically falling drops of water
2 Finger; objects greater than 12.5mm in diameter	2 Vertically falling drops of water with enclosure tilted 15 degrees
3 Tools or objects greater than 2.5mm in diameter	3 Spraying water
4 Tools or objects greater than 1.0mm in diameter	4 Splashing water
5 Dust-protected (Dust may enter during specified test but must not interfere with operation of the equipment or impair safety)	5 Water jets
6 Dust tight (No dust observable inside enclosure at end of test)	6 Powerful water jets
	7 Temporary submersion
	8 Continuous submersion

Example: IP41 describes an enclosure which is designed to protect against the entry of tools or objects greater than 1mm in diameter and to protect against vertically dripping water under specified test conditions.

Note: All first numerals, and second numerals up to and including characteristic numeral **6**, imply compliance also with the requirements for all lower characteristic numerals in their respective series (first or second). Second numerals **7** and **8** do **not** imply suitability for exposure to water jets (second characteristic numeral **5** or **6**) unless dual coded; e.g. **IP_5/IP_7**.

■ The IEC standard permits use of certain supplementary letters with the characteristic numerals. If such letters are used, refer to IEC 529 for the explanation.

NEMA/IEC EQUIVALENTS

NEMA 1 EQUIVALENT TO IP42
NEMA 4 EQUIVALENT TO IP66
NEMA 12 EQUIVALENT TO IP65

FIGURE 4-22 IEC designations as IP numbers. NEMA to IEC enclosure comparisons. *(Courtesy of Rockwell Automation)*

ENCLOSURES
IEC

ABRIDGED DESCRIPTIONS OF IEC ENCLOSURE TEST REQUIREMENTS — Cont'd

Tests for protection against access to hazardous parts (first characteristic numeral) — Cont'd

- **IP4_** — A test wire 1mm in diameter shall not penetrate and adequate clearance shall be kept from hazardous live parts (as specified on Page 18). Force = 1 N.
- **IP5_** — A test wire 1mm in diameter shall not penetrate and adequate clearance shall be kept from hazardous live parts (as specified on Page 18). Force = 1 N.
- **IP6_** — A test wire 1mm in diameter shall not penetrate and adequate clearance shall be kept from hazardous live parts (as specified on Page 18). Force = 1 N.

Tests for protection against solid foreign objects (first characteristic numeral)

For first numerals **1, 2, 3** and **4** the protection against solid foreign objects is satisfactory if the full diameter of the specified probe does not pass through any opening. Note that for first numerals **3** and **4** the probes are intended to simulate foreign objects which may be spherical. Where shape of the entry path leaves any doubt about ingress of a spherical object capable of motion, it may be necessary to examine drawings or to provide special access for the object probe. For first numerals **5** and **6** see test descriptions below for acceptance criteria.

- **IP0_** — No test required.
- **IP1_** — The full diameter of a rigid sphere 50mm in diameter must not pass through any opening at a test force of 50 N.
- **IP2_** — The full diameter of a rigid sphere 12.5mm in diameter must not pass through any opening at a test force of 30 N.
- **IP3_** — A rigid steel rod 2.5mm in diameter must not pass through any opening at a test force of 3 N.
- **IP4_** — A rigid steel wire 1mm in diameter must not pass through any opening at a test force of 1 N.
- **IP5_** — The test specimen is supported inside a specified dust chamber where talcum powder able to pass through a square-meshed sieve with wire diameter 50 μm and width between wires 75 μm, is kept in suspension.

    Enclosures for equipment subject to thermal cycling effects (category 1) are vacuum pumped to a reduced internal pressure relative to the surrounding atmosphere: maximum depression = 2 kPa; maximum extraction rate = 60 volumes per hour. If extraction rate of 40 to 60 volumes/hr. is obtained, test is continued until 80 volumes have been drawn through or 8 hr. has elapsed. If extraction rate is less than 40 volumes/hr. at 20 kPa depression, test time = 8 hr.

    Enclosures for equipment not subject to thermal cycling effects **and** designated category 2 in the relevant product standard are tested for 8 hr. without vacuum pumping.

    Protection is satisfactory if talcum powder has not accumulated in a quantity or location such that, as with any other kind of dust, it could interfere with the correct operation of the equipment or impair safety; and no dust has been deposited where it could lead to tracking along creepage distances.

- **IP6_** — All enclosures are tested as category 1, as specified above for **IP5_**. The protection is satisfactory if no deposit of dust is observable inside the enclosure at the end of the test.

Tests for protection against water (second characteristic numeral)

The second characteristic numeral of the IP number indicates compliance with the following tests for the degree of protection against water. For numerals **1** through **7**, the protection is satisfactory if any water which has entered does not interfere with satisfactory operation, does not reach live parts not designed to operate when wet, and does not accumulate near a cable entry or enter the cable. For second numeral **8** the protection is satisfactory if no water has entered the enclosure.

- **IP_0** — No test required.
- **IP_1** — Water is dripped onto the enclosure from a "drip box" having spouts spaced on a 20mm square pattern, at a "rainfall" rate of 1 mm/min. The enclosure is placed in its normal operating position under the drip box. Test time = 10 min.
- **IP_2** — Water is dripped onto the enclosure from a "drip box" having spouts spaced on a 20mm square pattern, at a "rainfall" rate of 3 mm/min. The enclosure is placed in 4 fixed positions tilted 15° from its normal operating position, under the drip box. Test time = 2.5 min. for each position of tilt.
- **IP_3** — Water is sprayed onto all sides of the enclosure over an arc of 60° from vertical, using an oscillating tube device with spray holes 50mm apart (or a hand-held nozzle for larger enclosures). Flow rate, oscillating tube device = 0.07 l/min. per hole x number of holes; for hand-held nozzle = 10 l/min. Test time, oscillating tube = 10 min.; for hand-held nozzle = 1 min./m² of enclosure surface area, 5 min. minimum.
- **IP_4** — Same as test for **IP_3** except spray covers an arc of 180° from vertical.
- **IP_5** — Enclosure is sprayed from all practicable directions with a stream of water at 12.5 l/min. from a 6.3mm nozzle from a distance of 2.5 to 3m. Test time = 1 min./m² of enclosure surface area to be sprayed, 3 min. minimum.

FIGURE 4-22 (cont.)

Controllers, Relays, and Timers   109

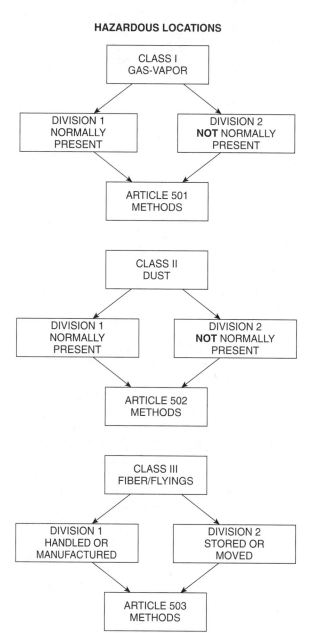

FIGURE 4-23 A simplified diagram to determine class and division of hazardous locations.

fibers, and flyings adopted from the IEC codes. These locations are classified as Class XX, XXI, or XXII. As was the pattern in Class I, Zones 0, 1, and 2, the concentration and likelihood of hazardous material are the key elements in determining locations. Class II, Zone 0 is a location where the combustible dusts or flyings are present continuously. Class II, Zone 1 is where the combustible dusts or ignitable flyings are likely to exist occasionally or when the equipment is under repair. Class II, Zone 2 is an area where the hazardous materials are not usually present or are present for only short periods of time. Each acceptable wiring method is described in Article 506.

Another form of safe operating procedures and enclosures for hazardous areas uses the idea of intrinsic safety (IS). This implies that the safety of the system is intrinsic to the design of the system. Article 504 of the NEC describes the uses of an intrinsically safe system. The concept is to create such low levels of energy for the circuit in the explosive locations that an arc or spark produced by the electrical connections will be so low that it does not have enough thermal energy to ignite the surrounding atmosphere. See Figure 4-24. By using zener diodes to control the voltage and resistors to limit the amount of voltage and also limit the amount of circuit current, the hazardous location can have standard Class I wiring used (without explosion-proof enclosures) and no explosion will occur. It is the electrician's responsibility to make sure the intrinsically safe wiring does not get connected to, or run, in the same piping as the normal higher voltage wiring.

ELECTROMECHANICAL CONTROLS, RELAYS, AND TIMERS

Relays and Timers

There are many different types and styles of relays, but they all provide the same basic function. The 2005 NEC defines a **device** as a unit of an electrical system that is intended to

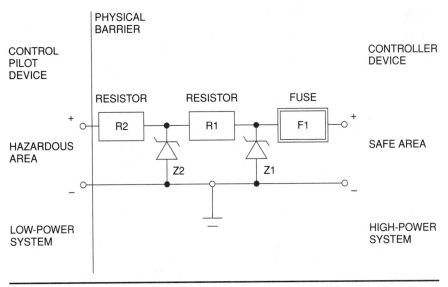

FIGURE 4-24 Intrinsic safety system concept.

carry or control, but not utilize, electric energy. As such, the relay is intended to carry or control a circuit but is not intended to utilize the electric energy. It may use power, as does a dimmer control for lighting control, but that is still considered a device. Most relays use some type of internal activating mechanism that receives a signal from one circuit and then opens or closes a circuit to another associated circuit.

Electromechanical Relay (EMR)

A different sort of relay is one where the actuating system is not part of the same device package. A relay that is often used for the presence detection (proximity sensor) system in an intrusion alarm system would be a reed relay. These relays are typically sealed in a glass or a plastic case. See Figure 4-25. The contacts are often plated contacts with gold or silver alloy to provide a very low resistance connection when the contacts are closed. They also carry small currents, approximately .25 mA. The contacts can be NO or NC or a three-point contact with NO and NC available with a common center connection. The contacts are pulled from their static position with an external magnet. This is a common type of actuation when a door or window with a magnet attached is closed and the reed relay activates to close the contacts. With the door open or if the reed relay is disconnected, the circuit is opened and indicates a failure of the protection system. Another use of the reed relay is detecting if a metal cover or safety shield is in place. If a magnet is used to close the contact with no safety shield in place, then when the shield is in place, the metal cover will come between the magnet and the switch. This effectively shunts the magnetic field and allows the reed relay to release, signaling "safety barrier in place." The reed relay can be used in control systems to turn solenoids off or on, or as a contact point for timer initiation. Because of the small current capabilities, they are used in control systems to control larger relays or contactors.

Relays are designed to make and/or break single or multiple contacts. You will often see specifications for different kinds of relays with different "form" contacts. To standardize between relay manufacturers, a standard marking

Controllers, Relays, and Timers

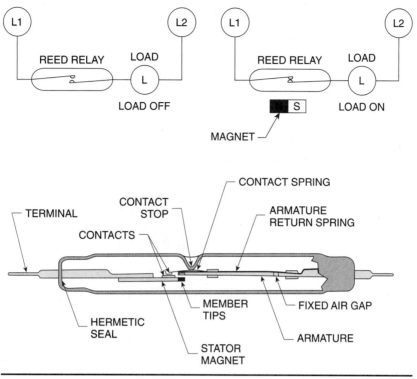

FIGURE 4-25 "Reed" switches operate by proximity to a magnet.

system has been developed to indicate how many, and in what sequence, the contacts respond to the relay coil. See Figure 4-26 for relay form standard designations. The most common type of relay is the form C relay, where a contact has two positions. The relay contacts are shown with no control power applied (or nonactuated). In form C, one contact opens (breaks) before the other contact closes (makes). This is sometimes referred to as a break-before-make contact. Notice that a form D is a make-before-break style contact.

Electromechanical control refers to the use of relays, timers, and so on that need to move mechanically in order to perform their function. Until recently, with the advent of electronic (solid-state) relays and timers, or the use of programmable logic controllers or PLCs, most control functions were provided by relay logic. In order to control multiple functions, control relays are used, as in Figure 4-27. The relays can also be operated at lower voltage and current levels than the motors they control, thereby providing a degree of protection to the operator.

Relays are used to transmit a small control signal to another circuit or to a larger coil. A typical electromechanical relay is shown in Figure 4-28 with the associated coil symbol. The contacts are separate from the electrical coil circuit just as they are in the magnetic starter.

Contacts are shown in the deenergized state. When no power is applied to the coil, the contacts are shown as normally open (NO) or normally closed (NC). Many relays have multiple sets of contacts with various combinations of NO and NC or convertible contacts. When ordering these relays for installation or replacement, the proper number and type of contacts must be specified as well as the coil

ELECTRIC MOTORS AND MOTOR CONTROLS

RELAY FORM IDENTIFICATION			
DESIGN	SEQUENCE	SYMBOL	FORM
SPST-NO	MAKE (1)		A
SPST-NC	BREAK (1)		B
SPDT	BREAK (1) MAKE (2)		C
SPDT	MAKE (1) BEFORE BREAK (2)		D
SPDT (B-M-B)	BREAK (1) MAKE (2) BEFORE BREAK (3)	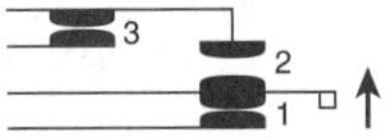	E
SPDT-NO	CENTER OFF	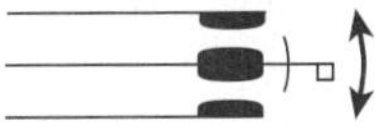	K
SPST-NO (DM)	DOUBLE MAKE (1)	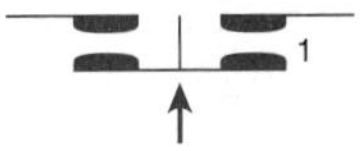	X
SPST-NC (DB)	DOUBLE BREAK (1)	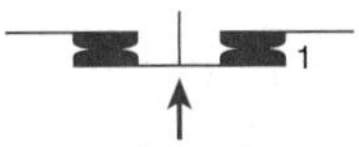	Y
SPDT-NC-NO (DB-DM)	DOUBLE BREAK (1) DOUBLE MAKE (2)	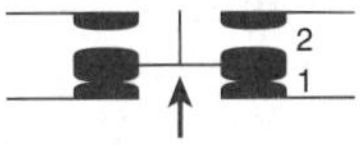	Z

FIGURE 4-26 Form letter identifies the relay contact design.

voltage rating and the contact current and voltage rating.

Convertible contacts can be changed from NO to NC by disassembly of the relay body and repositioning of the contact points. Many relays have contacts with three contact points. Figure 4-28 shows a relay with a shunt connection to a common point. The relay will pull the center contact from NO to closed. This style of *contact* is a single pole–double throw.

As mentioned, relays come in a variety of shapes and sizes. Some relays are mounted directly to a panel with mounting feet, as in Figure 4-29. This relay has multiple sets of contacts all operated by the same electromagnetic coil moving the contact assembly. The chart that accompanies the picture displays the combinations available with the four main contacts and the add-on contacts. The small L-shaped upper contact symbol indicates that

Controllers, Relays, and Timers

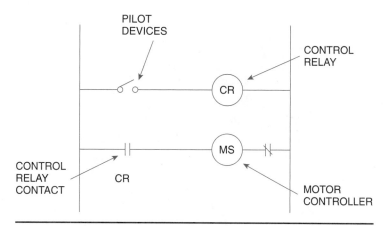

FIGURE 4-27 Control relay functions to operate another electrical load. Control relays are often labeled "CR."

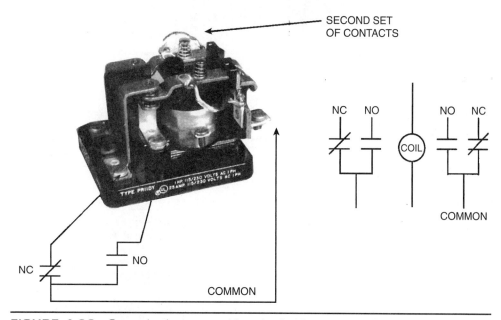

FIGURE 4-28 Control relay as used in relay ladder system. Shown with exposed contacts.

the contact is closed when deenergized. Some relays use a socket arrangement for mounting. The sockets allow easy wiring and removal of the relays. Typical sockets are shown in Figure 4-30. Often the type of relay that uses a socket is a plastic-encapsulated relay. The wiring diagram is printed on the outside of the relay case with the proper pin arrangements shown. These clear plastic-covered relays are sometimes called *ice cube relays* because they look similar to an ice cube. Ice cube relays allow you to see the relay in action to determine if the contacts are moving in response to the applied coil voltage.

ELECTRIC MOTORS AND MOTOR CONTROLS

Control Relays

Product Selection

AC RELAYS

	Relay Contacts		Overlapping Side Mounted Contacts		Relay Arrangement	Auxiliary	Cat. No. ❶
	NO	NC	NO	NC			
With 1 or 2 sets of overlapping contacts 4 Pole Control Relays IP20 660V Max.	4	0	1	1	A1 13 23 33 43 A2 14 24 34 44		700–FZ1510 ⊕
	3	1	1	1	A1 13 21 33 43 A2 14 22 34 44		700–FZ1420 ⊕
	2	2	1	1	A1 13 21 31 43 A2 14 22 32 44		700–FZ1330 ⊕
	4	0	2	2	A1 13 23 33 43 A2 14 24 34 44		700–FZ2620 ⊕
	3	1	2	2	A1 13 21 33 43 A2 14 22 34 44		700–FZ2530 ⊕
	2	2	2	2	A1 13 21 31 43		700–FZ2440 ⊕

FIGURE 4-29 Multiple-pole relay with contact diagram. *(Courtesy of Rockwell Automation)*

FIGURE 4-30 Eight-pin or eleven-pin bases used as connections for removable "ice cube" relays.

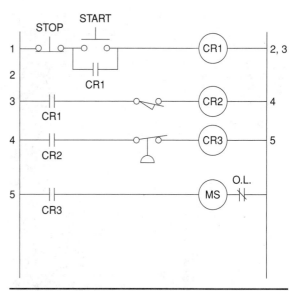

FIGURE 4-31 Typical multiline diagram of schematic ladder diagram used in motor control. Note how the numbers on the right runner correspond to the location of the contacts associated with the loads. These locate the contact on the proper rung (number to left of runner).

When installing relays in a motor control schematic, the practice is to place the coils in the rungs of the **ladder diagram** and label the relays as "CR" (control relay). The control relays are numbered as an identifier and the rungs of the ladder also are numbered. The NO and NC contacts are listed to the right to indicate where to find the contacts operated by the relay on the rung. See Figure 4-31.

Most relays are operated by electromagnetic coils. The coils should never be placed in series with each other as the voltage on the coils will be insufficient and cause the relays not to operate properly. Coil voltages, AC or DC, and proper frequency must be observed when ordering the appropriate relay, as well as contact sequence, number of NO and NC contacts. Whether it is single-pole–single-throw SPST or double-pole–double-throw DPDT needs to be analyzed.

The actual current the contacts will need to handle is also part of the information needed when ordering relays or replacement relays. The current ratings are essential. This determines the size of the contact surface. The larger the current, the more contact surface is needed. The voltage rating will determine contact movement and separation. Higher voltage-rated contacts must separate faster and farther to help extinguish the arc. Also, contacts that separate DC circuits are different from AC contacts. DC circuits are more destructive on separation.

Often, if a coil needs very strong pull-in and seal-in power, DC is used. DC voltage does not pass through zero each half-cycle and therefore provides a stronger, more consistent pull. There is no need for a shading ring on DC coils as there is on AC relays. The other feature not necessary on DC relays is the need for a laminated steel core. Because the voltage does

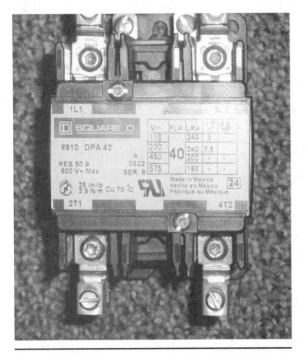

FIGURE 4-32 A relay with listed voltage and current ratings for the contacts.

not fluctuate, no eddy currents are induced in the iron. Laminations and low hysteresis steel are used to reduce the effects of eddy currents in AC systems but are not needed in DC coils.

Relays are available in many shapes and forms including relays used just for lighting, specified as lighting contactors, or large power relays with various number of poles and various control voltages. Figure 4-32 shows a two-pole, 40-A relay with a 120-V coil. The voltage rating on the contacts can be as high as 575 V and the contacts can carry different amounts of locked rotor amps (LRAs) based on the voltage that the contacts must break.

IEC relays also are available. These relays are compatible with IEC starters. They are generally smaller than standard relays and often come ready to mount on DIN rails for easy installation and replacement. DIN rails are racks that are mounted to the enclosure. These racks or rails stay in place and the relays snap into the mountings without the need for additional mounting hardware. Many of the same requirements are necessary when ordering IEC relays.

Solid-State Relays

Solid-state relays (SSRs) are designed to provide the same purpose as electromechanical relays. (See Figure 4-33.) The solid-state relay has no coil or moving contacts but is usually an SCR or TRIAC circuit controlled by separate voltage. Again, the purpose of the relay is to relay a control signal to another device. There are many factors to consider when applying an SSR to a particular situation.

Because solid-state relays have no coils, electromagnets, or moving parts, the control voltage that activates the relay and causes the load terminals to conduct can be a wide variety of signals. Relays can be TTL operated (digital logic levels) or operate on 3–30 V DC or 120 V AC. The input does not need to have inrush current ratings or pickup voltage ratings. In fact, many of the SSR are opto isolated devices. This means that the input operates a light emitting diode (LED). The light operates a photoreceiver, which causes the output to conduct. This has the effect of completely isolating the load voltage from the control signal voltage.

The SSR has no actual contacts to open and close so there is no spark created when opening a circuit under load. This advantage allows the SSR to be used in situations where an arc or a spark may be dangerous (such as hazardous locations). The drawback to not having contact closure is that solid-state devices produce a voltage drop even when conducting. The load voltage is usually large in comparison to the small drop at the relay (.7–1.4 V) so the relay drop is not critical (for example, 120-V source drops 1.4 V at relay, therefore 118.6 V remains at the load). (See Figure 4-34.) However, if the load voltage is small, the relay contact drop may be a critical loss and affect the load.

Controllers, Relays, and Timers 117

FIGURE 4-33 Solid-state relay has no moving parts. *(Courtesy of Carlo Gavazzi Company)*

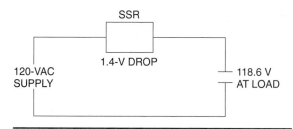

FIGURE 4-34 Voltage drops within the solid-state relay (SSR) may be objectionable at low voltages.

Another consideration when dealing with the fact that there are no contacts to separate, is the characteristics of solid state devices discussed in electronic starters. That is, even with the relay in the off state, or no control voltage supplied, there is some small leakage current. This small leakage current may be enough to activate some loads, or turn on output indicators such as neon lights.

SSRs usually have one set of NO contacts. Therefore, when the requirements are for multiple-pole switching or double-pole switching (common to two points) SSRs are not as versatile. The advantages of the SSR, other than no contacts, include the actual switching speed or the response time. Because no parts need to move, the speed of the switch for the SSR is many times faster than the electromechanical relay (EMR). Response time for EMRs in the hundredths of seconds is typical, but the SSR response is microseconds. If loads are to be switched continually and quickly, the SSR is a superior device.

Types of loads will affect the type of SSR you choose. Resistive-type loads would work best with a relay that is classified as a zero switch relay. This means the relay will turn on as the load voltage crosses zero. This is an advantage in resistive loads as it allows the current and voltage to build slowly and prevent spikes that may burn out resistive loads (such as incandescent lamps) faster. Inductive loads work better with an SSR that has an instant on relay. This means the relay may turn on at any time during the AC load waveform. This is an advantage to prevent current surges produced by inductive devices.

The third type of SSR is used for combination loads—the universal switch. This switching occurs as the AC load voltage has peaked and is moving toward zero cross.

Peak switching or universal switching is preferred when the load voltage and current are 90 degrees out of phase, the voltage is at peak, but the current is near zero. An analog switching relay has a feature that permits the output voltage at the relay output to ramp up over time. As the relay is turned on, the output voltage slowly rises to the full applied voltage available. As the control voltage is removed, the relay "contacts" stay closed until the load current crosses zero.

Comparison of Electromechanical Relay (EMR) and Solid-State Relay (SSR)

EMR advantages	SSR advantages
Multiple pole relays with NO/NC combinations	No mechanical wear or false relaying because of vibration or motion
Contacts are definite make or break	Switching modes variable
Contacts do not create voltage drop	No contact wear
Lower cost	Fast switching speeds
Easy to inspect and troubleshoot	Usually have wide range of control voltage
EMR disadvantages	SSR disadvantages
Contacts wear out and arc	Limited number of switching per relay
Slower switching repeatability times	Off state may have leakage current to load
Susceptible to vibration and electromagnetic interference	Susceptible to high heat

Finally, consider the need for isolation of the load. Does the load need definite turn off? SSRs do not provide definite open contacts. Will the relay operate in high ambient temperatures? SSRs do not work well in high ambient temperatures. Will the relay be operating in an electrically noisy environment? SSRs are susceptible to electrical noise and may have nuisance turn-on.

When deenergizing the power to inductive loads, there is a possibility of the collapsing magnetic field creating an inductive kick. The collapsing magnetic field generates a large spike or kick. This large voltage is impressed on the SSR load control. Sometimes, this is enough to destroy the SSR. One solution for this situation is to use a varistor designed for the normal line voltage. As the voltage applied to the varistor peaks above the rated voltage, the varistor's resistance decreases, thereby dissipating the inductive kick. Upon return to normal voltage, the varistor returns to a very high resistance and does not affect the load. See comparison list of EMR and SSR relays.

Timers

There are three categories of timers available. They are synchronous clock timers, dashpot timers (two general styles), and electronic timers (discussed later).

Synchronous clock timers use an electric clock motor driven from the AC power line to maintain a clock in sync with standard time. (See Figure 4-35.) As the clock drives a dial, tabs are often placed on the dial to trip cams to open or close contacts. Again the clock power is derived from one circuit and the contacts are used to control another circuit. The circuit is electromechanical as the tabs actually mechanically trip a contact mechanism. These types of timers are used to control motors or other devices where precise timing is not critical. Often, they are 24-hour or seven-day clocks that allow you to skip days or have multiple turn-on/turn-off cycles on each day. The contacts can be NO or NC and change state

FIGURE 4-35 Synchronous clock timer.

when tripped by the tab. They will stay that way until tripped back again by another tab. These timers do not provide on delay or off delay, only time of day control.

Timers also are an important part of control sequences. Electromechanical timers have different ways to create a time delay. The first major distinction between timers is an "on delay" or an "off delay." The different symbols are shown in Figure 4-36. The electromagnetic coil is again part of one circuit while the contacts are part of another circuit. One type of timer is an *off-delay* timer. The contacts change state from NO to closed or vice versa when power is applied. When power is removed from the timer coil, the contacts remain as is until time has elapsed, then change back to their normal deenergized state. The timed change of the contacts has occurred after the coil power has been turned off, or deenergized. Time delay on deenergization is referred to as TDD timers.

Another timing option is the *on-delay* timer. Timing occurs as power is turned on to the relay's coil. As power is applied, the

ON DELAY, TDE, TIME DELAY ON ENERGIZATION
NORMALLY OPEN, TIME CLOSING

 TIME DELAY ON ENERGIZATION "ON DELAY"
NORMALLY CLOSED, TIMED OPENING

 TIME DELAY AFTER DEENERGIZATION "OFF DELAY"
NORMALLY CLOSED, TIME CLOSING

 TDD (TIME DELAY ON DEENERGIZATION)
NORMALLY OPEN, TIME OPENING

FIGURE 4-36 Time-delay symbols used for timers.

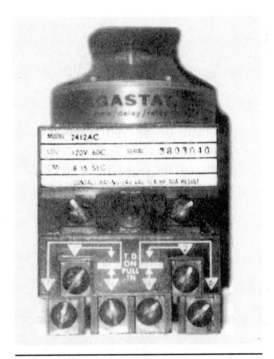

FIGURE 4-37 Pneumatic timing relay.

contacts do not immediately change state but rather stay in the original deenergized state, NO or NC until time has elapsed. At the end of the on delay, the contacts change position and remain that way until power is removed. Time delay on energization is referred to as TDE timers. When power is removed, the contacts immediately return to their normal deenergized state. The delay has occurred during the power on cycle, therefore the names *on delay* or *delay on energization*.

On- and off-delay timers found in control circuits can be dashpot timers. Dashpot refers to the method of control. Air *dashpot* timers are referred to as pneumatic timing. No special air pressure lines are required, only ambient air. As the electromagnet is energized, air is forced toward a controlled opening by a diaphragm. (See Figure 4-37.) If the air is allowed to escape slowly, the diaphragm takes a long time to move to its final position. When the diaphragm has moved completely, it moves a set of contacts and the delay has occurred on energization, *on delay*. By adjusting a needle valve in the orifice, the air escapes faster or slower, providing adjustable timing. Upon deenergization, a check valve allows the air to refill the chamber quickly and no delay is encountered.

Likewise, on *off delays* the air is allowed to escape quickly, thereby changing the contact's state quickly. The air allowed to refill the chamber is controlled by a needle valve in the opening, letting air in faster or slower.

Another style of dashpot timing is accomplished by an oil flow control, rather than an air flow control. The same principle applies, but the oil is held in a reservoir. By means of a control valve and a check valve the liquid is allowed to flow faster or slower to provide the same type of timing as in the pneumatic

Controllers, Relays, and Timers

timing. Electromagnetic coils are still used extensively for control relays and timers. However, the solid-state or electronic timers have become much cheaper and have a wider range of control and more precise timing.

ELECTRONIC TIMERS

Electronic timers (see Figure 4-38) have become much more rugged and are more versatile than many mechanical timers. The internal timing usually takes place with an RC time-constant circuit or with quartz timing. Many of the timers can be set for on or off delay, trigger timing, and they have the versatility of a very wide range of timing. As in other electronic controls, the switching is done electronically, not mechanically, so the same precautions that apply to other electronic switching also apply to the electronic timers. Timers often come with eight- or ten-pin bases.

Timers

A programmable timer is another form of an electronic timer, but with many more adjustable features. Features include on and off delays, instantaneous contacts, convertible contacts from NO to NC, and also the retentive timing feature. The retentive feature allows the timer to hold its accumulated time even if the control signal is interrupted. In other words, the timing will continue from the point of interruption when the control signal returns. A nonretentive time does *not* retain its timing memory.

One-shot timers are timers that provide one set of timing operations. The contacts may wait to change position, either open or closed, until a preset time has elapsed. The contacts then change and remain that way for another preset time. They will not change again until the whole sequence is reinitiated.

Recycle timers are used to recycle the timing mode. They are similar to one-shot timers in that they may wait to turn on a load for a predetermined time, then turn off again, but they recycle that time sequence over and over again at preset intervals until the control power is removed.

Multifunction timers are designed to be versatile in that they can be programmed to do different applications. This allows the electrical control designer to set it up to perform various functions. See Figure 4-39. To use the relay in various applications the electrical control designer must be aware of its options and how to read timing and contact charts. With two sets of single-pole–double-throw contacts the relay may have one set of NO and one set of NC contacts. See Figure 4-40. The relay used as an on-delay timer shows that the contacts switching after the preset time has elapsed, stay that way until power is removed,

FIGURE 4-38 Electronic timers with multiple and adjustable functions and wide range of timing.

Function Relays, Interfaces and Converters
Solid-State Time Relays

3RP20 / 3RP15

Overview

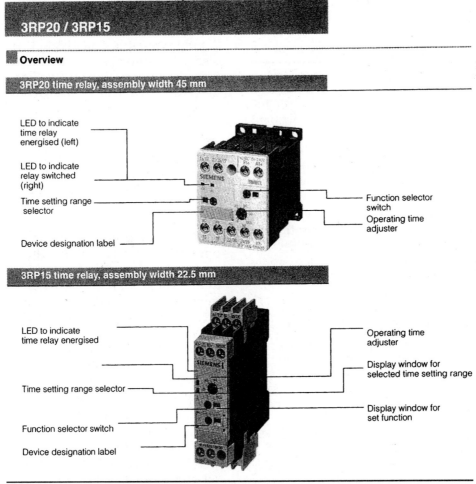

FIGURE 4-39 Solid-state multifunction timers.

and then instantly change back to normal. The same contacts can also be used, as in the next line down, where one contact is timed and the other contact is instantaneous. The same relay is used in the same manner but is now an off delay. The next scenario uses the timer contacts with both an on delay and an off delay. The next sequence shows one contact of the pair as an on delay and an off delay with the second contact operating as an instantaneous relay contact. The next sequence is called a flashing contact where the contact continuously changes contact close to open in a continuous cycle of preset times. The last sequence is a passing-make contact where the contacts change once for a timed interval, then return to their normal state for that "pass" of control signal. These timers are

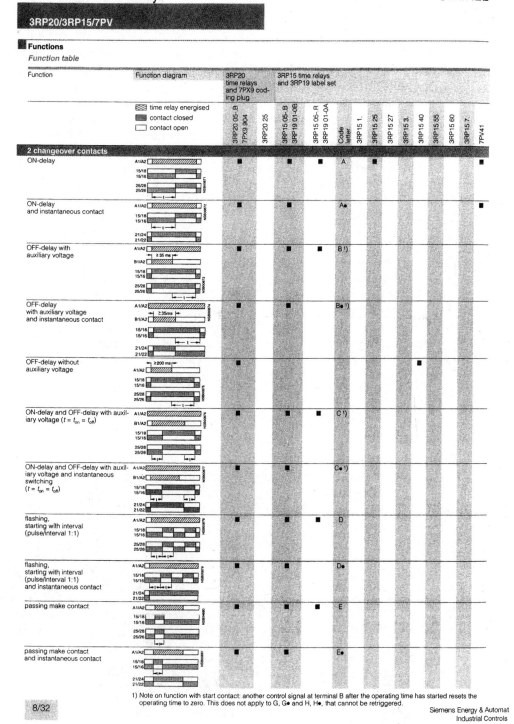

FIGURE 4-40 Timing contacts chart shows sequence of timing and contact closure.

adjustable from the front of the relay package. See Figure 4-39 for a legend of how the timer is set.

Programmable relays are available to be set up and programmed to operate under various modes and with adjustable parameters. One style of programmable logic module is made by Siemens, under the name LOGO. The basic model has four sets of output contacts that can be programmed to operate in many different ways, creating a very versatile relay for almost any use. The programs are entered by a computer program and then downloaded to the relay module via a connection cable. The computer is disconnected and the relay runs independently. The programs can be stored on a removable EEPROM so that the program can be transferred to other relays without the need for further computer programming. A system voltage that is always applied to the relay during operation powers these relays. There is class 1 for up to 24 V AC or DC. Class 2 is used for up to 240 V AC or DC. Figure 4-41 is a basic module with eight input connections and four output contact operations. Models also come with analog inputs and analog outputs.

The relay contacts can be individually programmed to any of eight basic functions. These functions are set up as logic functions. (1) All inputs are connected as AND functions so that all inputs are made to create a output contact closure. (2) All inputs can be a series in a NOT function with all inputs being made to give an open output contact. (3) All inputs are connected in parallel to create an OR function to create an output contact closure. (4) The inputs can create an exclusive XOR. (5) Inputs are configured as a NOT input which means a closed input yields an open output. (6) All the contacts can be in parallel to create a NAND-Not And. (7) Each input can be set with an edge-triggered pulse to create an AND gate with a retentive input pulse. (8) The last basic setting is a NAND gate using an edge-trigger. See Figure 4-42. These parameters can be adjusted using the curser keys on the front of the relay and using the LCD panel on the relay. For instance, the timing modes can have their time frames adjusted

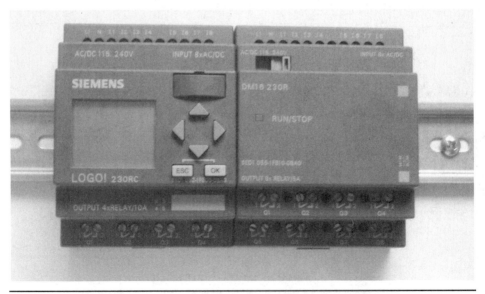

FIGURE 4-41 Photo of Siemens multiple-function electronic relay.

FIGURE 4-42 Solid-state motor controller.

without having to reprogram through the computer interface.

There are twenty-six special functions that can be programmed into the contact operation making this programmable relay similar to a small PLC. Some common modes are on or off delays, retentive timers, short time to twelve-month time switches, up/down counters, and a latching relay. The programs can be entered in two formats. This allows the electrical personnel to enter programs in whichever electrical language they are most familiar with, either Function Block Diagrams or Ladder Logic.

SOLID-STATE MOTOR CONTROL

Electronics has not been used extensively in motor control until recently. With the production of higher power handling capability and with the industrial hardening of the electronic circuits, an increasing number of electronic devices are being used to control large power systems.

Solid-state motor control is now available in a variety of forms. Solid-state controllers (as shown in Figure 4-42) can be used to soft-start a motor. This means that the applied voltage to the motor is ramped up to a full voltage level and allows the motor to gradually accelerate to rated speed. As will be discussed in Chapter 6, as a lower voltage is applied to a motor, the torque will also be reduced. This allows the motor to accelerate slowly to full speed, thereby reducing starting torque on gears, coupling, or belts. This effect also reduces the inrush current to the motor and reduces voltage fluctuations on the line.

A second form of starting is the current limit mode. The current limit mode only allows a preset maximum current flow to the motor during starting. (See Figure 4-43.) The current can be set so that a maximum current between 250 percent to 550 percent of full-load amps of the motor is delivered during the

SMC PLUS™ Solid–State Motor Controller

Product Description

MODES OF OPERATION

The controller provides the following modes of operation: soft start with selectable kickstart, current limit starting, or across the line starting.

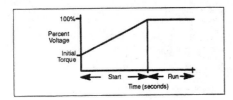

Soft Start
This method has the most general application. The motor is raised to an initial torque value. This is adjustable between 5 and 90% of locked rotor torque. The motor voltage is gradually increased during the acceleration ramp time, which can be adjusted from 2 to 30 seconds. These customer settings are set for the best starting performance over the required load range.

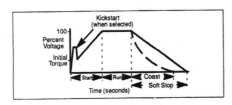

Soft Start *with* Selectable Kickstart
A kickstart or boost can be provided. This is intended to provide a current pulse of 500% of full load current and is adjustable from 0.4 to 2 seconds. This will allow the motor to develop additional torque at start for loads which may need a boost to get started.

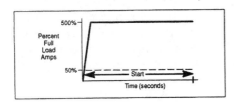

Current Limit Starting
This starting mode is used when it is necessary to limit the maximum starting current. This can be adjusted for 50 to 500% of full load amperes as shown.

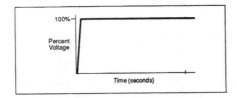

Across the Line Starting
This mode is used for applications requiring across the line starting. The ramp time is set for less than 1/4 second as shown.

FIGURE 4-43 Modes of operation that can be used with solid-state motor starters. *(Courtesy of Rockwell Automation)*

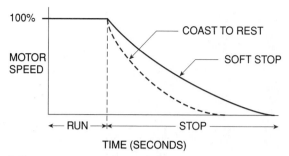

Soft stop extends stopping time to minimize load shifting or spillage during stopping.

FIGURE 4-44 Graph shows ramp for soft-stop control with electronic controllers. (Courtesy of Rockwell Automation)

starting time, which is also adjustable. This method provides better starting torque but still keeps the inrush current to reasonable levels. The solid-state starter can also provide full voltage to the motor as does a conventional "across the line" starter. Additional features on some starters provide for a soft stop control. This feature provides a continued voltage that is ramped down to allow the motor to come to a more gradual stop. (See Figure 4-44.)

Another type of AC solid-state controller is the adjustable frequency device (to be reviewed in Chapter 7).

A word of caution regarding electronic starters: As with all semiconductor devices, there is no perfect insulator or conductor. The controller is made of a semiconductor material. This means that even when the starter is turned off, there may be some current flow to the motor. This off-state current is the leakage through the semiconductors. Make sure you provide isolation switches that can definitely be opened before working on the power circuit.

Solid-state controls also are more susceptible to electrical noise compared to magnetic starters. Electrical noise could be in the form of fluctuations that are carried on the 60-Hz sine wave. These peaks and power fluctuations are delivered to the solid-state starter. Where there is a great deal of electrical noise caused by switching, and so on, the starters may be random in turn on or turn off characteristics. Also, semiconductors are more sensitive to overvoltages and surge currents. Surge protection may be required to prevent inadvertent operation.

Failure of solid-state control often occurs in the "on" mode. This is crucial if the starter is controlling critical safety-related systems. Many times, electromechanical devices are better suited for critical systems where an "on" failure could result in costly damage.

CONTACT MAINTENANCE

For proper operation of contacts there must be solid contact between the conducting surfaces. If there is not good connection, there is a voltage drop because of high resistance; therefore, less voltage is delivered to the load. Also, heating occurs at the contacts, which may cause further damage. One of the main reasons contacts wear is the arc damage that occurs when the contacts are opened. As the contacts open, under load, there is still current flow between the contact surfaces. The path for the continued flow is through the ionized air that creates the arc. If allowed to continue, the hot arc will begin to melt the contact surface. (See Figure 4-45.) One way to remove the arc from the contact surface is to provide *arc horns*. The horns, as shown in Figure 4-45, allow the hot arc to expand and rise up the horns, as heated

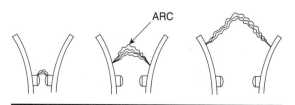

FIGURE 4-45 Contacts with arc horns are used to move arc off contact surfaces.

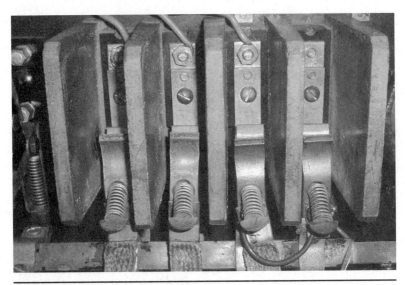

FIGURE 4-46 Arc barriers are used to contain and direct the arc.

air rises. Notice the arc horns are tipped away. This causes the arc to expand and will eventually break apart. In the meantime, the arc has heated the tips of the arc horn and not the contact surface.

Another way to remove the arc from the contact surface is to use arc chutes. As seen in Figure 4-46, the arc horns actually deliver the arc to the arc chute. The segmented sections split the arc into many small arcs that are allowed to rise into the insulating material. Each arc is then cooled and allowed to break apart without damage to the contacts.

For larger arcs in heavy-duty contactors, magnetic blowout coils may be used. Figure 4-47 illustrates the appearance and connection of the magnetic blowout coils. The large turns of wire are placed in series with the contacts, so that main motor current passes through the coil and the contacts. As the contacts open, current still continues to flow. This current flow also continues through the blowout coil which produces a magnetic field. This magnetic field reacts with the arc, which is a current carrying path and magnetically blows the arc out of the way. As the arc is stretched, it cools and breaks. As the arc is extinguished, the circuit is broken and the current flow is interrupted; therefore, there is no longer current through the blowout coil.

Often, when contacts become too damaged to properly conduct current, they can be replaced. See the manufacturer's recommendations for replacement procedures. Figure 4-48 shows a progression of arc damage.

SUMMARY

This chapter introduced you to the basics of motor controllers and the difference between control circuits and power circuits. There are differences between NEMA and IEC starters, and reasons were presented for different applications.

Different pilot devices were explained by using the various forms of the sensors. Relays, photoelectric, proximity, and timers are all examples of possible pilot devices that can provide control signals to a motor control

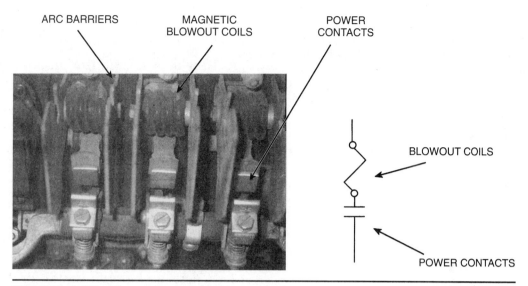

FIGURE 4-47 Photo of magnetic blowout coils that magnetically force arc away from contact surfaces and help extinguish the arc.

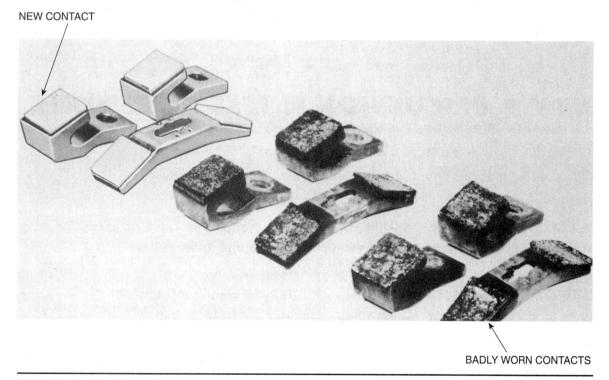

FIGURE 4-48 The left contacts are new, the center contacts are beginning to wear, the right contacts are worn out and need replacing.

circuit. The basic ladder diagram or schematic diagram was introduced to enable you to see the electrical sequence of controls. Differences between electromechanical controls and solid-state controls were presented to allow you to begin determining the proper control for an application.

Finally, if you do need to maintain electomechanical controls, contact maintenance becomes important. You should be able to analyze the problems with contacts and determine if they need replacing. Also, arc control was explained to help you decide if all the arc control techniques were working properly to control the arc.

QUESTIONS

1. What does the acronym NEMA stand for?
2. What part of the contactor is used to prevent chatter and unwanted armature release during AC operation?
3. If an electric starter is to be opened and closed frequently, which contactor design would you use? Why?
4. Explain why starter inrush current is higher than normal operating current.
5. At what percent of normal voltage will contactors deenergize or drop out?
6. Give two reasons why contactor coils may become hot.
7. How does IEC utilization philosophy differ from NEMA philosophy?
8. Explain what is meant by pilot devices.
9. Draw the symbol for an NC pressure switch.
10. Explain two styles of photoelectric sensors.
11. What is the purpose of a NEMA 1 enclosure?
12. Where are hazardous locations specified in the NEC®?
13. When wiring control relays for motor control, why not place relay coils in series?
14. Explain why voltage and current ratings are both needed for contact rating.
15. Draw an on-delay NOTC contact symbol.
16. Name one undesirable feature of electronic starters.
17. Give one advantage of SSRs.
18. Name one way to draw the arc away from contacts.

CHAPTER 5

THREE-PHASE MOTORS, CONTROLS, AND FULL-VOLTAGE STARTING

OBJECTIVES

After completing this chapter and the chapter questions, you should be able to

- Identify different types of three-phase controllers
- Connect three-phase motors to manual or magnetic starters and connect them as reversing starters
- Begin designing and troubleshooting motor control schematics
- Recognize and use sequence control circuits
- Use multispeed motor control techniques in the power and control circuits
- Understand the difference between wiring diagrams and schematic diagrams
- Compare the differences between relay control and programmable logic control
- Recognize the block function of variable frequency electronic drives
- Make power and control connections for wye–delta starting

- Identify the benefits of motor control centers
- Work safely and follow the Occupational Safety and Health Administration (OSHA) requirements for lock out/tag out

KEY TERMS IN THIS CHAPTER

MCC: A motor control center is used to house a number of motor controllers in one location. It has the advantage of using a common feeder and grouping controllers for easier control interaction.

Open or Closed Transition: This is a term that indicates how a motor is transferred from the starting mode to the running mode of operation. If power to the motor is interrupted during the transition, it is open transition. If there is no interruption of power to the motor, it is closed transition.

Part Winding Start: A method of applying full-line voltage to a motor during start, but initially only to part of the windings, therefore reducing line current during starting.

PLCs: Programmable logic controllers are used to control complex control systems. These controllers are the electronic equivalent of an electromechanical relay control system.

Schematic Diagram: A diagram that represents the electrical relationship of each of the control functions, laid out in a logical sequence.

Sealing Circuit: The electrical circuit that is established in a motor controller designed to keep the magnetic coil energized once a momentary push button has been reopened.

Single Phasing: A condition that exists where one of the three-phase wires has been disconnected from a three-phase motor.

Starting Torque: The percentage of normal running torque that a motor will develop during starting. This is also referred to as locked rotor torque.

Variable Frequency Drives: A motor speed control system that uses electronics to vary the applied frequency to an AC motor and therefore control the speed.

Wiring Diagram: A diagram that is laid out in a manner to show physical location of components and how the wiring is to be run from one to another.

Wye–Delta: A method of starting three-phase motors that applies line voltage to the motor but reconnects the coils of the motor to reduce the line current during starting.

INTRODUCTION

As you know, motors are used for many applications in providing mechanical horsepower. Sizes range from fractional horsepower to thousands of horsepower. Control systems used to start, stop, and vary the operations of these motors are sometimes quite complicated. The essentials of the type of control and the desired outcome are essential in installing, maintaining, and troubleshooting the motors and the control system. Some basic considerations are whether the motors will have full voltage applied to them immediately or whether the voltage will be reduced in some manner. Another decision to make is whether the motor will be controlled automatically or manually. Most motors will be controlled automatically.

MANUAL CONTROL

Occasionally, motors will be installed that have a local control requirement and the motor can be controlled manually. This means that the equipment operator will provide the physical energy needed to close contacts in the motor circuit. For three-phase motors, these

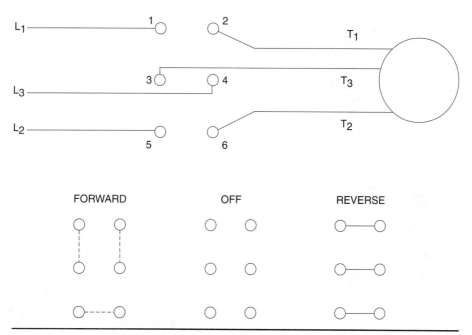

FIGURE 5-1 Drum-switch control used for a three-phase motor.

are typically drum controllers. As referred to in Chapter 3, for single-phase motors, the drum-switch handle is used to forward-stop and reverse the motor. The three-phase motor can be controlled in the same manner. (See Figure 5-1.)

Note on the connection diagram there are only six leads. Three of the lead connections are for the input voltage L_1-L_2-L_3 and the other three act as the power connections to the motor. Even though the motor has nine leads, the motor connections are assumed to be connected per the normal connection patterns and three power leads brought to them. Some manual drum controllers have eight lead connections for forward and reverse. (See Figure 5-2.)

Other types of manual controls resemble a magnetic starter. Figure 5-3 illustrates a manual controller. As with the drum controller, the contacts are manually or physically pushed closed to close the contacts supplying power to the motor. These controllers have overload

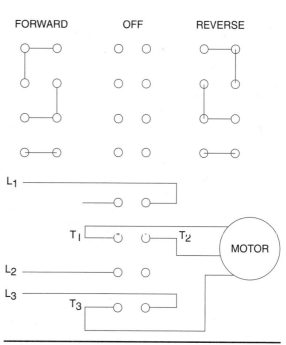

FIGURE 5-2 Eight-lead drum controller for three-phase motor.

ELECTRIC MOTORS AND MOTOR CONTROLS

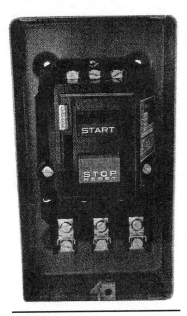

FIGURE 5-3 Three-phase manual starter without low-voltage protection. (Courtesy of Square D/Schneider Electric)

circuit and the power circuit. One method of control is governed by two-wire control, where a pilot device is the on–off decision maker. The second style of control is the three-wire control where a stop/start station or a momentary contact switch is used, and the auxiliary holding contact of the controller also is used.

New starters are available that provide different protections compared to the basic starter described in Chapter 3. Some starters provide extra protection because of the addition of electronics or solid-state components to the basic starter design. Although the starters are monitoring mechanisms that will release the power contacts if there is an overcurrent to the motor.

In most manual controllers there is no control circuit involved. Some manual starters make use of a holding coil or magnetic coil as described in Chapter 3. This coil will hold the contacts shut after they are physically closed. If a magnetic coil is used on a manual controller, there can also be a low-voltage protection system available. If the voltage goes below standard limits, the coil will release the contacts and remove power from the motor.

MAGNETIC CONTROL

Magnetic control devices are the most common type of control for motors in commercial and industrial settings. As discussed in Chapter 3, distinctions must be made between the control

FIGURE 5-4 Magnetic starter with electronic overloads.

also NEMA rated, they can provide extra protection such as phase loss protection and self-protection against short circuits. The overloads in the new type starters may be of the heaterless design. This type of starter (Figure 5-4) uses current sensors and solid-state overload, adjustable trip point overload relays.

The overload section has an adjustment to control the actual trip point of the overload mechanism. If the trip point is not available on the overload relay you have, you can loop the T_1 leads going to the motor through the overload section to increase the apparent amps through the overload current sensor. The method is described in Figure 5-5. The concept is the same as passing multiple turns of a conductor through a clamp-on-type ammeter. One conductor passing through the jaws will read the actual magnetic field of one conductor. If you pass two conductors (one loop) through the jaws, you double the magnetic field and will read twice the actual single conductor current.

Other starters are available that use mechanical contacts for the power circuit but are also a hybrid between the electromechanical and the electronic sensing relays. (See Figure 5-6.) The control voltage is monitored by a microprocessor chip. The chip monitors the control voltage to determine if there is sufficient voltage to energize the magnetic coil and close the power contacts. If there is sufficient voltage, the contacts will close. The chip will open the contacts to prevent chatter and the "kiss" of the contacts under varying control voltage. Kiss of the contacts means that the contact faces alternate between touching and separating; this creates arcing of the contacts. Arcing of the contacts creates heat and wear and also may weld the contacts together. This particular starter (contactor and overload) also has increased contact pressure to reduce heating and reduced contact bounce to lessen the physical and electrical wear that occurs on contact closure. The overload relay section is also an electronic sensor design. Many considerations are made to reduce the amount of heat within the starter. The advantage to reducing heat in all areas is to physically reduce the size of the starter.

This particular starter (Figure 5-6) also has flexible style of overloads that can be changed

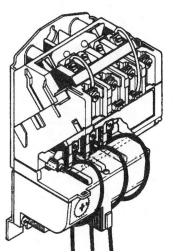

LOOPING OPTION

One loop is an easy procedure and multiple looping gives you maximum flexibility in situations where motor, horsepower, and voltage are unknown. Imagine the advantage, for example, of an overload with a current range of 9 to 18 amps that could be used on 460V motors from 1½ to 10 horsepower. The following table demonstrates how the looping process reduces the current setting of the overload by the number of times the wires pass through the windows of the overload thereby extending the basic full-load amp range of the overload.

	Overload current range	# of loops	# of times wire passes through window
Current range shown on label	9–18	0	1
	4.5–9.0	1	2
	3.0–6.0	2	3
	2.25–4.50	3	4

All current values are expressed in amps.

FIGURE 5-5 Looping the motor conductor through the current sensor extends the overload sensor range. *(Courtesy of Furnas Electric Company)*

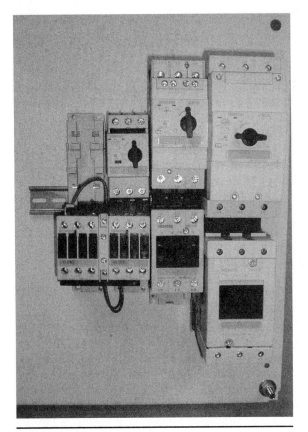

FIGURE 5-6 Solid-state motor starters.

from class 10 to 30 to 20 by changing a dual in-line package (DIP) switch. The DIP switch can be switched to provide three classes of protection. Class 10 protection is used to protect hermetic refrigeration motors, submersible pumps, or IEC-designed motors that are set up for quick starting. Class 30 protection is intended for motors that have long acceleration time, so they may have extended high-current loading. Class 10 overloads are designed to take the motor off-line in 10 seconds with 600 percent of full-load current; likewise, Class 20 trips within 20 seconds, and Class 30 trips in 30 seconds. Class 20 protection is the normal design and is used for all other applications. In addition, these overload sections can provide ground fault protection of equipment. Ground fault protection prevents damage to equipment from line to ground faults. It opens the power circuit at a lower current level than the short circuit protection on the supply circuit. (See NEC® Article 230.95 for requirements.) It also can monitor for phase loss or imbalance and open the power contacts when an imbalance occurs.

Because control of motors has become so critical to industry and business, close monitoring may be essential. The last starter discussed also has a communications module that allows the starter to transmit data to a computer network. The starter receives information on start, stop, and reset control and can provide data on the current in the three phases, phase imbalance, control voltage, status of starter, overload trip cause and trip data, and DIP switch setting. (See Figure 5-7.)

Wiring of this type of starter differs slightly from the usual electromechanical type, as described in Chapter 3. See Figure 5-8 for typical three-wire connection of start/stop switches.

Up to this point, all the power to the motor has been on or off in a simple, single-motor circuit. The control has been either two- or three-wire control used to operate a starter. The starter may be electromechanical (where the power contacts actually move) or electronic (where there is no physical contact motion).

Electrical control manufacturers have been implementing control schemes that incorporate the large-scale motor control systems, the programmable controller systems, the electronic drives, and variable frequency drives (VFDs) along with the Internet to provide control and monitoring from local control to control from any location in the world. Several worldwide companies have cooperated to create a system of communication protocols that interface with each other. One such system is called Profibus. Profibus is an open (nonproprietary) fieldbus connection platform that is used to allow devices and controllers to talk to each other in industrial communications. Fieldbus is a system that manufacturers use as a proprietary system

FIGURE 5-7 Profibus-rated equipment.

that can connect sensors, drives, controllers, transducers, actuators, and so forth on a digital communication system through wire, fiber optics, or wireless communication. Through this evolving open system, various manufacturers have collaborated on a system that they can all use to connect their equipment and have it function in a factory infrastructure or in process automation.

To be able to see the functioning of the fieldbus and the status of devices or information about system components such as starters or timers, a method to convert the data to usable text is often used. This equipment is known as a human machine interface (HMI). These devices can either be in the form of a small control panel screen that indicates the status of a particular piece of equipment (see Figure 5-8) or the HMI can be a computer screen that monitors the total operation and provides data about the system flow or operations. Sometimes the HMI uses touch-screen controls to not only view operational data but also to control the functions of various components of the system. See Figure 5-9 as an example of how local area networks (LANs), proprietary industrial networks, HMI, and device communications evolved through collaboration on an international basis.

As can be seen in Figure 5-10, the integration of the industrial control network to the equipment on the manufacturing or automation floor can be accomplished through a fieldbus system. The communication to the office and management information systems (MISs) is typically though an Ethernet system. The connection of these two systems is called vertical integration. Another system called PROFINET is available to integrate these systems to provide factory real-time data to office managers. This system is again an open, cross-vendor system that allows various manufacturers to connect to the Internet and provide information and control.

ELECTRIC MOTORS AND MOTOR CONTROLS

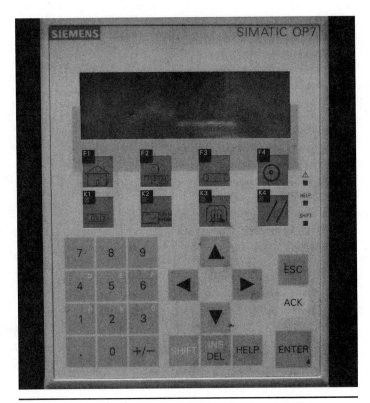

FIGURE 5-8 LCD panel used for human machine interface (HMI).

Network Level	Examples	Examples of Devices or Systems Connected
Management Level	Ethernet	Computers Running Management Information Systems (MIS)
Control Level	Ethernet	PLCs, Industrial Computers
Field and Process Level	PROFIBUS FMS PROFIBUS DP PPOFIBUS PA	PLCs, CNCs, Variable-Speed Drives, HMIs, Industrial PCs
Device Level	Actuator Sensor Interface (ASI)	Actuators (Output Devices), Sensors (Input Devices)

FIGURE 5-9 Chart of levels of components connected to a vertically integrated system.

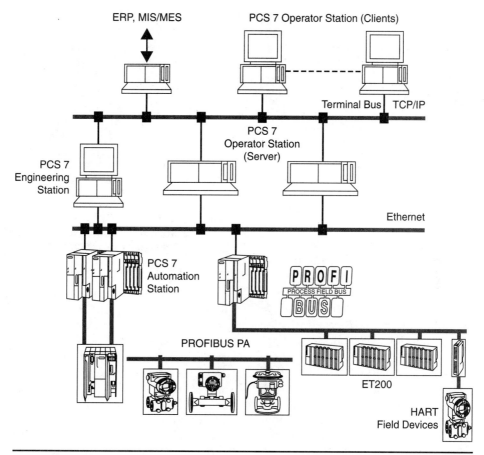

FIGURE 5-10 Example of vertical integration from factory floor to office information systems.

Electronic Control Devices

Control functions are now controlled over a wide range of capabilities and functions. Relays and control devices are constantly changing to reflect the current trends in automation and modular systems. Control relays are available with a wide variety of functions built into the components. The trend is to reduce inventory cost by stocking fewer base relays and then adapting them to specific functions. For instance, one model of control relay is designed to be the contact block (see Figure 5-11). To that contact block a snap-on module can be added, such as a timing module, auxiliary contacts, or surge suppressors. Because programmable logic controllers (PLCs) are so popular as controllers, a special module can be added to the line voltage rated contacts to interface with a PLC. This module uses the PLC output and creates a 24-V DC low-power consumption coil to operate the relay.

Contactors are also more flexible and often smaller, based on the IEC standards rather than NEMA. For example, a 75-HP-rated motor contactor is only 70 mm wide (about 2.75 inches

FIGURE 5-11 A multifunction relay that is adapted by add-on components.

wide). The overload section for this starter uses bimetal operation based on heaters in the overload section. The heaters heat the bimetal strip directly based on the motor current and are not affected by harmonics, or nonsinusoidal waveforms. The overload section also has another bimetal element that is sensitive to the ambient temperature that prevents nuisance tripping if the controller is at a higher temperature than the motor. In the event of a phase loss, another function of the overload section causes the contactor to trip out sooner if a phase loss is detected. The overload can be manual or automatic reset controlled in the field, or can have a remote reset added to the section.

Combination Starter

Motor starter protectors provide the functions of the required disconnect and the short-circuit protection needed in motor control applications. The motor starter protector (MSP) can be used by itself as a manual motor starter. Because it has adjustable overloads, it provides the short-circuit protection through the magnetic trip protection and also provides running overload protection through thermal overloads. By adding a magnetic contactor to the bottom of the MSP, the system becomes a combination starter. See Figure 5-6.

REVERSING THREE-PHASE MOTOR STARTERS

One of the most used control circuits that is slightly more complex than a single-direction motor is the three-phase reversing starter. As shown in Figure 5-12, the power circuit for the motor is composed of a starter (with overloads) and a contactor (without overloads). The process used to reverse a three-phase motor is to interchange two line leads to two motor leads. Typically, L_2 is the line lead that is connected to the T_2 motor lead in both forward and reverse. L_1 and L_3 are interchanged on the top side of the contactors. If you purchase the reversing controller as a prebuilt unit, the top and bottom power leads are usually already connected. Notice that there is only one overload relay unit. This must sense the current in both the forward and reverse direction.

A simple reversing control schematic circuit is shown in Figure 5-13. Notice that in the ladder diagram, the rungs of the ladder are numbered on the left-hand side. Numbers on the right-hand side are a listing of the lines where the contacts for the associated relays will be found. It is a good idea to label your drawings this way. Notice that some of the right-hand numbers are underlined. This means that the contacts on that numbered rung will normally be closed contacts. When the relay on that rung has no power, the associated contacts will

Three-Phase Motors, Controls, and Full-Voltage Starting 141

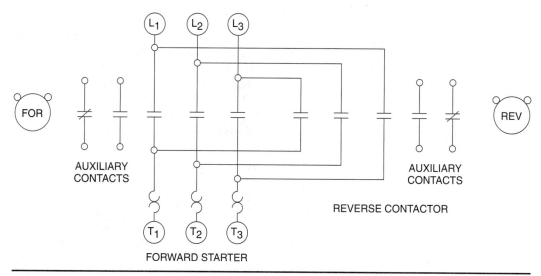

FIGURE 5-12 Power circuit for a three-phase reversing starter.

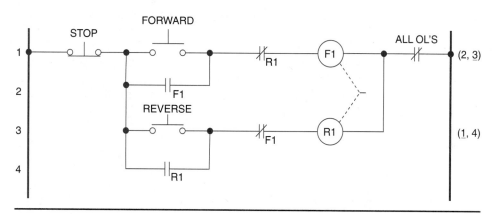

FIGURE 5-13 Simple reversing control circuit with forward and reverse mechanical interlocks. Note that the right-hand number indicates on which rung the contacts associated with the coil are located.

be closed. The stop button must always interrupt power to the control relays and the motor starters.

As shown in Figure 5-13, pressing the stop button prevents any current flow to the F1 or R1 (forward or reverse) coils on the starters. By pressing the forward momentary contact, voltage is supplied up to the normally closed R1 interlock. If the reverse starter is not energized, this R1 contact will be closed and voltage will be supplied to the F1 coil.

Assuming the OL contact is closed, the F1 coil will pick up (energize) and all F contacts on the power circuit will close (as shown in Figure 5-12) connecting the line leads to the motor terminals. Also, the F auxiliary contacts close to provide a **sealing circuit** around the momentary forward push button as illustrated in Figure 5-13. The F1 normally closed auxiliary contact will open, creating an electrical interlock and preventing power from accidentally energizing the R1 coil simultaneously.

This electrical interlock is good practice but is not a code requirement. In addition to the electrical interlock, there is a physical or mechanical interlock that is connected to the two contactors. If one contactor is energized, a push rod or bar prevents the other contactor from closing. If you make your own reversing controller out of two single starters, you have to add the mechanical interlock. Again, this mechanical interlock is not required, but it is good practice. If both contactors should close simultaneously, you would have a direct short between L_1 and L_3 and an explosion could result.

A variation on this simple interlock control system is a double set of interlocks as shown in Figure 5-14. In this example, push buttons are ordered with both NO and NC momentary contacts. A set of forward and reverse limit switches have also been added to this drawing. The push button for reverse has two sections. The NC reverse section is on line 1 and interrupts power to the forward coil. The NO reverse section is shown on line 3 and connects power to the reverse coil. The dashed line between them indicates that they are mechanically linked and that both contacts move when the push button is pressed.

Also, notice that even though they are physically mounted together, as in Figure 5-15, the electrical function is shown on the schematic diagram on two separate lines. The contactor electrical interlocks are in place as well as the mechanical interlocks. With this type of control, you can change motor direction without first pressing stop. With the simpler control in Figure 5-13, you must first press stop to change direction.

Figure 5-14 also has travel limit switches. The location in the circuit allows the one direction of travel to be stopped if the motor is driving a device that has limits to its travel (for example, a crane that has only so much cable). The opposite direction of travel is not affected by one travel limit being opened. As soon as the motor is reversed and the actuator is no longer holding the limit switch open, it will return to the NC position.

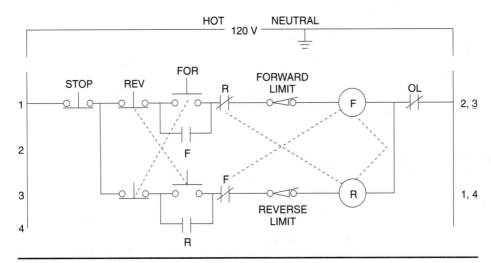

FIGURE 5-14 Reversing control scheme for three-phase motor control includes push-button interlocks, electrical auxiliary interlocks, and limit switches.

FIGURE 5-15 Set of push-button contacts with normally open and normally closed set.

Troubleshooting Control Circuits

Although this technique can be used for any type of circuit, the explanation will be applied to the reversing control circuit in Figure 5-11. It is critical that you understand the general objective of each of the control circuits you will troubleshoot. This allows you to choose the testing procedure. As mentioned, each control circuit may be slightly different, but the method of finding the problems essentially will be the same.

Be careful! You may be measuring live voltage as you work through the troubleshooting procedure. The control voltage may be only 120 V, often the power circuit is much higher and the possibility that you may come into contact with live voltage is very dangerous. *Do not* push the contactor's armature (movable piece) closed to test the power circuit. If the motor will not start, find out why! The motor control circuit may be disabled for some reason (such as a person inspecting the equipment). If you push the starter contacts closed, you have just eliminated many of the safety features. Also, you must be aware of OSHA requirements on lock out and tag out (discussed later in this chapter).

Before attempting to troubleshoot any control circuit, be sure you know where the power for the motors is disconnected and whether this also disconnects the control circuits. The NEC® allows these two voltages to be supplied from different disconnecting means as provided in NEC® Article 430.74. The first step is to disconnect all power from both the power circuit and the control circuit, to visually check the components. Test the overload relay by resetting it. If it has tripped, you usually can hear a reset ratchet click into place. Try to determine why the motor overheated to trip the overload (OL) relay. Is there a mechanical overload on the motor? Are the motor bearings operating smoothly and without drag? If the OL was not tripped, test the starters for free movement. They should move without binding or catching. Are there signs of overheating? If all the mechanical operations are functioning, you can now reapply power. You need a voltmeter to check with live power. Be careful not to ground yourself and keep your hands off live parts or bare ends of test leads.

Use Figure 5-11 for the sequence to follow in the control circuit example.

1. Test for control voltage and for motor power on L_1, L_2, L_3—test all three phases.

2. Keep one lead of the voltmeter (VM) on the neutral conductor of the control circuit, or on the right side runner of the ladder.

3. If the control voltage is 120 V, measure to the opposite runner of the ladder diagram, ahead of the stop button for 120 V AC. If the control voltage is not available, check the control fuse or the main power from the lines.

4. If control power is available at the left runner, move the left VM lead and measure to the right side of the stop. There still should be 120 V on your meter if the stop is closed.

5. Measure to the right of the NC reverse push button on line 1. There should be 120 V on your meter.

6. Measure to the right of the NO forward push button on line 1. Your meter should read 0 volts until you press the forward push button. Press the push button to see if the contacts make connection to conduct the power.

7. Measure to the right of the NC reverse interlock. If you hold the forward push button closed, there should be 120 V on your meter. If not, the reverse interlock auxiliary contact is open.

8. Measure to the right of the forward travel limit. If 120 V is present, double-check the OL relay contact or measure directly across the coil. If you have 120 V across the coil, but it does not pick up, it is a faulty coil or the starter is jammed.

9. If the starter picks up but does not stay energized, check the auxiliary contacts in the holding circuit. These auxiliary contacts must close when the forward contactor closes. The reverse auxiliary interlock contacts are in the forward circuit and vice versa. Check to be sure these are functioning properly.

Refer to Appendix D for more information.

SEQUENCE CONTROL/ COMPELLING CONTROL

As the name sequence control implies, there will be a definite sequence to the starting and stopping of the motors involved in the control of a piece of equipment. Timers are usually involved and may be electromechanical, synchronous clock, or electronic timing (see Chapter 4's section on timers). The control may have many variations, but the idea is to delay the starting of some motors until others are operating. A typical sequence control is shown in Figure 5-16A.

The operation of a basic sequence control is simple. By pressing the start button on line 1, you energize the M_1 coil. This M_1 operates the contactors to supply power to motor 1 as shown in Figure 5-16B. At the same time the M_1 coil is energized, timing relay 1 (TR_1) is also energized on line 2. The M_1 auxiliary (aux) contacts on line 2 provide the holding circuit

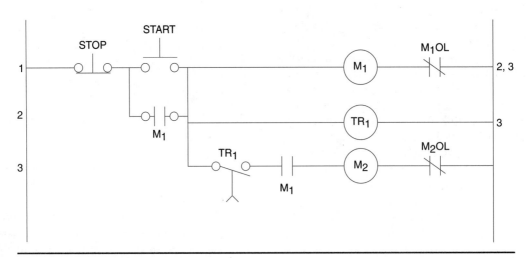

FIGURE 5-16A Sequence control scheme.

Three-Phase Motors, Controls, and Full-Voltage Starting

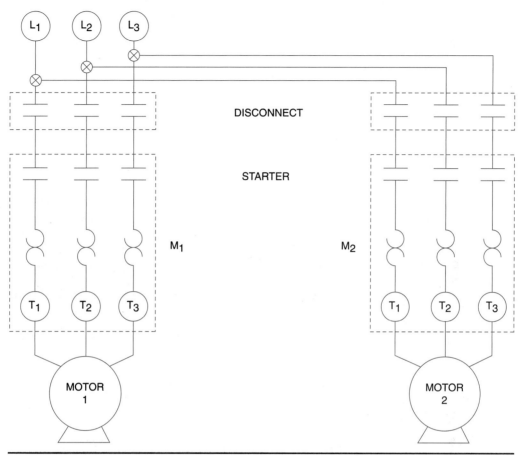

FIGURE 5-16B Power circuit used with sequence controller in Figure 5-16A.

for the M_1 coil and the TR_1 coil. The timer is a time delay on energization (TDE) (see Chapter 4) and will close the contacts on line 3 after a predetermined time. If the M_1 coil is also still energized, the M_1 contacts on line 3 will also be closed and will provide control power to M_2 to start motor 2. The stop sequence for this control scheme is that all motors will stop by pressing a single stop button. Any number of motors can be added with timers or other conditions that need to be met before the next motor will start. The power circuit is illustrated in Figure 5-16B.

A variation to the simple sequence control is shown in Figure 5-17. This figure uses a control transformer to bring the control voltage to 120 V AC. The addition of a timing relay with instantaneous contacts and timing contacts allows the motors to run in automatic with the normal sequence: M_1, then M_2 start. The motors can also be run independently by placing the hand/off/auto selector switch into the "hand" position. Using this technique allows you to start motor 1 in hand position without changing the condition of motor 2. Note also that a single stop button is master and will stop both motors regardless of the hand/off switch position. Likewise, either motor can be left in the off position without affecting the other motor in hand.

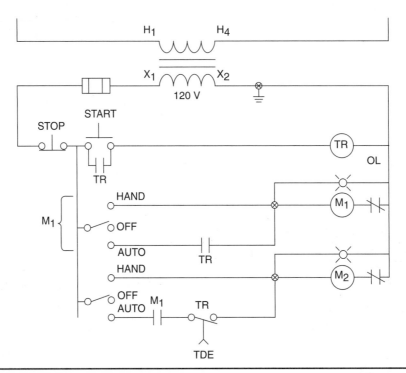

FIGURE 5-17 Definite sequence controller with independent hand/off/auto controls.

Each motor is treated separately with respect to the OL sizing. Additionally, each may have a separate disconnect as shown in Figure 5-16B. NEC® Article 430.110C allows you to use a single disconnecting means for all the motors of the group as long as you size it according to Article 430.110C.

MULTISPEED CONTROLLERS

The principles of multispeed motors were discussed in Chapter 2. There are two basic styles of multispeed motors. You need to know which motor type you are attempting to control before you can develop the control circuit. Is the motor a consequent pole motor (speeds are multiples of each other based on double the number of poles) or are the motors alternate pole motors where there are two separate windings for each speed?

If the motor is a two-speed consequent pole motor, the typical wiring will look like Figure 5-18. The **wiring diagram** is different from the **schematic diagram.** The wiring diagram shows the approximate physical location of the components with the wires connected to the approximate real physical location. Schematic diagrams show electrical relationships without regard to real physical location. Notice that there are additional power contacts for the slower-speed power connection. The connection with the larger number of poles requires T_4-T_5-T_6 to be tied together.

The two-speed starter schematic, Figure 5-19, allows the motor to start in either slow or fast

Three-Phase Motors, Controls, and Full-Voltage Starting

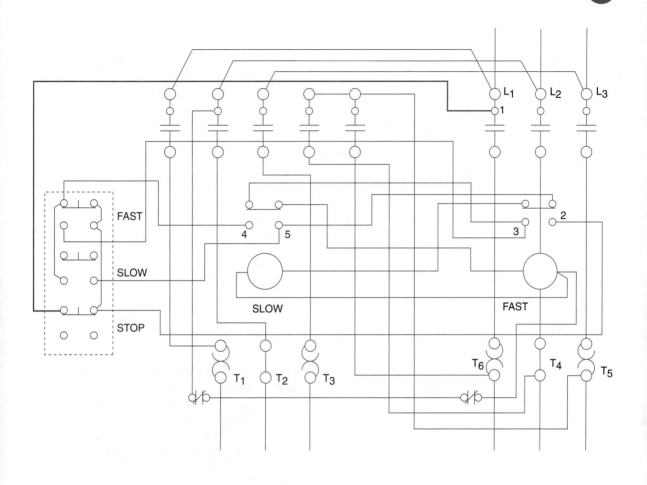

SYNOPSIS OF MOTOR CONNECTIONS				
SPEED	SUPPLY LINES L₁ L₂ L₃		OPEN	TOGETHER
SLOW	T₁ T₂ T₃		NONE	T₄, T₅, T₆
FAST	T₆ T₄ T₅		T₁, T₂, T₃	NONE

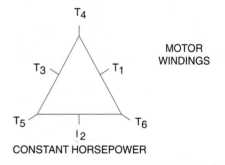

MOTOR WINDINGS

CONSTANT HORSEPOWER

FIGURE 5-18 Two-speed motor control wiring diagram.

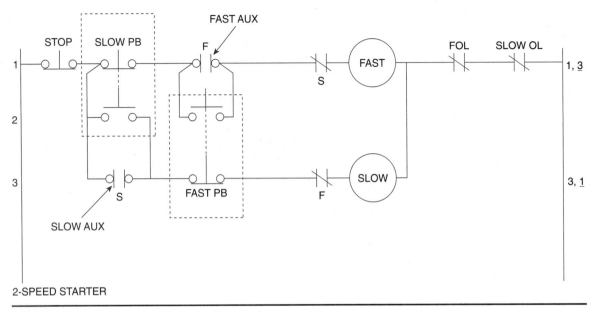

FIGURE 5-19 Two-speed motor ladder diagram.

speed. However, the electrical interlocks, normally closed fast or normally closed slow, prevent both slow and fast contactors from being closed at the same time.

There are two sets of electrical interlocks. One set is built into the push-button stations. As you press the slow push button, one section of the switch breaks a set of contacts before the normally open momentary section makes contact. This is the same concept used in Figure 5-14. Assuming the stop button (which must be master) is not pressed, the slow contactor will pick up on line 3. As the slow starter picks up, the normally closed contact on line 1 breaks to an open position and also prevents power from reaching the fast coil. The same is true for the fast coil circuit.

If the fast push button is pressed, it will break the circuit to the slow coil on line 3 before it makes the circuit to the fast coil on line 1. To ensure that the slow coil deenergizes and disconnects the power to the slow motor terminals, the slow auxiliary NC interlock is also used in series with the fast coil on line 1. Usually the starters will have mechanical interlocks located within the starter enclosure. If you buy the two contactors as an assembled two-speed starter, there will be a mechanical piece on each starter that prevents the other starter from pulling shut after one starter is pulled in.

With the addition of a control relay (CR) (Figure 5-20) on line 4, the motor sequence is changed. This circuit is sometimes called the compelling control scheme with the addition of the CR relay. This means that the system is compelled to start in a particular sequence. Notice that the CR relay has contacts in series with the fast coil. The fast contactor cannot pick up until the CR relay is energized. The CR relay can only be picked if the slow push button is pressed first. This requires the control systems to be started in slow. Then when the fast push button is pressed, the slow coil is deenergized, but not the CR relay, because of its own holding contacts on line 4. Now the circuit works the same as the two-speed

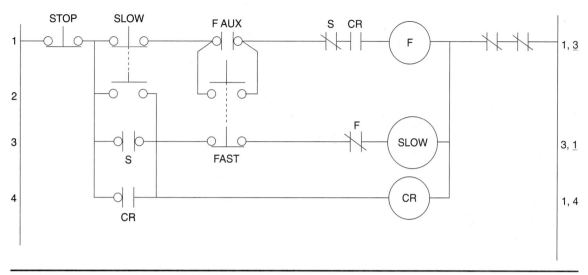

FIGURE 5-20 Compelling control sequence used to ensure the motor starts in slow speed.

controller mentioned earlier. This circuit allows you to go back to the slower speed if necessary.

An important consideration is to carefully connect the leads as indicated on the motor nameplate or in the connection diagram in Figure 5-18. Notice that T_1 and T_6 are connected to the L_1 leads, and so on. It is critical that you test each speed connection separately for direction of rotation before connecting the mechanical load.

The NEC® requires you to protect each winding or connection against overloads and shorts. Article 430.32(A)(1) requires separate overload protection. Article 430.32B-1 indicates that for overload protection for multispeed motors, each winding is considered separately. As shown in the wiring diagram in Figure 5-18, each OL relay senses the current to each of the appropriate winding connections. These connections are usually of different amperages. The actual overload contact is in series with both of the starters on line 1 of Figure 5-19, so that if either OL trips, both speeds are disabled.

NEC® Article 430.83 requires the starter to be capable of controlling the horsepower rating of the motor. If your motor changes horsepower while changing speed, make sure the starters are sized for the proper horsepower. Usually the sizing is dictated by the larger nameplate horsepower rating and each starter is the same NEMA rating. NEC® Article 430.52A refers to the motor short-circuit and ground fault protection. For multispeed motors, a single protective device (for example, fuse) can protect two or more windings of the motor provided the ratings do not exceed the values listed in Article 430.52(4) for the smallest winding protected. See the exception to this rule when using separate overload protection as is required and the branch circuit conductors are sized according to the highest full-load current.

Two-speed motors with separate windings are easier to connect. They require only three poles per starter. The power circuit is simpler as it is connected like two different motors with typical lead markings as shown in Figure 5-21. The control circuits are identical to the two-speed, one-winding control circuits.

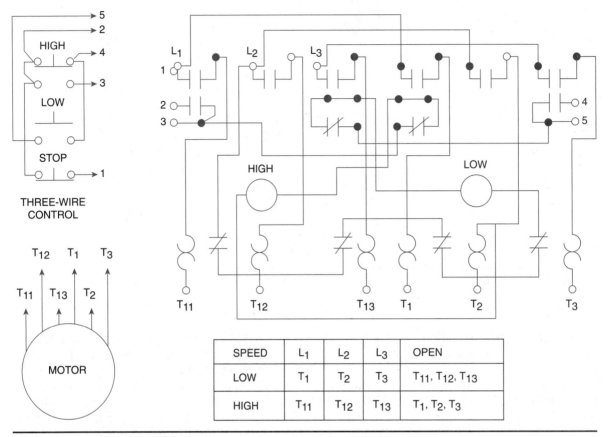

FIGURE 5-21 Two-speed, two-winding motor speed controller.

PROGRAMMABLE CONTROLLERS

Previous descriptions of motor control have used relay logic. This means that the process is controlled by turning relays or timers on or off in a logical sequence. Although there is still a great deal of relay logic control being used, as the control sequences become more lengthy and sophisticated another type of control system may be used. Programmable controllers or **programmable logic controllers (PLCs)** are electronic equivalents to the relay logic control system.

Some electromechanical relays were replaced with solid-state relays, but the control systems had to be physically wired to each relay and the control sequence was not easily changed. With the advent of more industrially hardened integrated circuit technology, a microprocessor system is now used in an industrial environment to electronically replace electromechanical controls.

Because the microprocessor can be programmed to provide many different functions, the control sequence can be altered, changing the control of the system. The control can be altered without rewiring all the physical pilot devices or the output points. PLCs are used increasingly where controls need to be changed or altered frequently. The PLC also offers a greater range of functions that would have

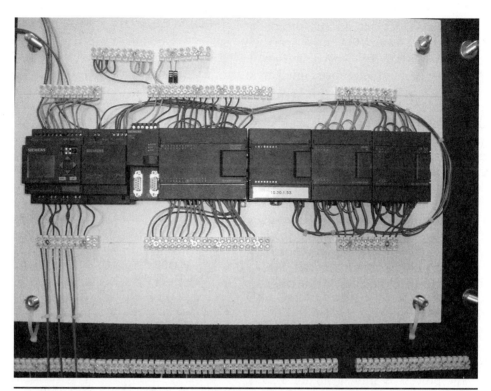

FIGURE 5-22 PLC being used for prototype control testing.

required extensive hardware, space, and power to operate. An example would be where many relays, timers, sequencers, counters, and so on are needed. A PLC has all those functions located within the processor.

Programmable controllers are microprocessor systems that need to be programmed to provide the function you desire. Because electricians were used to ladder logic, the programs are displayed as relays, timers, and so on as in ladder diagrams. The programs are entered into the processor by means of a program panel. Some panels are specific to the controller; with other PLCs the programs are written on a computer and then loaded into the processor unit of the PLC.

The actual input events, such as push buttons, flow switches, proximity switches, or others, are delivered to the processor through input cards in the PLC rack. The output of the processor is delivered to the real-world output devices such as motor controllers, solenoids, lights, or analog signals, through output cards in the PLC rack.

This book will not try to cover all the details of programmable controllers. A thorough study of the particular manufacturer's manual will allow you to program the controller for your specific use. However, a thorough knowledge of relay control theory will help you understand programming the controller. Several PLCs are shown in Figure 5-22.

ELECTRONIC SPEED CONTROL

Unlike the speed-changing control where the number of poles of the AC motor is changed, another option is to change the applied frequency. There are many advantages to controlling the frequency on an AC motor. The speed of the motor has a wide range. Speed can vary from 0 RPM to full speed and in some cases, higher than full-load speed rated at 60 Hz.

The speed can be changed quickly and in very small increments. The speed control can be used to bring up the motor torque slowly, thereby reducing the mechanical stress caused by a sudden change of inertia, such as suddenly starting a heavy load from zero to maximum speed. Another advantage with most controllers is the ability to match the speed of another motor based on an input signal from the master motor. The master–slave control scheme is used where one motor sets the pace and the other motors are matched to the original motor.

The devices being used to control the frequency are one of three types. The variable voltage input (VVI), the pulse width modulation (PWM), or the current source input (CSI) are all used. The basic process is similar. Figure 5-23 shows a block diagram of the process involved. The three-phase AC input is rectified as it enters the controller. The rectifier block uses either a three-phase bridge rectifier using SCRs or a simple diode rectifier for three phase. The object is to create DC to be fed into the inverter to be converted back into three-phase AC.

When changing the frequency to an AC motor, the internal impedance of the windings also changes. For example, if you reduce the applied frequency, the motor impedance will be reduced because of lower XL. (Remember, $X_L = 2\pi FL$.) The applied voltage must also be reduced to prevent overheating of the motor. Similarly, if the frequency is raised above 60 Hz, the motor impedance will increase and the applied voltage must rise to provide the proper current for the motor.

Variable Voltage Input

Variable voltage input (VVI) drives use a three-phase converter to deliver a controlled voltage

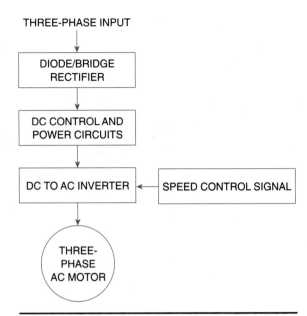

FIGURE 5-23 Block diagram of typical AC variable-frequency electronic drive.

DC output. The converter section may be an SCR bridge circuit where the firing of the SCR is controlled to provide the desired level of DC output, or the method may be a diode bridge with a chopper circuit to rectify and adjust the DC voltage output out of the converter section. The DC voltage is adjusted by a regulator or control chip in the DC control section to provide the proper level of DC voltage to correspond to the desired output frequency. As mentioned, the output voltage must change along with the output frequency.

The variable DC voltage is then inverted (or changed back into AC). A microprocessor controller controls the firing of output devices to control the frequency and the voltage to the output. The power switching takes place through a set of six bipolar transistors, MOSFETs (metal oxide semiconductor field effect transistors), IGBTs (insulated gate bipolar transistors), or thyristors. See Figure 5-24. Because the voltage has already been adjusted—hence the name "variable voltage"—the

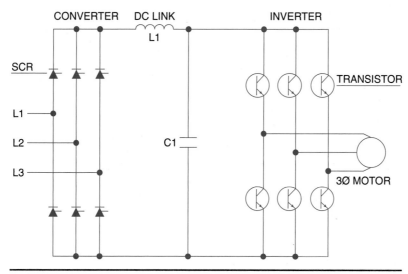

FIGURE 5-24 Schematic diagram of three main sections of VVI speed drive.

inverter just produces on/off pulses in a six-step pattern to simulate a sine wave. Each pulse lasts 60 degrees, so six steps produce a 360-degree waveform. Although not a true sine wave, the current pattern is close to a sine wave shape. See Figure 5-25. The speed of the motor is dependent on the frequency and the voltage is based on the volts per hertz ratio for the motor. Motor torque is dependent on the voltage delivered to the motor, which also affects the delivered current. This type of control has the disadvantage of pulsations at slow speeds. Because of the stepping of the AC waveform, the motor will appear to pulse or produce cogging at a very slow speed. Because the input is not a true sine wave, there is more heating of the motor coils, and the motor cannot provide its full-rated horsepower without overheating and must be derated.

Pulse Width Modulation

Pulse width modulation (PWM) is another method to electronically control the frequency and the voltage to an AC motor for speed control. This is the most common method in practice. The rectification of AC to DC on the input to the drive is accomplished by diodes, which produce a standard level of DC to the DC bus section. This process is simpler than the other types of drives as there is no special control needed on the input. The DC is smoothed, or filtered, and yields about 1.33 times the input RMS value of the AC. In other words a 480-V AC drive would have about 650 V of DC on the DC bus. *Use caution when taking readings.*

The output section (inverter) typically uses bipolar transistors or insulated gate bipolar

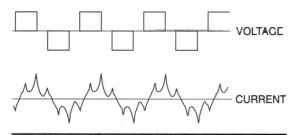

FIGURE 5-25 Representative output waveforms from VVI adjustable-frequency drive.

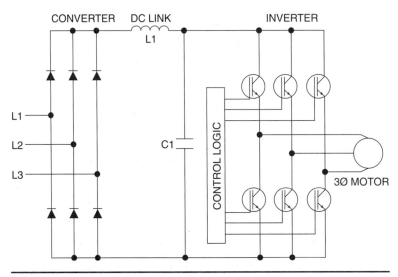

FIGURE 5-26 Example of blocks of PWM drive.

transistors (IGBTs) to produce the variable-frequency AC output. See Figure 5-26. IGBTs are capable of the very high switching speeds required of the output devices to turn on/off thousands of times per second. Because the DC voltage delivered to the invert is a constant value, switching the DC to stay on for either longer or shorter intervals produces the AC waveform. The associated microprocessor controls or *modulates* the *width* of the DC *pulse* (PWM) to the motor. The output current waveform to the motor is very close to a sine wave at all frequencies and therefore the motor does not have the cogging effects at low speed. See Figure 5-27 for current waveforms and Figure 5-28 for the waveforms that are produced as the motor speed and therefore voltage are altered. As discussed earlier, the voltage to the motor must increase as the frequency and the speed increase.

Pulse width modulation uses the same basic principle of rectification, then inversion back to an AC approximation of a three-phase sine wave. The difference between PWM and VVI is that the DC is always delivered at the same value to the DC inverter section. Thus, the amplitude of the output AC will also be fixed, but the average AC voltage output can be changed.

The process used to control the average output voltage is to adjust the width of the output pulse. By adjusting the time the output devices are on, the amount of time the voltage is at peak and at zero voltage is adjusted. See Figure 5-26. If the time that the output is turned off is extended, the zero voltage point is extended. This reduces the frequency of the output waveform and reduces the average AC output voltage. Figure 5-27 illustrates how the pulses are varied to produce an approximation of the AC sine wave. The current waveform is a closer approximation than is produced by the VVI controller.

Current Source Input

The current source input (CSI) again starts with a converter/rectifier. This rectifier block is similar to the VVI module in that the DC output voltage amplitude is controlled at this point. The DC is smoothed by a filter network and sent to the inverter. The inverter section of the CSI uses SCRs to provide the switching as in Figure 5-28. A current sensor is used to monitor the

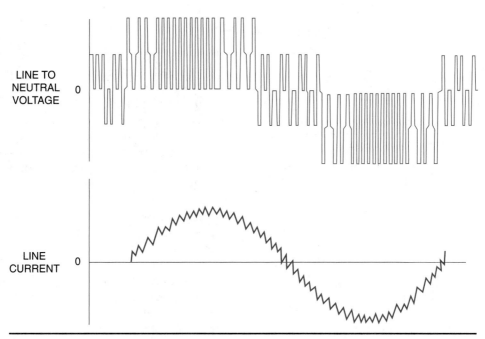

FIGURE 5-27   PWM output waveforms. Each phase has the same waveform.

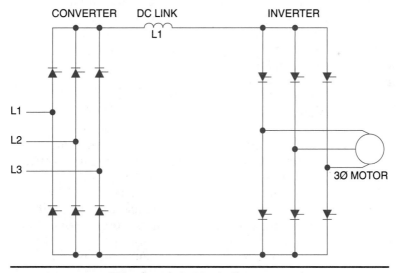

FIGURE 5-28   CSI schematic block diagram.

current between the converter and the inverter. The current sensor is compared to the output power requirements and the speed/frequency control. The current controller then adjusts the DC voltage to provide desired current at the desired frequency to the motor. Figure 5-29 shows the output voltage and resultant current delivered by a CSI variable-frequency drive.

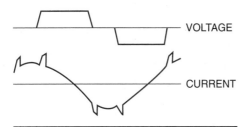

FIGURE 5-29 CSI variable-frequency drive: representative voltage and current waveforms.

Flux Vector Drive

The flux vector drive is a version of the PWM drive technology with added capability. The flux vector drive has the ability to hold speeds very constant over long periods or with varying loads. If exact speed control is not necessary, the flux vector controls are not necessary for general speed control or general energy management. The advantage of the flux vector controller is produced by the feedback system that provides information to the drive from the actual motor performance. It can produce maximum torque at all speeds, and will respond to changing loads with speed and torque control almost instantly. The terminology *vector* comes from the mathematical notation to indicate both magnitude and direction. The *flux* of the motor can be calculated to determine the magnitude and relative direction or position of the magnetic fields in the motor. The ideal displacement between rotor flux and stator flux is 90 degrees to produce maximum twisting effort (torque). By adjusting timing and quantity, the magnetic flux of the stator can be adjusted to provide optimum relationships with the rotor flux.

In a normal squirrel cage induction motor the rotor slips behind the stator speed by 3 to 5 percent at full load. The slip can be measured and converted to a vector-type equation as angular velocity. As the mechanical load on the motor increases, the slip increases and the angular velocity increases (known as the slip-hertz). By monitoring the stator field flux magnitude and position, a computation can determine speed and torque of the motor. By adjusting the PWM electronic output the speed and torque can be adjusted.

By sensing the current to the motor, current feedback information is gathered. By monitoring speed, a speed feedback system is created. The speed of the motor is actually measured by a tachometer, a resolver, or a shaft encoder. The speed signal is processed by the microprocessor in the controller and the input to the motor is adjusted to bring the speed back to the set point. See Figure 5-30 for a sensor-type flux vector drive.

Another type of vector drive is the sensorless drive. This drive does not directly measure the output speed but instead measures the input parameters and adjusts the flux vector relationship for optimum performance. The current measurements differentiate between the current that creates fundamental flux and the resultant current that creates torque. Figure 5-31 shows a block diagram of a complicated process.

Variable-Frequency Drive Considerations

Different mechanical loads require different settings on the **variable-frequency drives.** Some loads remain constant (such as friction loads). This type of load requires constant torque. The torque controls on the adjustable frequency drive would be adjusted to provide constant torque over the speed range. Constant horsepower loads are found on machine tool spindles, extruders, and so forth. The horsepower remains constant; as speed increases, torque decreases proportionally. (See Figure 5-32.) Remember the horsepower formula is a product of torque and speed. The variable torque loads are used in large centrifugal pumps and fans where the load is added as the speed increases. In this variable torque

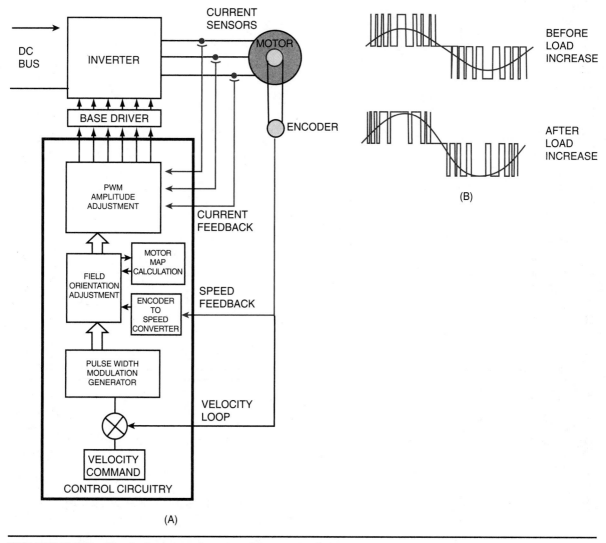

FIGURE 5-30 Block diagram of vector-drive controller.

control style, speed and torque are increased together.

Electronic Drives and NEC®

The NEC® refers to electronic drives in several places. Initially, Article 430.2 refers to adjustable speed drives. The feeder rating to the controller shall be based on the rated input to the power conversion equipment. If overload protection is provided by the conversion equipment, no further overload protection is required. Article 430.6C specifies the ampacity and motor circuit ratings. Use the maximum ratings marked on the motor controller for sizing the motor circuit components. If there is no maximum current marked, use 150 percent of the values listed in the NEC® motor

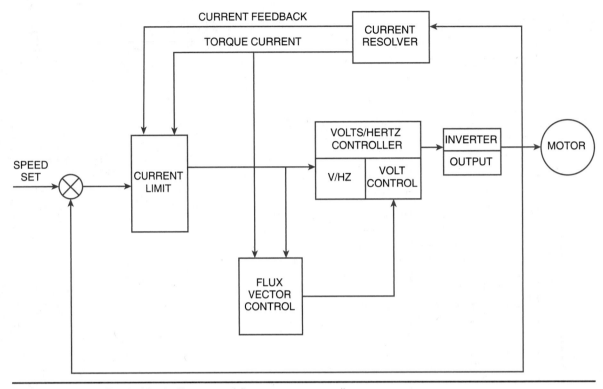

FIGURE 5-31 Basic block diagram of flux vector controller.

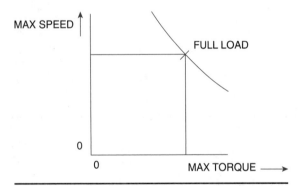

FIGURE 5-32 Constant horsepower speed/torque.

Tables 430.249 and 430.250. Article 430.52(c)(5) refers to fuses inserted by the manufacturer. These fuses are designed for the electronic control requirements. You may use these fuses (recommended) as long as the manufacturer indicates proper fuse replacements next to the fuses.

Alternate Full-Voltage Starting Methods

Applying full-line voltage directly to the motor may cause problems when starting large motors. The method known as "across the line" starting would provide line voltage to the motor at full value without changing the motor configuration. This method can cause extreme electrical and mechanical stress on the motor when starting from a stopped position to full speed. The power system that delivers the locked rotor current to the motor (sometimes referred to as locked rotor amps (LRA)) also suffers stress. There are large currents carried along the system, which create large voltage drops along the supply lines and subsequent low voltage delivered to other loads in the system.

The starting methods presented in this chapter are actually full-voltage starting. **Wye–delta** starters and **part winding starters** actually change the motor connections to provide for reduced stress on the motor and electrical system by reducing the locked rotor current. The control of these starters is presented (in this chapter) with full-voltage starters.

WYE–DELTA STARTING

Wye–delta (Y–Δ) starting is also called star–delta starting. Although technically the actual voltage to the motor is not reduced, it is a method used to lessen the inrush current and reduce the **starting torque**. This type of starting is used frequently on large centrifugal air-conditioning systems where the inrush current needs to be limited. (See Figure 5-33.)

In this system the motors are designed to be run in the delta-connected pattern. However, enough leads are brought out to connect the motor in either the wye or the delta pattern. The motors are typically single voltage so that only six leads are needed. The leads are numbered as coil 1—T_1 and T_4, Coil 2—T_2 and T_5, Coil 3—T_3 and T_6 connected in a wye pattern. T_4, T_5 and T_6 are connected to form a wye (as in Figure 5-34A). The motor is started in this pattern by closing a set of contactors to bring T_4-T_5-T_6 together. When this is done the full-line voltage will be connected from T_1 to T_2 to T_3 in a three-phase star pattern.

Remember, in a star or wye system, the voltage line to line is 1.73 times the line to wye point voltage. (Phase voltage is 58 percent of line voltage; 1/1.73 = .58.) With this constant in mind, only 58 percent of the normal line voltage is available to the wye-connected motor coils. If there is only 58 percent of the normal

FIGURE 5-33 Centrifugal compressor for a building's chiller.

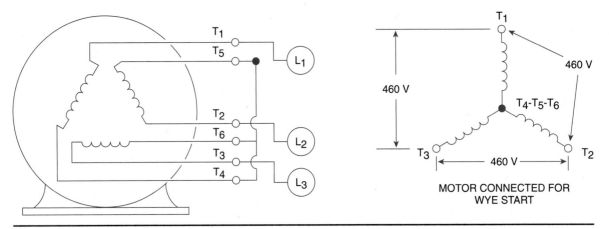

FIGURE 5-34A Motor connections for wye–delta starting.

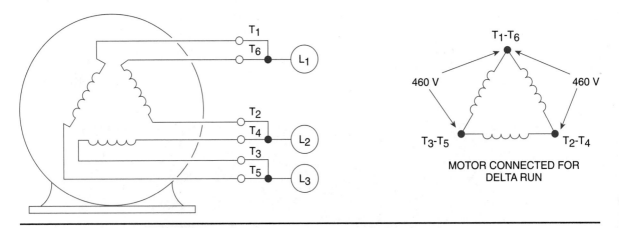

FIGURE 5-34B Motor connection for delta run in wye–delta starter.

voltage available to the coils, only 58 percent of the normal starting current will be available. Torque is directly proportional to the square of the applied voltage.

Because the torque depends on the voltage and current, if one reduces the voltage and that causes a similar reduction of current, the torque will be affected at the squared rate. Reducing the voltage to .58 times the normal coil voltage will reduce the torque to $(.58)^2$ or approximately 33 percent of normal starting torque. Starting torque is the normal starting torque if the motor is connected to full voltage and the normal connection patterns are used initially.

The motor is now operating at 58 percent of the normal voltage applied to the coil, but only 33 percent of the normal inrush current was drawn from the feeder. To illustrate this large reduction in starting current drawn from the line, consider Figure 5-34A. Assume a voltage of 460 volts available at the line terminal L_1, L2, L_3. For example, assume an impedance for each coil at 460 ohms. Connected in a wye pattern only .58 amps would be drawn from line L_1. Reconnect the same coils into a delta pattern (Figure 5-34B) and reapply the 460 V. The line current from line 1 is now 1.73 amps. The wye-connected line current is only 33 percent of the normal delta line current.

OPEN/CLOSED TRANSITION

Normally a timer is set to make the transition from Y to Δ after a preset time. Figure 5-35 shows a typical control circuit used to transfer

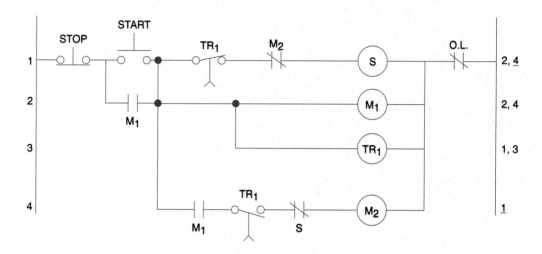

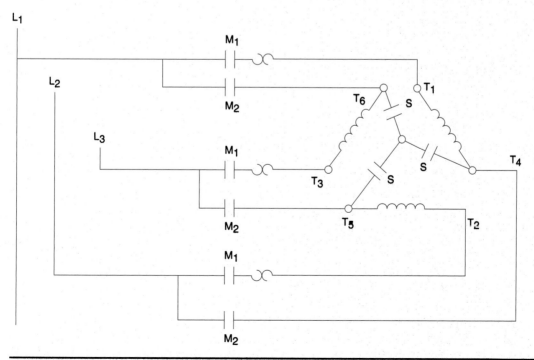

FIGURE 5-35 Wye–delta motor starter control schematic.

the motor connections from the wye pattern to the delta pattern. To do this, the power can be disconnected from the motor coils, the lead connections changed (quickly), and the power reapplied. This disconnection and reconnection of power is referred to as **open transition** because the power circuit is opened during the transfer of motor connections. This is accomplished by breaking the star point.

To analyze the circuit, review the concept of connecting in wye and then reconnecting after a preset time into a delta pattern. Press the start button and energize S (star) which closes the M_1 contactor and begins timing on TR_1. The M_1 contactor provides line power to T_1-T_2-T_3. Notice that the overloads are in this circuit. The S contactor closes the connection to T_4-T_5-T_6 to form the star point. After TR_1 times out (TDE) the S coil circuit is broken (line 1) and the M_2 coil circuit is completed (line 4). This closes the connections T_1 and T_6, T_2 and T_4, T_3 and T_5 after opening the star point. The motor is now reconnected into the delta run position and will operate as a normal delta-connected motor at full voltage.

This particular design is an open transition because the actual power circuit is broken at the star point so that (for a period of time) no current flows in the coils. If you study the motor performance graph (Figure 5-36), you will notice a slight dip in the speed torque curve occurs at the open transition of a motor under load. To compensate for this dip, wye–delta starters can use a form of **closed transition** starting. Closed transition means that there is never a time during starting circuit operation that no line current flows through the coils. With the addition of high-power resistors and additional contactors, closed transition can occur. Resistors are inserted into the circuit during the transition time to continue current flow to the coils. These resistors are high-watt, low-ohm resistors and are only in the circuit for a few seconds during transition.

This closed transition method may be useful if the motor is under heavy load and a

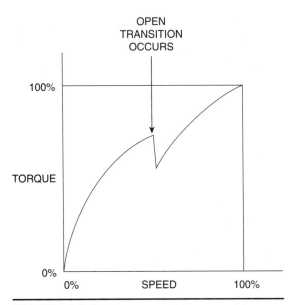

FIGURE 5-36 Open transition speed/torque curve shows a dip when transition occurs.

slight dip in torque cannot be tolerated. If the motor is started under light load, the expense of closed transition may not be necessary. Figure 5-37 shows a controller with closed transition power resistors used in wye–delta starting.

NEC® and Wye–Delta

For most considerations on sizing the motor components, you will need to base your calculations on the running current of the motor. The running current will be based on the delta connection. The type of starting is still considered full voltage according to NEC®.

The other consideration is in the sizing of the overload protection. Article 430.32A(1) allows you to provide protection based on the location of the overload sensor. If it is in the motor coil circuit as in Figure 5-35, then the current will be only 58 percent of the marked delta current and the heater should be sized accordingly. Another consideration may allow you to size the conductors from the contactors windings may be sufficient. An example of this application is starting air-conditioning or refrigeration compressors that are unloaded during startup. If the start time is under approximately five seconds, then no damage to the motor will occur. Do not allow the motor to run on only part of the windings.

The concept is to use a dual-voltage motor and connect only one-half of the parallel coil windings to the line voltage. The line voltage applied must be the lower rated value. If you connect one-half of the coils (the other half remains open), then the line current should be about 50 percent of normal starting current. This is not true because all the motor losses and magnetizing current are still needed. The actual line current is closer to 65 percent of normal starting current. Again, although the starting current is reduced, the actual voltage at the motor terminals remains full-line voltage.

A simple control circuit is used and only two starters are needed for this inexpensive starter. Figure 5-38 illustrates how the first wye point is connected to form the low-voltage, first star group of coils. T_4-T_5-T_6 are connected in the same manner as the low-voltage connection on a dual-voltage, wye-connected motor. M_1 coil contacts deliver line power to T_1-T_2-T_3 while M_2 remains open.

Caution: There will be induced voltage into the open coils and it will be present at the M_2 starter leads. Be sure to make the proper motor connections on T_7-T_8-T_9 so the second set of coils will attempt to turn the motor the same direction as the T_1-T_2-T_3 connections. In other words, L_1 needs to be connected to properly marked T_1 and T_7, and so on. As you press the start button, M_1 contacts close and apply power to the first wye-connected coils. TR_1 begins timing and after a preset time, TR_1 contacts energize M_2 coil, which applies power to the second set of coils.

Some part winding starters use four contacts on the first starter and only two on the run contactor. This enables the current to be

FIGURE 5-37 Closed transition wye–delta starter.

to the motor leads at 125 percent of the 58 percent of full load current. Because the conductors from the load side of the contactors carry only coil current and not line current, your local authority would have to grant you permission to reduce the size of these conductors. The feeder leads or L_1-L_2-L_3 are still based on the delta running current times 1.25 or 125 percent.

PART WINDING STARTERS

When the situation allows large dual-voltage motors to be started without the need for high torque, using only part of the available

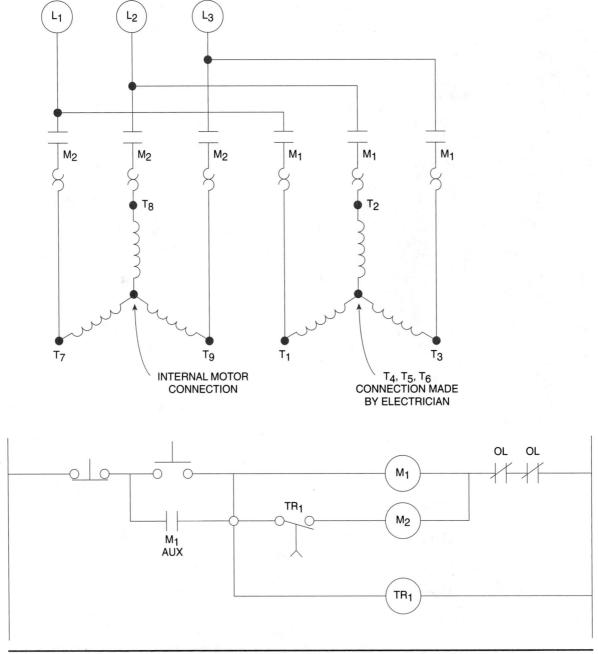

FIGURE 5-38 Diagrams for a part winding starter.

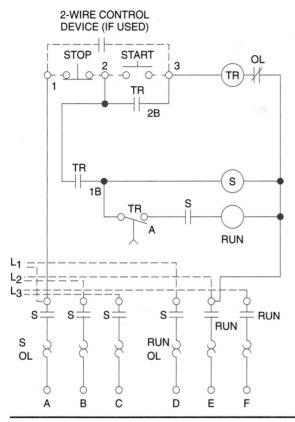

MOTOR LEAD CONNECTIONS TABLE						
PART WINDING SCHEMES	LETTERED TERMINALS IN PANEL					
	A	B	C	D	E	F
1/2 WYE OR DELTA 6 LEADS	T_1	T_2	T_3	T_7	T_8	T_9
1/2 WYE 9 LEADS (1)	T_1	T_2	T_3	T_7	T_8	T_9
1/2 DELTA 9 LEADS (2)	T_1	T_8	T_3	T_6	T_2	T_9
2/3 WYE OR DELTA 6 LEADS	T_1	T_2	T_9	T_7	T_8	T_3
2/3 WYE 9 LEADS (1)	T_1	T_2	T_9	T_7	T_8	T_3
2/3 DELTA 9 LEADS (2)	T_1	T_4	T_9	T_6	T_2	T_3

(1) CONNECT TERMINALS T_4, T_5 AND T_6 TOGETHER AT TERMINAL BOX
(2) CONNECT TERMINALS T_4 AND T_8, T_5 AND T_9, T_6 AND T_7 TOGETHER IN THREE SEPARATE PAIRS AT TERMINAL BOX.

FIGURE 5-39 Options for a part winding starter connection.

balanced better on delta-connected motors and also provides the possibility of creating a two-thirds wye or delta start to improve starting torque. Figure 5-39 illustrates these options. This creates a reduced load by connecting an open Y- or open Δ-type pattern.

The part winding starter is a closed transition starter. No coils need to have the current flow interrupted during transition from start to run. Under normal operations, the line current is reduced to 65 percent of normal start current as previously mentioned. The torque is reduced as a square of this rate. Therefore, $(.65 \text{ percent})^2$ is approximately 42 percent of normal start torque.

Review of Article 430.4 is necessary to size the components in this system. Not all dual-voltage motors are suitable for part winding starting. Consult the manufacturer's specifications to determine if part winding starting is acceptable. Part winding start motors are designed so that both halves of the winding carry equal current. Note that hermetic refrigeration compressors are not considered standard part winding motors.

Typically the motor starters are sized to carry one-half of the motor rating. This allows you to use a NEMA or IEC size that is smaller than the full-load value of the motor. Also the overload devices are sized and installed to protect each half of the motor windings individually. This protection is based on the manufacturers nameplate current rating and Section 430.32. If any of the overload relays sense an overload, they must open the contacts and break the circuit to both starters.

Article 430.4 also states that each motor winding shall have short-circuit protection based on one-half the value specified by Article 430.52. See the exception to this rule. If you use a single dual-element fuse for both windings, the rating shall not exceed 150 percent of the motor full-load current. Motor conductors are not mentioned so you will have to obtain permission from your local authority having jurisdiction to reduce the motor conductors from the starters to the terminal housing.

PHASE FAILURE RELAY

As with any of the three-phase motor systems, it is critical that the phase rotation be known and that all three phases be present. If one of the three phases is lost (for example, blown fuse) a three-phase motor may continue to operate, but with severe difficulty. The overloads should trip and take the motor off the line, but there may already be damage to the motor due to overheating. At starting, the motor typically will not start to rotate and will simply begin to burn the windings from excess current and heat buildup. When one phase is lost the condition is called **single phasing**. Phase protection relays are discussed in Chapter 2.

MOTOR CONTROL CENTERS

Often in large facilities the motor starters are grouped for ease of installation or maintenance. If the motors to be controlled are in the same area, a **motor control center** (**MCC**) is often used. Figure 5-40 illustrates an MCC.

SUMMARY

This chapter explained some of the control systems used to operate three-phase motors. Basic manual and magnetic control was

FIGURE 5-40 Motor control centers are used to group controllers for ease of installation and maintenance.

analyzed to show how the power circuit and the control circuits were separated. Methods of reversing three-phase motors were presented. Different types of control were explained (such as sequence control and multispeed controllers). Basics of troubleshooting longer, ladder diagram–type, schematic drawings were explained.

Electronic speed control was introduced in this chapter. Included in the descriptions was the introduction of PLCs, which are used to replace electromechanical relay systems. Also, the basic concept of variable-frequency AC motor speed control was introduced.

Different methods of full-voltage starting were explained. Wye–delta, part winding starters, and associated control circuits were presented. Related motor system equipment was introduced (such as the motor control center).

QUESTIONS

1. Give a brief explanation of what is meant by manual motor control.
2. Explain the difference among class 10, 20, and 30 overloads.
3. On ladder diagrams, why are some right-hand numbers underlined?
4. What is meant by a mechanical interlock on reversing starters?
5. What is meant by electrical interlocks on control circuits?
6. Provide a brief description of a sequence control circuit.
7. How is the branch circuit protection sized for two-speed motors?
8. What is meant by PWM in variable-frequency drives?
9. What are the special motor considerations for wye–delta starting?
10. How much is line current reduced using wye–delta starting compared to normal full-delta starting?
11. Explain what is meant by open transition starting.
12. What are some conditions that would allow the use of part winding starting?
13. What is meant by "single phasing"?
14. Where would you use an MCC?
15. Is it necessary that you put your own lock on the electrical equipment on which you are working? Explain.
16. What is a PLC and how is it used?

CHAPTER 6

MOTOR ACCELERATION AND DECELERATION

OBJECTIVES

After completing this chapter and the chapter questions, you should be able to

- Understand the need for reduced voltage starting
- Calculate the line current and starting torque for various reduced voltage starting methods
- Identify and use primary resistor starting techniques
- Use the NEC® to determine requirements for reduced voltage starting
- Identify and use autotransformer starters
- Understand the applications for electronic soft start controllers
- Determine the type of synchronous motor control that is used, and begin troubleshooting the system
- Identify and use different electrical and mechanical brakes

KEY TERMS

AC Synchronous Motor: A type of motor that is driven by an AC rotating magnetic field but is designed to run at synchronous speed.

Autotransformer Starting: A form of reduced voltage starting where an autotransformer is used to reduce the line current drawn from the primary lines.

DC Braking: The method used for electrical braking when a DC field is applied to the stator of an AC motor.

Dynamic Braking: A form of electrical braking that uses the spinning rotor as a generator and temporarily applies an electrical load to the generator output.

Plugging: A style of electrical braking where the line voltage to the motor is reversed momentarily.

Primary Resistor Starter: A form of reduced voltage starting where a resistor is inserted into the primary line in order to reduce the starting current and reduce the voltage at the motor terminals.

Reduced Voltage Starting: A method of starting a motor by reducing the line voltage that is delivered to the motor terminals. This method is used to reduce line current draw and also to reduce the starting torque of the motor.

Soft Start Controller: A controller that is used to reduce the line current delivered to a motor in order to reduce the starting torque and create a smooth acceleration up to full speed.

INTRODUCTION

Starting and stopping motors may be a simple or complex process. Starting a motor may involve a mechanical, electromagnetic, or electronic control. You may use full-line voltage (as presented in Chapter 5) or a variation of the full value of voltage and current.

An electrician must also be aware of how to stop a motor. Many cases exist where the motor should not be allowed to freewheel to a stop. Examples of such cases are elevator cable motors, hoists, or cranes. Mechanical brakes may be applied, but electrical braking may also be required.

Acceleration circuits are used to bring a motor up to speed without drawing excessive amounts of current from the line. As you have probably experienced, when a motor starts it may cause lights to dim or other devices to have a brownout or reduction of voltage. This brownout is caused by the large current drawn during motor startup. As discussed earlier, the motor may draw six to ten times the normal current upon starting. Large motors draw large amounts of current. This large current demand causes a voltage drop in the conductors that bring power to the motor. Many times the feeders supply not only the motor, but also other equipment, including lighting. The large voltage drop in the conductors means that there is less available voltage at the motor terminals and at the branch panels that supply lighting.

As the motor accelerates, the line current begins to drop back to normal and the line drop is not as great in the line wires. The voltage is again available at the circuit branches and normal light levels and voltage levels return.

Another less obvious effect of large starting currents is the electrical stress placed on the power delivery equipment. Although motors typically have a short duration start, the high currents can cause heating of the transformers, contacts, and circuit breakers or fuses. The electrical stress may eventually cause premature failure in supply equipment.

Another factor to consider when starting large motors is the mechanical stress placed on the driven machinery. When starting under full power, there is an enormous value of starting torque developed. This torque is much larger than the required running torque. This large and immediate twisting effort may be enough to physically damage the driven equipment.

If acceleration methods are important in bringing a motor up to speed, so are the methods used to bring a motor speed to a stop. These circuits are called *deceleration circuits.* This chapter includes discussion of mechanical brakes and electrical braking provided by deceleration circuits.

REDUCED VOLTAGE STARTING

Most of the time you will be using reduced voltage starting techniques to reduce the locked rotor amps (LRA) on the motor nameplate.

"Locked rotor" refers to the time that the rotor has not yet begun to move even after power is applied. It does not mean that the rotor is actually held in place. After a period of time, the motor accelerates and the LRA decreases. As the motor reaches full load with full voltage, the current reaches the nameplate value of full-load amps (FLA).

Starting torque is the twisting effort delivered by the rotor to the mechanical load. With full-voltage motor starting, and with ample line current available, motors may start with 200 percent (two times) normal running torque. This higher torque is due to the effects of a large amount of stator current and also the larger amount of rotor current. These two currents are factors in the torque calculation. (See Chapter 5.)

Reducing the amount of inrush current available to the motor also reduces the amount of rotor current proportionally. For example, reducing the line current to 65 percent of normal means the torque is reduced to $(.65)^2$ or 42 percent of normal starting torque. If a motor has 10 foot-pounds of torque while running and normally starts at 200 percent with full voltage, the reduction to 65 percent voltage yields $(.65)^2 \times 2.0 \times 10 = 8.45$ ft-lbs. This torque may be too low to move the rotor and a different method of starting may be required.

PRIMARY RESISTOR STARTING

Figure 6-1 shows a typical reduced voltage starter using resistors to reduce the line voltage. The resistors are placed in the primary power circuit, hence the name **primary resistor starter**. The resistors are high-wattage, low-ohm resistors such as those shown in Figure 6-2. The resistors are sized to provide about 35 percent drop in voltage at the resistors when the circuit is first energized (locked rotor current). It is essential that you know the nameplate data of the motor when ordering or sizing these resistors. Typically, the motor receives 65 percent of the initial line voltage at the initial closure of the starter.

Referring to the comparison chart in Figure 6-3, note that the starting voltage available at the motor is 65 percent. These values are approximations and vary by manufacturer. By reducing the motor voltage to 65 percent, only 65 percent of the normal starting current is delivered to the motor. The starting torque is reduced at the percent reduction squared. For example, if a 30-HP, 230-V, three-phase motor, Code H is started across the line, it draws 6.3 KVA per horsepower. That equates to 475 amps line draw. By using 35 percent voltage reduction, 65 percent of the line voltage is available at the motor and 65 percent of the 475 amps locked rotor current will also be drawn from the line. Therefore, 65 percent of 475 A is 308 amps starting line current. The starting torque is also reduced from normal start torque. If the same 30-HP motor operates at 1750 RPM, the normal running torque is 90 foot-pounds (see torque formula in Appendix B). Starting torque for this motor may be 160 percent of running torque, or 144 foot-pounds. By reducing the applied voltage to 65 percent, the starting torque is reduced to 42 percent of 144 foot-pounds or 60 foot-pounds. You need to be sure that this is enough torque to start the load.

To change the applied voltage or if you change motors, the ohmic value of the series line drop resistors must be changed. If a larger motor is installed, or if the code letter changes to a higher locked rotor KVA per horsepower, the ohmic value is reduced to maintain the same 65 percent voltage. The watt ratings are high, but they are not the full rating you would expect by using $I^2 \times R$ watt calculations. The starting current is a short duration and the resistors do not heat to the maximum level. As the motor accelerates, the line current decreases and the amount of voltage dropped on the series resistor also decreases. This means that there is more voltage available at the motor and torque also is increased. This gives the motor smooth acceleration.

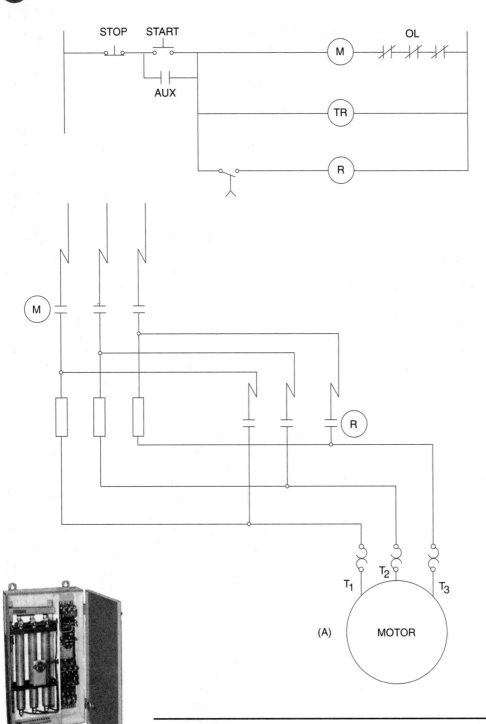

FIGURE 6-1 (A) Control schematic and power circuit for a resistance starter. (B) Photo of 25-HP two-point primary resistor starter. *(Courtesy of Rockwell Automation)*

FIGURE 6-2 Resistors used in primary resistor starters.

Figure 6-4 shows a simple control circuit for a resistor starter. The power circuit is also shown. By pressing the start push button the "M" contactor is energized and the TR_1 timer begins timing. At the end of the predetermined time, TR_1 contact on line 3 closes, and the by-pass contactor "A" is also energized. This acceleration contactor now shunts the line current around the resistors and delivers full voltage to the motor. The timer setting is based on the motor load and motor size. Typically, the timer is adjusted to shunt the resistors when the motor reaches 75 percent of full-load speed. The "M" and "A" contactors remain closed until the stop button is pressed. This contactor sequence means that the motor acceleration is a closed transition between start and run. There is no interruption of line current to the motor. Note that this can be used with two-wire control by inserting the dashed portion instead of a three-wire start/stop.

This method is considered the least expensive of any of the methods that actually reduce the level of voltage to the motor. Resistors are relatively cheap and the control circuit, as well as the power circuit, is not expensive. This method provides for smooth acceleration because as the motor speed increases, the motor voltage also increases gradually. The closed transition method makes very little line disturbance as it shifts to the full-voltage position. If tapped resistors are used, more than one step of starting reduction can be accomplished. One incidental advantage to this starter is that the power factor of the motor circuit is high during starting because resistors are added to the power circuit.

This advantage also creates a disadvantage. By adding resistors to the power circuit, the watt loss is increased. Watt loss is in the form of heat and the heat must be dissipated. Usually the resistors are in a ventilated portion of the starter. The watt loss means that the starter itself consumes power and is therefore considered less efficient than other reduced voltage starters. These starters also provide lower torque

Type of Starter	Starting Characteristics			Advantages	Limitations
	Voltage at Motor	Line Current	Starting Torque		
Full Voltage	100%	100%	100%	• Lowest cost. • Less maintenance. • Highest starting torque.	• Starting inrush current may exceed limits of electrical distribution system. • Starting torque may be too high for the application.
Auto Transformer	80 65 50	64 42 25	64 42 25	• Provides most torque per ampere of the line current. • Taps on auto transformer permit adjustment of starting voltage. • Suitable for long starting periods. • Closed transition starting.	• In lower HP ratings, is most expensive design. • Heavy, physically largest type. • Low power factor. • Most complex of reduced voltage starters because proper sequencing of energization must be maintained.
Primary Resistance	65	65	42	• Least complex method to obtain reduced voltage starting characteristics on low-capacity systems because interlocking of contactors is unnecessary. • Smoothest acceleration of electromechanical types. • Improves starting power factor because voltage current lag is shortened by putting a resistance series with the motor. • Less expensive than auto transformer starter in lower HP ratings.	• Additional power loss in resistors compared to other types of starters. • Low torque efficiency (decreases as voltage is decreased). • Starting characteristics not easily adjusted after manufacture. • Duty cycle may be limited by resistor rating. • High initial inrush current.

FIGURE 6-3 Comparison chart for reduced voltage starters. *(Courtesy of Furnas Electric Company)*

Motor Acceleration and Deceleration

Type of Starter	Starting Characteristics			Advantages	Limitations
	Voltage at Motor	Line Current	Starting Torque		
Part Winding	100	65	42	• Starter less expensive than other types of reduced voltage control. • Closed circuit transition. • Most dual voltage motors can be started part winding on lower of two voltages. • Control smaller than other types.	• Torque efficiency usually poor for 3600 RPM motors. • Possibility of motor not fully accelerating due to torque dips. • Unsuitable for high inertia, long-standing loads. • Requires special motor design for voltages other than 230 V.
Wye Delta	100	33	33	• Low torque efficiency • No torque dips or unusual winding stresses occur as in part winding starting.	• Requires special motor design. • Starting torque is low. • Usually not suitable for high inertia loads. • Control more complex than many other starter types.
Solid State	Adjust	Adjust	Adjust	• Includes constant current, ramped current, or tachometer type starting. • Adjustable current limit and starting time. • Increased duty cycle compared to electro-mechanical types. • Power factor controller and line voltage limiting included. • Multiple adjustable points over wide range. • Smoothest acceleration.	• Specialized maintenance required. • Shorting contactor is required for NEMA 4 and 12 enclosure. • Ventilation required. • Higher priced. • Isolation contactor may be required.

FIGURE 6-3 *(cont.)*

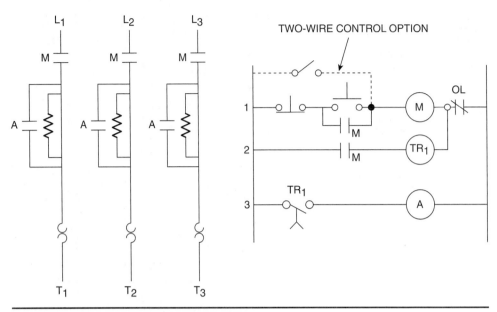

FIGURE 6-4 Primary resistor starter diagrams.

per line current compared to the transformer starters. Primary resistance starters are limited in application to lower horsepower motors although they are available up to NEMA size 7.

Resistance Starters and NEC®

Short-circuit protection is based on NEC® Article 430.52 which refers you to Table 430.52. The protection is based on the same values as full-voltage starting. Resistor or reactor starting refers to the primary resistor starting or the same style starter where the resistor is replaced by a reactor (coil). In the reactor starter, a reactance replaces the resistance. There is less power loss and this method can be used on larger motors than the resistors. The drawbacks to reactor starters are that they are more expensive and they reduce the power factor compared to resistor starters.

Autotransformer Starters

Using transformers to change the voltage delivered to the motor terminals provides a different set of characteristics than does the primary resistor starter. The concept of using a transformer or **autotransformer** provides the benefits of reducing current draw from the line without reducing the motor current at the same ratio. Therefore, the motor current is not affected as much as the input line current to the transformer. In other words, the autotransformer allows more motor torque per line current than a resistance starter.

The transformer principle is used to keep the primary current low and the secondary current high. At the same time the secondary voltage is reduced. A step-down transformer system is used. When a transformer steps down the voltage, it will step up the current so that voltamps in equal voltamps out. Transformers will be discussed in more detail in Chapter 9. Figure 6-5 shows a single autotransformer that is tapped for different voltages. The autotransformers used for starting are often tapped to provide different levels of starting voltage. For example, in Figure 6-5 the primary voltage may be 230 V and the VA rating may be 2300 VA. This tells you that the primary current could

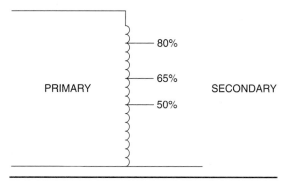

FIGURE 6-5 Autotransformer tapped for different voltages.

reach 10 amps without damage. If you use the 50 percent secondary tap, the secondary, or motor voltage, would be 115 V, but the secondary current could be a maximum of 20 amps. By using the transformer, 20 amps can still be delivered to the motor without drawing 20 amps from the primary line.

Autotransformer starters can be either open or closed transition. As seen in Figure 6-6, using the 65 percent tap will limit the starting current to less than 200 percent of normal full-load amps. This provides 42 percent of the normal starting torque, which equates to nearly 100 percent of normal running torque. Also note that transition occurs at approximately 85 percent of synchronous speed and very little line fluctuation occurs. A larger fluctuation would occur if open transition were used.

A typical power circuit and control circuit for closed transition–autotransformer starting is shown in Figure 6-7. In this circuit the start button is pressed and a timer TR_1 is energized. The timer has instantaneous contacts that immediately close on lines 2 and 3, and timing contacts <u>T</u>ime <u>D</u>elay of <u>E</u>nergization (TDE) that operate on lines 3 and 4. The instantaneous contacts on line 3 allow the "S" coil (start coil) to energize through the NC run electrical interlock. When the S coil energizes, it also closes contacts on line 5, energizing S_1. The power contacts of S and S_1 close in the power circuit to form a wye transformer pattern with line voltage applied to H_1. H_2 terminals are all connected in a wye. The motor already is connected to the tap chosen. In this example the 65 percent tap is used. Now 65 percent of the line voltage is delivered to the motor. If the line voltage is 220 volts, only 143 volts are actually

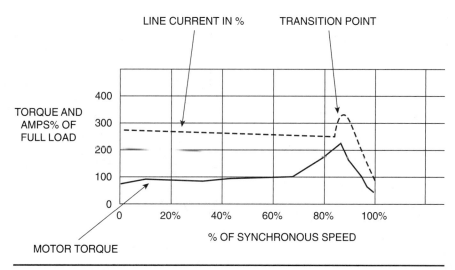

FIGURE 6-6 Graph shows line current and motor torque with a 65 percent tap on an autotransformer starter.

ELECTRIC MOTORS AND MOTOR CONTROLS

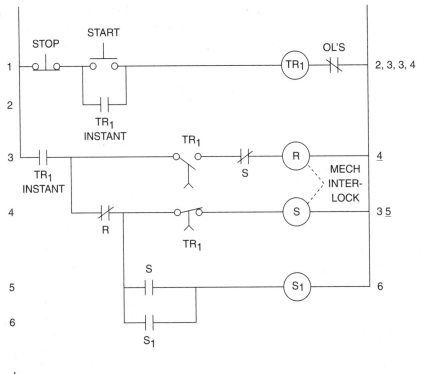

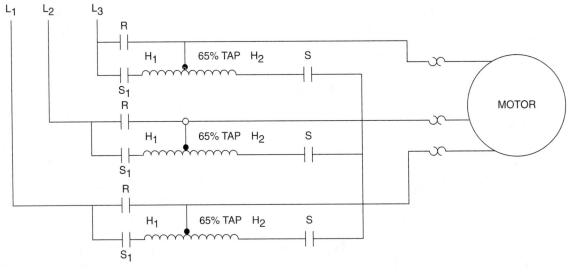

FIGURE 6-7 Autotransformer starter control diagram.

delivered to the motor terminals or a ratio of 1:.65. When 65 percent normal voltage is supplied, only 65 percent of normal motor starting current flows in the secondary of the transformer. The current ratios in the transformer are .65:1—primary to secondary, respectively. Therefore, only 65 percent of the 65 percent secondary current will be drawn from the primary line. This means only 42 percent of the normal locked rotor current will be drawn from the primary line even though 65 percent of locked rotor current is delivered to the motor by the transformer secondary.

After the TR_1 timer finishes, the on-delay contacts on line 4 open, deenergizing the S coil but not the S_1 coil. The S_1 coil has a self-sealing circuit on line 6. Now the TR_1 contacts on line 3 close to energize the Run (R) coil. R power contacts also close to provide power directly to the motor. The closed transition has occurred because power was never completely disconnected from the motor. Only the S contacts were broken, disconnecting H_2 leads, but current flow to the motor still occurred through H_1 to the 65 percent tap connector. S_1 contacts open when R contacts close. See Figure 6-3 for values of line current and starting torque at other tap points.

The advantage of the autotransformer starter is that it provides more motor torque per ampere of line current than any other starter. These starters are used for larger motors where line current disturbances are more noticeable. They typically range in size from 15 to 400 HP at 460 V and up to 200 HP at 230 V. The transformers have low watt loss in the starter and are usable for extended starting periods. They have closed transition starting available and usually have taps to provide for field changes to compensate for different motor characteristics.

One disadvantage of the autotransformer starter is that it is expensive. In addition, it is usually large and heavy. It has low power factor during the starting cycle. This type of starting is used for motors that have long run times. It is not used for frequently started motors (more than five times per hour). The transformers may have thermal protection to guard against overheating and this thermal cutout is wired into the control circuit.

AUTOTRANSFORMER STARTERS

NEC® Article 430.52 governs the ratings of motor branch circuit, short-circuit, and ground fault protection. The general rules refer to Table 430.52. Notice that the ratings are different on the non-time-delay fuse and the instantanous trip breaker setting. The ratings are nearly the same for the dual-element fuse and the inverse time breaker. Most of the other requirements for motor installations remain the same because the motor will run on the normal line voltage and require the same considerations as full-voltage motor installation.

SOLID-STATE REDUCED VOLTAGE STARTING

Prior to the industrial hardening of many solid-state devices and the ability to handle higher currents, solid state was used only for control. The purpose of the other reduced voltage/reduced torque starting systems (such as, part winding, primary resistor, autotransformer, or wye–delta) was to reduce the high starting current drawn from the line. The other consideration was to reduce the starting torque on large motors to avoid physical stresses on the motor and driven machinery.

Electronic **soft start** or **reduced voltage starters** are available from various manufacturers. Figure 6-8 shows an electronic starter in a NEMA one enclosure. The electronics of a soft start or an electronic control are discussed in Chapter 2. The process is similar to speed or torque control. The controller rectifies the three-phase AC to DC, then uses a controller to

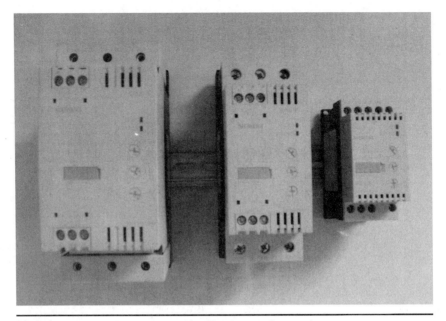

FIGURE 6-8 Solid-state reduced voltage electronic starter.

fire electronics to invert to an AC waveform. The amplitude of the AC is controlled to cause a reduced voltage to be applied to the motor. The motor current may be monitored to provide a precise current control. Current limit control is only recommended on systems that have constant motor load such as constant volume fans and flywheels.

Voltage controls can be set to determine the amount of voltage applied to the motor at turn-on. This control will adjust the starting torque just as other mechanical starters would. In other words, the torque is still dependent on the percent of voltage applied squared. Another adjustment is the time that elapses to bring the motor to full voltage. This is referred to as a *dual-ramp system*. As mentioned earlier, electronic controls usually have the added feature of a deceleration control. This feature is mostly used on motors with friction loads such as cranes or conveyors where a quick stop with a brake may cause equipment or product damage. Because the circuit is monitored electronically for the control function, other protections normally are available. Undervoltage, overvoltage, and phase failure protection is available.

Other parameters to be aware of on solid-state controls are factors that the installer, or the motor specialist, can help program. These parameters are field adjustable and can be used to accelerate, decelerate, or monitor the function of the motor. The motor can be programmed locally, or in newer installations can be set and monitored from a network connection. Motor full-load amps (FLA on nameplate) is a main parameter to be entered into the controller data. From this data, the proper service factor can be programmed to allow the motor to perform over its desired range. The motor acceleration curves and the time of acceleration are programmed to allows the motor to reach full load over a prescribed time, or at a prescribed rate of increase. When the start command is generated, the motor should start to rotate. The starting current can be programmed as a percentage of the motor FLA. This feature allows for the soft start; however, it needs to be high enough to turn the motor if it is loaded. Too much start current could produce too much torque and a motor with a heavy mechanical load will damage the motor or the load.

The maximum current control is adjusted and is coordinated with the ramp-up time setting. As suggested, the time allowed for acceleration must match the current maximums needed for acceleration. Too little time with a maximum current limit will not allow the motor to reach full speed before the time is finished. This may create a motor stall. A stall time parameter determines the time between when the end of the acceleration ramp time occurs and the time the motor must meet designed speed.

Likewise, the motor will decelerate over a timed period or a ramp-down slope to allow a gradual stop. If the motor must stop quickly, the motor deceleration time is minimum. Braking may occur to stop the motor. Again, be aware of the mechanical load on the motor and what the consequences of quick stops or long-time stops can do to the motor shaft and the driven load.

By monitoring the current to the motor, phase imbalance, phase loss (single phasing), or improper phase rotation can be monitored. The class of the overload, 10, 20, or 30, can be entered to protect the motor properly, based on motor application. Voltage monitoring allows the user to select under- or overvoltage protection, to take the motor off-line if conditions are not correct.

Troubleshooting electronic drives know as variable-frequency drives (VFDs) or adjustable-speed drives (ASDs) or simply inverter drives for AC motors usually involve the information provided by the diagnostics built into the drive. This information is accessed through the human machine interface (HMI), a display screen, also known ac a man machine interface (MMI), or a simple LED indicator on a fault indicator. Many of the VFDs are compatible with a computer interface. Either AC or DC drives have four areas where trouble is likely. As we analyze the "system," we can decipher the individual condition of the (1) power supply and the DC bus, (2) the drive controller and inverter section, (3) the motor and connections, and (4) the feedback system that provides signals back to the controller. As discussed in the previous chapter, the incoming power must be the correct voltage level and have the proper phases present. A phase imbalance or missing phase will cause serious problems. If the incoming AC power is correct, make sure the rectifier section is providing the proper voltage level and the capacitor filters are delivering the proper DC voltage to the DC bus or DC link.

The next step to verify proper conditions is to check the motor and the connections from the drive to the motor. Be sure the motor is operational under full-voltage conditions. Check that all the proper voltages and phase are being delivered to the motor and that it is free to turn. Verify that the motor nameplate provides the correct information and is a proper match for the connected VFD. Lead length and control wiring run with the power wiring often cause problems. The information provided with the drive will usually address this issue.

If these first items are correct, the next step is to check for the feedback system. In closed loop control, there needs to be a way for the motor to communicate what is actually happening and compare that with the commands sent from the controller. This feedback can be through a tachometer for speed control or some sort of encoder to other information as well as current and voltage monitoring. (See previous chapter for details.) If the feedback signals are received at the controller and the motor and the power supply are all correct, then the last step is checking the actual processor and the output electronics—or the inverter section. A double check on all the set-up parameters is needed to be sure all specifications for the desired outcome have been properly programmed. If it is determined that the problem is in the inverter section, the quickest way to troubleshoot is to replace the components suspected, or replace the entire board. Some controllers have the power electronics available so that you can replace individual components; many have protected components so they are not field replaceable. If you suspect failure has occurred, again check the DC bus voltage and the heat dissipating capability of the power electronics. Make sure the heat sinks are properly attached to the electronic controls and the fans for cooling the cabinet are operational. Check the vents and airflow obstructions around the controller to be sure there is plenty of clean air available—sometimes filtered.

SYNCHRONOUS MOTOR STARTING

Synchronous motors are not normal AC induction motors. Remember that induction motors are motors that must slip behind the rotating magnetic field in order to induce rotor current. Synchronous motors, by definition, do not slip behind the synchronous speed of the rotating magnetic field and do rotate at synchronous speed.

The **AC synchronous motor** has an induction motor stator but has extra windings in the rotor. The large synchronous motor should be distinguished from a small AC clock motor that also runs at synchronous speed, but the small clock motors are not designed to carry a heavy load.

Large AC synchronous motors are designed to carry heavy loads at the synchronous speed of the rotating magnetic field (determined by the number of poles and the applied frequency). There are two general types of synchronous motors. One style uses an externally generated DC generator to apply DC to the special rotor in order to establish magnetic rotor poles. The second style uses an electronic rectifier assembly mounted on the rotor to provide DC to the rotor, much like the brushless generator system described in Chapter 2. Figure 6-9 shows a nameplate for a 30-HP synchronous motor.

Sample Synchronous Motor Nameplate			
HP	30	KVA	
Volts	230/460	AMPS	83/41.5
PF	90	RPM	1800
Phase	3	Hz	60
Temp. Rise	Armature Field	40°	
DC Excitation 125 V @ 4.8 A			

FIGURE 6-9 Typical data found on synchronous motor nameplate.

Synchronous motors are used where they can start unloaded or with light mechanical loads. The motor is started using a conventional starting system that is used for induction motors. The rotor of the synchronous motor has a small AC squirrel cage winding called an *armortisseur winding* wound on the rotor. This small AC winding is used during starting to get the rotor moving and increase it to nearly 95 percent of synchronous speed. The winding is not designed to carry a load because it is relatively small and cannot carry load current without overheating. As the rotor approaches synchronous speed, DC is applied to the rotor to create a fixed set of poles in the rotor so that current does not have to be induced to create poles.

The number of rotor poles matches the number of stator poles and the rotor will lock into synchronism and follow the rotating magnetic field of the stator. At the start, the DC field is disconnected from the rotor. At synchronous speed, the DC provides the rotor poles and no current is induced into the squirrel cage winding because there is no slip.

One style of control still in use is the polarized field frequency relay shown in Figure 6-10.

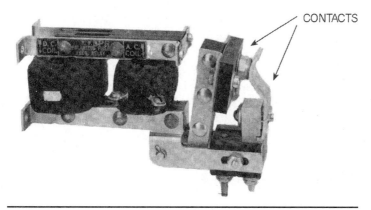

FIGURE 6-10 Field frequency relay for synchronous motor starting. *(Courtesy of Rockwell Automation)*

The field frequency relay is an electromechanical device used to monitor the frequency of the rotor circuit. As the motor starts, high frequencies are induced into the rotor. As the rotor speed increases, the rotor frequency drops. Refer to Figure 6-11. A reactor (or coil) is connected in series with the DC field windings, the field discharge resistor, and the out-of-step relay coil. As the rotor starts, high frequency causes the inductive reactance of the reactor to increase; consequently it drops a higher percentage of voltage in the series circuit. This AC voltage dropped is also applied to the center winding of the polarized field frequency relay (PFFR). Notice that the left-hand coil is connected to the source of DC to be used to magnetize the rotor. The larger AC voltage produces a magnetic flux in the iron core that holds the armature at the right side in the open contact position (pulled tight to the iron core).

The voltage in the middle coil decreases as the rotor frequency drops, due to less voltage drop across the reactor. As this center coil voltage drops, it reaches a point where the magnetic field is cancelled by the DC coil field on the left end. The hinged armature will be released and fall to the closed contact position. This contact will close the DC contactor to apply DC to the rotor. The rotor now has a DC field applied and pulls the rest of the way into synchronism.

Notice also that as the field contactor closes to apply power to the rotor DC windings, the field discharge resistor circuit is opened, and the out-of-step relay is removed. The field discharge resistor is standard on DC magnetic fields. The purpose is to connect across the field coils when the coils are deenergized from the source of supply. As a DC field collapses, it induces a large inductive kick if it cannot discharge through a resistor. The large inductive kick produced by the collapsing magnetic field can be large enough to destroy the field windings. The out-of-step relay is also in the circuit to provide protection for the field.

The relay is an overload device to monitor the field. If there is an overcurrent situation caused by a pullout of synchronism, or if the starting period is too long, the relay will cause the line power to be opened to the stator windings.

The PFFR is used where there is a separate source of DC. For most sync motors of this style, the DC is provided by a small DC exciter generator mounted on the same shaft with the rotor. As the rotor turns, it also drives a DC generator to provide the DC for the rotor field at synchronous speed.

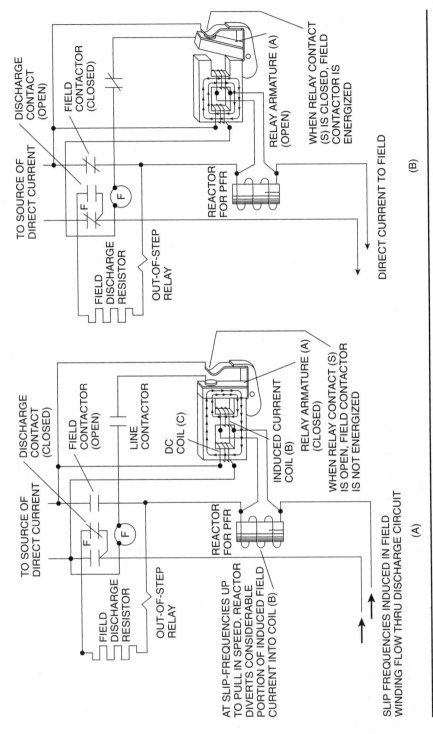

FIGURE 6-11 Synchronous motor polarized field frequency relay. *(Courtesy of Dresser Rand Electric Machinery)*

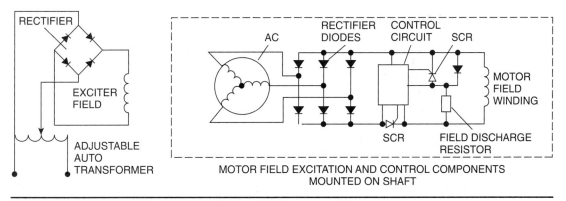

FIGURE 6-12 Brushless excitation used with synchronous motors.

Another style of synchronous motor uses the brushless exciter system to control power to the DC rotor fields. Figure 6-12 shows a simplified diagram of the brushless system. The rotor design of this motor is more complex, but it does not require the use of slip rings or brushes. There is no arcing of the brushes and the maintenance is greatly reduced. The principle is the same as used on brushless generators. The system uses a DC field embedded in the stator to induce voltage into a rotor AC armature. By controlling the DC stator field, the rotor armature voltage is controlled. The rotor armature AC voltage is rectified on the rotor by semiconductor diodes and fed to the DC rotor windings.

The electronic controls on the rotor have the same function as the PFFR described earlier. They monitor the frequency/speed and apply the DC to the rotor windings at the approximate times to achieve synchronisms.

A big advantage to using synchronous motors, other than the fact that the speed remains constant, is that they allow you to correct the power factor of the power system. By adjusting the DC field excitation current you can control the power factor of the motor from lagging to leading. The DC supplied to the rotor changes the operating characteristics of the AC stator supply. If the rotor is excited at the exact level needed to carry the load at synchronism, the stator will operate at 100 percent power factor. By overexciting the DC field, the power factor of the stator can be made a leading power factor. When connected to a feeder system, the power factor of the feeder is also affected. This system can be used to offset the lagging power factor of other motors on the system. Therefore, these motors can be used to help correct the system power factor and also drive large equipment at a synchronous speed.

NEC® Requirements for Synchronous Motors

Article 430.52 again refers to the protection needed for motor branch circuit, short-circuit, and ground fault protection. General motor applications refers you to Table 430.52. Synchronous motors referred to in Table 430.52 also refers to footnote 3. Synchronous motors are used to drive low-speed compressors and pumps. These are typically large motors and the fact that they can correct the power factor of the system is an added advantage. Referring to footnote 3, if these motors are started unloaded, the rating of the short-circuit protection does not require a fuse or circuit breaker setting to exceed 200 percent of the full-load current.

Electrical and Mechanical Brakes

Many different methods have been discussed to start the motors and make them rotate at different operating speeds. There also are many methods of stopping the motor. The simplest method is to disconnect line power from the stator circuit and allow the load to coast to a stop. This method requires no electrical connections or mechanical equipment. Sometimes the motor needs to be stopped quickly to accommodate the driven load. Obvious examples of this situation include motors that drive electrical elevators, hoists, and cranes.

ELECTRICAL BRAKES

Electrical brakes are distinguished from mechanical brakes in several ways. Mechanical brakes use a mechanical method to apply friction to the motor shaft mechanism to physically stop the motor. These are typically either drum and shoe brakes or disk brakes, similar to ones used on motor vehicles. Electrical brakes are electrical connections that are made to affect the electrical operations of the motor. Several methods are used, including **plugging**, **dynamic braking**, **DC braking**, and regenerative braking.

PLUGGING CONTROLS

"Plugging" refers to the method of bringing the motor to a quick stop by momentarily reversing the direction of rotation of the rotating magnetic field. Through the use of a reversing controller you may either manually push the reverse push button to apply power in the opposite direction or use a plugging switch as illustrated in Figure 6-13. The plugging switch is connected to the same motor shaft as the loaded motor. Inside the plugging switch are NO and NC contacts that can be operated through centrifugal force or through a magnetic assembly. The object of the switch is to

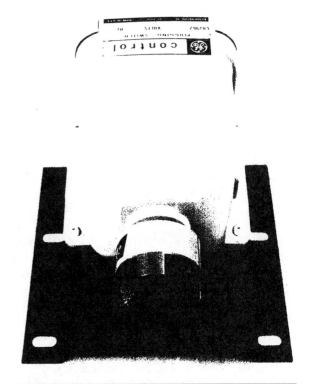

FIGURE 6-13 Plugging switches used when stopping motors quickly.

provide closure to the reversing circuit after the stop button is pressed but remove the control after the motor has nearly stopped. Without this type of mechanism, the automatic reversal may stay on too long and the motor could turn in reverse.

The use of a plugging switch in the control circuit is shown in Figure 6-14. In this circuit the motor is used for a reversing situation, so the plugging switch must be able to change the direction of the rotating field to apply countertorque to the rotor, no matter in which direction the motor operates. To operate in the normal forward direction, press the forward push button on line 1 to pick up the CR relay. Contacts on line 3 close to hold the relay energized. Line 4 contacts also close to energize the

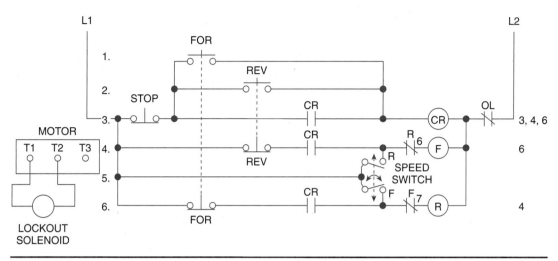

FIGURE 6-14 Plugging switch or zero-speed switch circuit.

forward contactor. Line 6 CR contacts close, but the push button has opened the circuit to the reverse coil. The motor power circuit is energized in the forward direction and the motor begins to turn. The bidirectional plugging switch (sometimes called the zero-speed switch) closes after a preset speed to the F contact on the lower part of the switch on line 5. No electrical circuit is established yet as the NC (F) contact on line 6 is now open. To stop the motor, press the stop push button. This de-energizes the CR coil and also breaks the circuit to the F coil. The forward power circuit is removed, but the plugging switch remains closed. As the F contact on line 6 recloses, the control circuit to the R coil is completed. The reverse power circuit is now complete and the motor receives connections to send a reversed magnetic field around the stator. The rotor will try to follow the stator field and the forward speed will be slowed or stopped. As the motor speed slows, the plugging switch opens the F portion of the contacts and breaks the circuit to the R coil, thereby disconnecting the power circuit from the motor. Typically, there are adjustments to the differential switching. This refers to the speed point where the contacts close and the lower speed where they reopen. If the contacts do not open soon enough, the motor may actually start to turn in the opposite direction when you try to stop it.

Lockout Relay for Plugging

Because the plugging switch creates a path directly to the forward and reverse coils, precautions may be necessary to prevent accidental closure of the contacts in the switch. If the motor shaft were to be turned accidentally, you wouldn't want the switch to energize one of the motor starter relays. The addition of a lockout solenoid is recommended to prevent accidental turn on of the switch and starters. The lockout solenoid is connected so that the plugging switch contacts will not change state unless power is applied to the motor leads. The solenoid coil is rated so that it can be connected electrically to T_1 and T_2 on the three-phase motor. When the motor is powered in one direction, the contacts will set into the plugging contact position. When the motor is stopped, or reversed, the contacts will remain as is until the power is reapplied to the motor in the opposite direction. Now

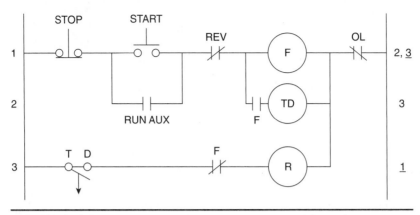

FIGURE 6-15 Plugging circuit that uses a time-delay relay.

the plugging contacts are again free to change and they will reopen as speed is reduced to a very low value. At this point, all power is removed from the motor and the contacts are again inoperable.

An important note in these diagrams is that the stop button always remains master in the circuit. At any time when you press the stop button, you must be able to disconnect the normal path to the forward and reverse coils. This is done through the control relay (CR) contacts.

Plugging Using Time Delay

Instead of using a plugging switch connected to the motor shaft on the driven equipment, timing relays may be used. Figure 6-15 shows how a time delay on deenergization (TDD) relay might be used. In the circuit, the timer is set to reopen the contacts on line 3 after a preset time. Pressing the start button energizes the forward contactor and the TD coil. The TD contacts on line 3 close, but no circuit is established as the NC "F" contacts are opened. The motor runs normally until the stop button is pressed, at which time the NC "F" contacts on line 3 reclose. The TD contacts remain closed and complete the circuit to the R coil. This circuit stays energized until the timing contacts reopen after the time delay.

This circuit needs careful evaluation to adjust the TDD timer. Although cheaper than the plugging switch, this circuit is not responsive to actual motor or shaft speed or conditions.

Antiplugging

This term refers to the way the plugging controls are applied. According to NEMA, antiplugging protection is achieved when a device prevents the motor torque from being reversed until the motor speed has slowed to an acceptable level. Figure 6-16 shows the schematic diagram application of this type of switch. The contacts on line 3 are open until the motor forward speed has slowed after pressing the stop button. As the motor forward speed slows, the F speed switch on line 3 recloses. This sequence prevents an operator from completing the reverse circuit until the motor plugging switch has reclosed the reverse circuit on line 3. The same switch application is used to prevent the forward contactor operation when the motor has been operating in reverse.

DC BRAKING

This method of braking uses the principle that the rotor will try to follow the stator field. By applying DC to the stator fields after the AC is removed, the stator field will become stationary. The magnetic poles on the rotor will try to

Motor Acceleration and Deceleration

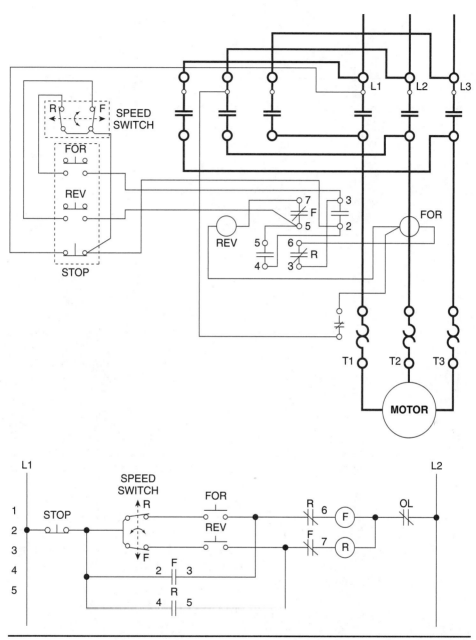

FIGURE 6-16 Antiplugging circuit. *(Courtesy Rockwell Automation)*

align themselves with the stationary stator fields and then try to also become stationary.

Figure 6-17 shows the power diagram as well as the control circuit. Notice that the DC is applied to T_1 and T_2 or T_1 and T_3. Only two motor leads are used. The DC usually is supplied by a bridge circuit in the controller. The braking relay can either connect the DC after it is rectified or it can be used to control the AC to the transformer ahead of the rectifier. The

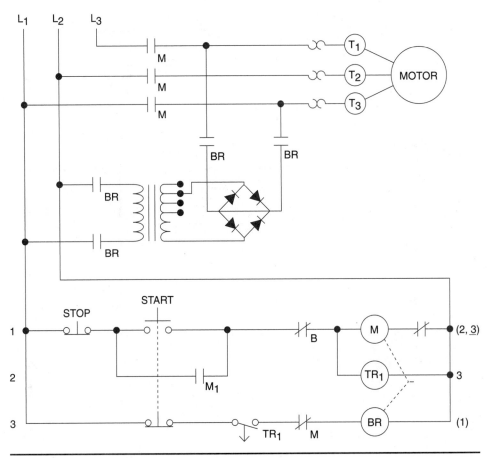

FIGURE 6-17 DC power applied to the motor provides electrical braking.

DC is a relatively low level because there is no inductive reactance to the motor coils and a lower value of DC voltage than AC voltage must be used.

The control circuit uses a TDD timer much like the plugging system that uses a timer. It is recommended to provide interlocks to prevent the AC and DC from both being applied. Therefore, the start push button has an NC contact in the Brake (BR) circuit as well as the M coil NC interlock. If possible, also use a mechanical interlock between the BR and the M coil. Pressing the stop button breaks the circuit to M_1 and also begins the timing sequence on TR_1. The M contacts on line 3 reclose and energize the brake relay thereby applying the DC power to brake the motor. At the end of the timing, TR_1 contacts on line 3 open, deenergizing the brake.

Dynamic Braking

Dynamic braking uses the generator principle that is in existence on every spinning motor. That concept states that the spinning magnetic field also generates a voltage back into the stationary member of the motor. In normal operation, this is counter electromotive force (CEMF) and is the effect that helps keep line current low. Dynamic braking is usually applied to DC motors because the rotating member has leads brought out (see Chapter 10). However, this principle can be used on some AC motors that have rotor lead connections, such as wound rotor motors.

The process of applying dynamic braking is to apply a resistor load to the rotor leads after the motor normal power has been disconnected. As the rotor spins in a magnetic field, from momentum, the rotor will generate an EMF. If the generator (rotor) is connected to a heavy electrical load (low resistance), then the mechanical drag on the spinning rotor will increase. (See "Generator Principles," Chapter 2.) The heavy electrical load in the form of a dynamic braking resistor can remain connected as long as the motor is at rest. (For dynamic braking circuits, see Chapter 10.)

MECHANICAL BRAKES

Often motor shafts need braking and need to be held in place once the shaft has stopped. Mechanical drum and shoe brakes, as in Figure 6-18, are often used to keep the motor shaft from moving. The brake drum is attached to the motor shaft and the brake shoes are used to hold the drum in place. In most applications, the brake is clamped in the closed position by a heavy spring when there is no power on the motor or brake.

This system is a fail-safe system that applies the brake in case of an electrical failure. When the power is applied to the motor, the brake solenoid coil is also energized and pulls the brake shoes away from the drum. Figure 6-19

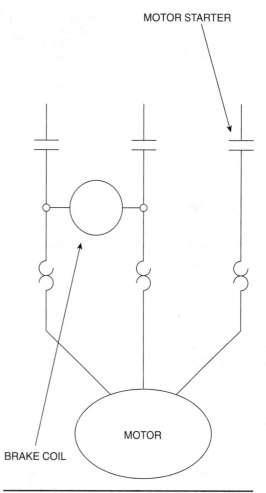

FIGURE 6-18 Drum brake applied to motor drum. *(Courtesy of Eaton Corporation)*

FIGURE 6-19 Electrical brakes are connected on the load side of the motor controller. It is best to connect them ahead of the motor overloads.

FIGURE 6-20 Disk brakes used as mechanical brakes. *(Courtesy of KEB AMERICA Inc.)*

shows how the brake solenoid is connected into the electrical circuit.

Disk Brakes

Disk brakes may use two different methods to apply the braking action. The brakes are shown in Figure 6-20. One style shows the brake disk and the brake pads mounted on the side. As the disk rotates, the pads are pulled off via electrical solenoids. As the power to the motor is interrupted, the pads will be pressed against the rotating disk and stop the motor. Some disks are applied to the rotor by electromagnetic action. This means that the braking action can be controlled. In some models, the higher the DC current to the brake, the harder the pads are pressed against the disk. In other styles a fail-safe system is used. "Fail-safe" means that permanent magnets are trying to force the pads to maximum contact pressure. Electric current must be applied to the brake to pull the pads away from the disk. With the fail-safe system, maximum braking will occur if there is an electrical power failure or a failure of the DC brake supply voltage.

FIGURE 6-21 Photo of brakes installed on motor housings.

Figures 6-21A and 6-21B show mechanical brakes applied to motors. Figure 6-21A has a brake installed on the same end as the shaft so that a load can be attached at the same end. Figure 6-21B has a brake attached to the opposite end of the motor from the drive pulley.

SUMMARY

The need for reduced voltage starting was presented to allow you to determine when it might be required or needed in a large-motor situation. The effects on the motor when using reduced voltage starting were also presented. Calculations on the resultant currents and motor torque when reduced voltage is applied were also explained.

Comparisons of different types of conventional reduced voltage starters included primary resistor, reactor starters, and autotransformer starters. Solid-state reduced voltage starting was introduced as another way to soft start a motor. Synchronous motor control was included in this chapter because of the style of starting in one mode and then switching to another mode of operation.

Deceleration or soft-stop controls, as well as electrical braking, are presented as ways to stop motor rotation. Mechanical brakes were discussed as methods to slow, stop, or hold a motor in place.

QUESTIONS

1. Why would you use reduced voltage starting?
2. How is torque affected by reducing the applied voltage to 70 percent?
3. Explain why line current drops as the motor accelerates.
4. What are some advantages when using the primary resistor starter?
5. Explain why an autotransformer starter provides more torque than other starters for the same line current.
6. What is meant by deceleration controls?
7. How do synchronous motors differ from induction motors?
8. What is the advantage to using synchronous motors rather than constant speed?
9. How is plugging used to stop a motor?
10. How is a DC electric brake applied to an AC motor?
11. Show in the diagram in Figure 6-22 how mechanical brakes are electrically connected.

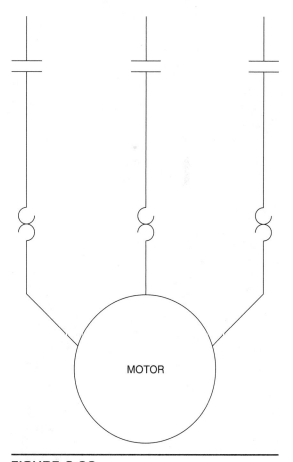

FIGURE 6-22

12. Explain why some electronic starters use a bypass contactor during the running stage.

13. Give at least three different parameters that you need to know when programming an electronic drive.

14. Give an example of where mechanical brakes and electrical brakes might be used.

15. When starting a synchronous motor, how does it transfer from an induction to a noninduction motor?

CHAPTER 7

MOTOR MAINTENANCE AND INSTALLATION

OBJECTIVES

After completing this chapter and the chapter questions, you should be able to

- Determine types of motor bearings and diagnose problems
- Understand how to remove and relubricate bearings
- Identify and mark motor leads that have no markings
- Mark the motor leads on different three-phase motors
- Calculate the horsepower output of motors
- Determine motor efficiency
- Recognize basic motor installation requirements and use the NEC® to verify safe installations

KEY TERMS IN THIS CHAPTER

Delta-Connected Motor: One method of connecting the coil windings of three-phase motors. It is a style used in larger motors to get more torque per ampere input.

Eddy Current: Current that circulates within a conductive material, caused by induced voltage.

Hysteresis: The lag that occurs when magnetic polarities are reversed in a magnetic material. It is a form of energy loss that occurs as the molecules of magnetic materials are reversed.

Magnet Wire: The wire that is used to make up the magnetic coils inside the motor. This wire has a varnish-type insulation.

Megohmmeter: A test instrument that is used to measure the resistance quality of magnet wire insulation in the motor winding.

Motor Bearings: Any of various means that are provided to support the motor shaft and allow it to spin with a minimum of friction (for example, roller bearings, ball bearings, sleeve bearings).

Motor Controller: The electrical equipment used to control the operation of the motor in some predetermined manner by controlling the connection to the line.

Wye-Connected Motor: A method of connecting the coil windings of three-phase motors. It is a style used on motors with higher voltages.

INTRODUCTION

Electrical personnel will be involved in the installation, troubleshooting, preventive maintenance, and replacement of motors.

As you know, single-phase motors are used in a wide variety of applications. The installation of the proper motor for the intended purpose is of utmost importance. Many times a motor has been installed to do a job, but it has been misapplied.

If the motor has been properly installed according to the NEC®, many of the safeguards that are installed will help determine if the motor functions according to specifications, or if malfunctions have occurred. Other tests and measurements will have to be performed to determine if it is a defective motor or a misapplication. A proper motor installation has built-in safeguards to prevent the motor from causing damage to surrounding equipment by overheating. The material in this chapter will help you determine the proper protections to install for a safe installation and should help you determine if there is a problem with the motor or the power system.

See Appendix D for motor components.

REPLACEMENT MOTORS

As mentioned in Chapter 1, motor manufacturers publish a chart of physical motor dimensions. The frame size of a motor allows you to replace a manufacturer's motor with one from another manufacturer if the frame number matches and both manufacturers subscribe to NEMA standards.

Some motors in the United States are not manufactured by NEMA member manufacturers. Motors are also manufactured according to international standards, or IEC specifications. IEC stands for the International Electrotechnical Commission. In this case, you will have to make accurate measurements of all the physical parameters before an exact replacement can be made. In fact, some non-NEMA motors do not list all the nameplate specifications that will allow you to directly size components from the NEC® or U.S. manufacturers' guidelines. If this is the case, you may have to consult the local electrical inspector in order to approve the safe installation.

BEARINGS—TYPES AND MAINTENANCE

Another item to consider in motor installation or replacement is the motor mounting and the need for the proper type of **motor bearings**. Some motors are designed for vertical mounts with the bearings taking the weight of the motor and load without adverse wear. These motors' bearings are installed for heavy thrust loads and are designed to carry loads that are applied parallel to the axis of the bearing. (See Figure 7-1A.)

Examples of axial loads would be motors that are suspended so that the motor shaft is

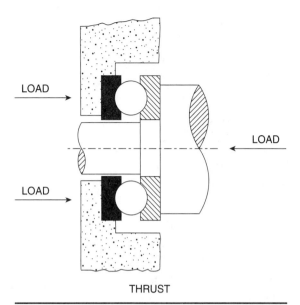

FIGURE 7-1A Thrust bearings are designed to carry loads that apply pressure parallel to the shaft.

FIGURE 7-1B Cooling tower fan is an example of a thrust bearing load.

vertical. The mounting of the motor requires the bearing to hold the weight of the motor and the driven device as the motor shaft spins. One such application might be a fan for a cooling tower on a cooling system, or the motors that drive a large mixer or centrifuge. See Figure 7-1B.

Other motor bearings are designed to carry radial loads. Loads that act perpendicular to the axis of the motor shaft require radial bearings. (See Figure 7-1C.)

Radial loads can be supported by several types of bearings. Sleeve bearings can be used to support shafts. Generally, these bearings are cylindrical sleeves that slide over the shaft. They often use an oil reservoir or oil wick system in the housing to maintain the oil required for lubrication. If the bearings' lubrication fails or there is too much misalignment, the steel shaft will wear away the softer bronze bearing surface. To replace these bearings, remove the lubrication mechanism and then tap off the bronze sleeve.

Many motors use ball bearings to carry the radial load. Deep groove bearings frequently are used. Other styles include self-aligning bearings; double-row bearings for heavier, larger motors; and angular contact bearings that will support heavy thrust loads in one direction. (See Figure 7-2.)

Other bearing styles include roller bearings. Roller bearings have higher load ratings than ball bearings of the same size and often are used in heavy-duty, lower-speed applications. (See Figure 7-3.)

As you check for bearing wear, you will need to listen and feel for suspected wear. Hot

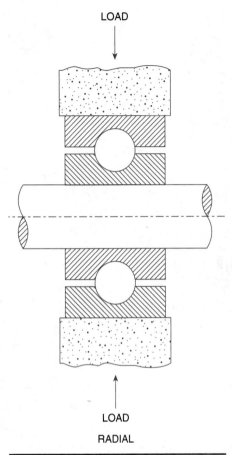

FIGURE 7-1C Radial bearings are bearings that carry a load that is perpendicular to the motor shaft.

spots on the motor housing near the bearing race could indicate worn or failing bearings. Likewise, by listening to an operating motor, bad bearings can be heard as a grinding or gritty sound. After the motor is stopped and disconnected from power, hand-turn the rotor. It should turn freely and easily. There should be imperceptible movement when moving the shaft in a radial direction and very little movement in the thrust direction. Oftentimes there will be a thrust washer installed with the radial bearing to compensate for minor thrust loads that the radial bearings might encounter. The thrust washer looks like a twisted washer, so it can spring with the thrust movement. If there is excessive movement or unusual noise in either direction, further inspection may be necessary. If the bearings are worn or damaged, you will be able to measure the air gap around the circumference of the rotor. Use a feeler gauge to check the air gap between the rotor and the stator. Ideally, the top and bottom air gaps measurements should be identical.

Replace the bearings with similar types. Replacement with the exact bearing manufacturer is not necessary as most bearing manufacturers adhere to consistent standards. Bearing pullers may be needed to remove the bearings. (See Figure 7-4.) Use lubrication on the end of the cleaned shaft to pull the old bearing off the shaft. Pull on the inner race and avoid scarring the shaft.

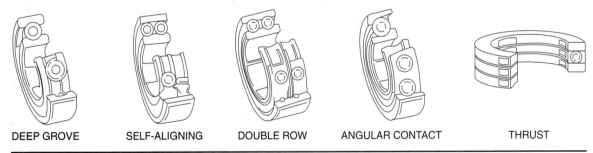

FIGURE 7-2 Various styles of ball bearings used in various applications.

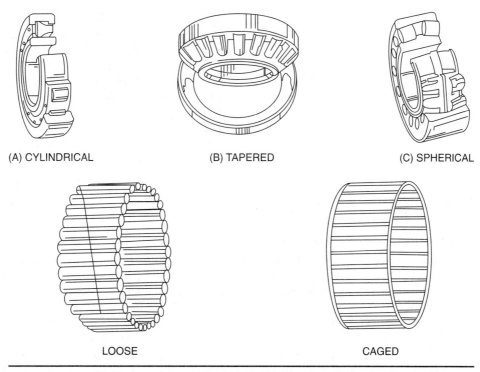

FIGURE 7-3 Roller bearings are used for heavier loads or slower speeds.

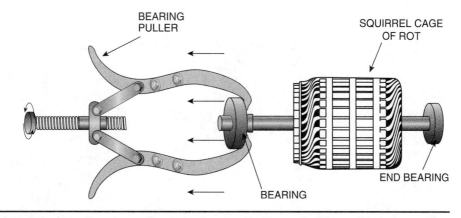

FIGURE 7-4 Bearing pullers use jaws and screw shaft to pull bearings off motor shaft.

To install the new bearings, reseat the bearing by driving the bearing inner race down the shaft's length. Be sure the race is aligned so as not to mar the shaft, and drive the bearing only with a soft metal tool against the inner race. *Do not* push against the outer race. Often the bearing can be heated slowly in an oven to expand the metal; it will then slip more easily over the shaft. As the bearing cools, it will contract around the shaft to create a firm bond. Similarly, the outer race should fit snugly into the bearing holder. The outside race should not move in its holder.

Sometimes bearings can be cleaned and relubricated to extend their lives. Bearings should be removed for most effective cleaning, but if that is not possible, cleaning and relubricating can take place on the shaft. Clean the bearing as much as possible with clean rags. To loosen old grease and hardened oil deposits, try using warmed kerosene. *Be careful not to ignite the kerosene.* Use low-pressure compressed air to blow the kerosene out of the bearing. *Do not* spin the ball bearings with the compressed air. Check the bearing for smooth operation. If there is no damage, the bearing can be relubricated and reinstalled.

Different bearings require different types of lubrication. Small fractional horsepower motors often use sleeve bearings that are lubricated by oil. Spring-loaded caps cover the oil reservoir. Typically, these motors are oiled on an annual basis with an oil grade SAE 10-20. Larger sleeve bearing motors use an oil well and a loose ring called a slinger ring that slings oil from the well up to a groove in the brass sleeve. Usually the oil becomes contaminated during the course of a year. The oil should be drained and refilled annually if the motor is used as a standard duty motor.

Ball and roller bearings require lubrication as part of a preventive maintenance program. Check to be sure the bearings are not sealed bearings. Do not try to force grease into sealed bearings. (See Figure 7-5.) If the bearings use grease as a lubrication method, the grease often can be applied using a grease gun. Clean the area around the grease inlet. (Often a zerk quick-attach grease fitting is used.) Remove the plug on the outlet side. If possible, run the motor. Add new grease to the inlet, forcing the old grease out and into a catch pan. Run the motor to allow the new grease to provide fresh lubrication to the motor. When no more old grease is forced out of the outlet hole, replace the outlet plug. Clean the motor and

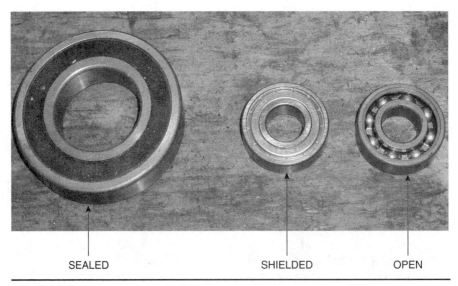

FIGURE 7-5 Examples of sealed bearings, shielded bearings, and open bearings.

housing of excess grease. Be sure to use a grease that has adequate temperature ratings for the intended location of the motor. High-temperature greases may be necessary to keep the lubricant from separating and leaking out from the surrounding medium. Grease also helps provide better seals against grit and dirt, than oil, if the proper grease is used for the application.

Remember to check bearings and ease of rotation regularly. As bearings lose lubrication, the motor does not spin as freely and the motor heats up and may cause more damage to the bearings, as well as the motor windings.

Maintenance Checks

Other maintenance checks can be performed. As mentioned, check for overheating and vibration. The motor housing and any air vents must be clean to allow cooling air to flow. Excess heat also may be caused by poor bearings.

Voltage and current measurements can be performed to be sure the motor is operating within the nameplate rating. Use a multimeter or a clamp-on meter as shown in Figure 7-6. Use the voltmeter function to verify that the voltage delivered to the motor is within 10 percent of the motor rated value. Use the clamp-on ammeter function to determine if the current draw is close to the nameplate rating. (Refer to Chapter 1 to verify other nameplate data.) If the current is out of safe limits or if the current is unbalanced on a three-phase motor, further investigation is required. (See Chapter 6.)

Motor windings are insulated from each other by a film of varnish. Wire used in the motor winding that has a varnish insulation is called **magnet wire**. As a motor ages, the heating and cooling of the windings as well as moisture, dirt, and oil all work to break down the varnish insulating qualities. The windings are also electrically insulated from the motor frame. Use of a **megohmmeter** will help to determine if there is a problem in the motor windings.

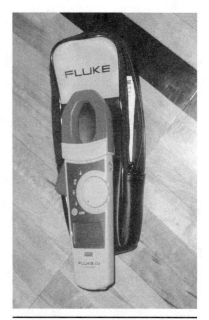

FIGURE 7-6 Clamp-on meters used to measure electrical circuit quantities.

MEGOHMMETERS

A megohmmeter is actually a small generator set that supplies a voltage to the test leads. As the voltage is applied to the motor leads, a current will flow. The current is measured on the insulation tester and calibrated in megohms (1,000,000) ohms. Be careful not to touch open leads of the insulation tester, because of a relatively large voltage output. Also, separate the open leads on a motor when testing. Voltage may be induced into adjacent windings and therefore voltage is available at the winding leads. Connections for motors are between windings to check for breakdown and between the windings and the frames. (See Figure 7-7.)

Test the megohmmeter first. With the leads open (not connected), energize the voltage source. This can be done by turning a hand crank or using a battery-powered model. See

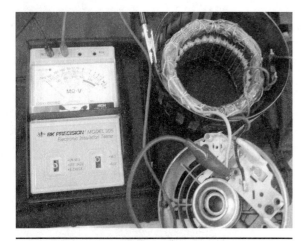

FIGURE 7-7 Analog megohmmeter connected for testing winding to frame resistance.

Figure 7-8 for a hand-crank megohmmeter and Figure 7-7 for a battery-operated model. The initial reading should be infinite resistance. Short the leads together and the resistance should drop to 0 ohms. When testing the motor coil to coil, or coil to frame, the readings should be near the meter's infinity range. Eighty megohms or more is an acceptable reading. If the readings are 1 megohm or less, insulation breakdown has occurred or is very probable. There is a rule of thumb called the 1-megohm rule. It states that there should be at least 1 megohm of insulation resistance for each 1000 V of the equipment's voltage rating. You should maintain at least 1 megohm for equipment under 1000-V rating.

Testing Motor Coils

Other checks can be made to determine if coils are open or shorted. These tests are done more easily in a motor shop or at a test bench. When analyzing single-phase motors, a little knowledge of motor type is needed. For instance, check to see if there is a starting switch involved. The switch is often located within the motor but may be mounted remotely. (See Chapter 1.) If the motor is of the universal

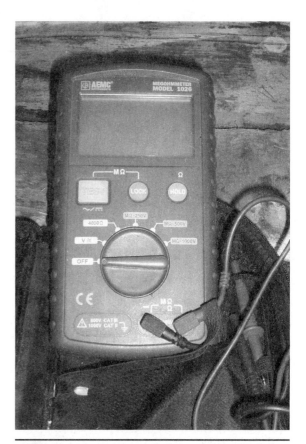

FIGURE 7-8 Digital megohmmeter for motor winding insulation testing.

style, then connections are made through the rotor via carbon brushes; these must also be checked for continuity.

GROWLERS

If these connection points are creating a solid connection, then the windings can be checked for opens by simply using an ohmmeter to check for end-to-end continuity of the coil. To check for shorted windings within a coil of the stator, an internal **growler** is used. One style of growler uses a vibrating armature. As the growler is placed against the inside of the stator and energized, an AC voltage is induced

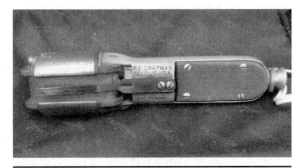

FIGURE 7-9 Internal growler used to detect shorted stator coils.

FIGURE 7-10 External growler used to test squirrel cage rotors.

into the coil. Be careful of the loose coil ends, because voltage is being induced into the winding. Because the coil is not connected, no current will flow in the windings of a normal coil and the vibrating armature on the internal growler does not vibrate. However, if there are shorted windings of the coil, a current flow in the stator winding will be generated (within the shorted windings). This induced current sets up a magnetic field and causes the growler to growl or vibrate. This growling takes place over the shorted coil. (See Figure 7-9.)

Rotor Testing

Squirrel cage rotors rarely are defective, but checks can be made to determine if there are broken or disconnected rotor bars. A typical symptom of a bad rotor is that the rotor will not start in one position of rest, but turning the rotor to a new position will allow the motor to start. To test the squirrel cage rotor, an external growler is used. (See Figure 7-10.) In a squirrel cage the bars are supposed to be shorted so that an induced voltage from the growler should cause an iron detector (blade) to vibrate or growl. When testing with the blade, if there is no vibration, an open rotor bar is the probable cause.

Another test can be performed to indicate an open rotor bar. This test can be performed on the motor while it is still assembled. Use a reduced voltage source applied to a single phase of the stator. (The voltage should be approximately 25 percent of normal nameplate voltage.) Monitor the stator current, then slowly turn the rotor by hand. If the stator current shows an obvious decrease, an open rotor bar is indicated. This is because there is no current being induced into an open rotor, therefore an associated drop in stator current is measured.

LEAD IDENTIFICATION

Lead identification of motors is necessary when lead markings have been lost or destroyed. To mark the leads of the motor, a basic understanding of motor coils is required. As described in Chapter 1, single-phase motors come in a variety of styles. The most common styles are the split-phase and the capacitor-start motor. Remember that these motors use the principles of inductance and resistance to create a phase split (shift) for the current. This allows the single-phase voltage to act like a two-phase system when energizing the coils.

If the motor is apart, the coils can be identified by the resistance and the position in the stator slots. The starting winding is the winding with higher resistance and is closer to the inside surface of the stator. The measurement is only a few ohms, so accurate measurements with precise meters are required. These starting winding leads are marked T_5 to T_8. If only one starting winding is used, it is marked T_5 and T_8. If a dual-voltage motor uses two windings, then the leads are marked T_5 and T_6, and T_7 and T_8. (See Figure 7-11.) Starting winding lead markings for split-phase and cap-start motors are the same.

The main (running winding) leads are deeper in the slots of the stator and have less resistance, even though there are more turns. The main windings must carry the running current and therefore use larger conductors which have fewer ohms per foot of wire. These leads are marked T_1 to T_4. On dual-voltage, single-phase motors (either split-phase or capacitor start) the leads are marked T_1 and T_2 for one coil, then T_3 and T_4 for the second coil. (See Figure 7-12.)

Simple continuity tests to determine the starting and ending of each of the coils is necessary. To verify the proper markings, the magnetic poles created must correspond to other poles inside the motor. Connect T_1 and T_2 to a low-voltage source, and connect T_5 through T_8 to the same voltage service. Allow the motor to run unloaded for a direction of rotation check. Now disconnect the motor and reconnect T_3 to the same line wire that T_1 was connected to, and T_4 to the same line wire where T_2 was connected. Make sure the starting winding remains connected as it was for the first test. Now reenergize the motor to verify direction of rotation. The motor must turn the same direction as the first test if you have the lead markings correct. If it does not, T_3 and T_4 lead markings are incorrect. With the leads labeled correctly, connect the motor for the intended voltage of the circuit. The

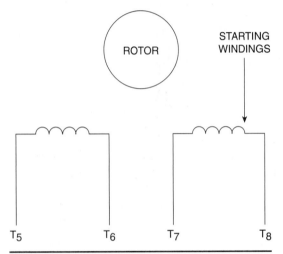

FIGURE 7-11 Two starting windings are marked T_5-T_6 and T_7-T_8.

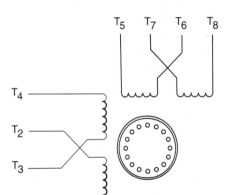

VOLTAGE RATING	L_1	L_2	TIE TOGETHER
115 VOLTS	T_1, T_3, T_5, T_7,	T_2, T_4, T_6, T_8,	—
230 VOLTS	T_1, T_5	T_4, T_8	T_2 AND T_3, T_6 AND T_7

FIGURE 7-12 Connection pattern for a dual-voltage motor with two run windings and two start windings.

Motor Maintenance and Installation

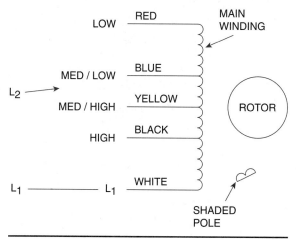

FIGURE 7-13A Motor lead labels for adjustable-speed shaded pole motor.

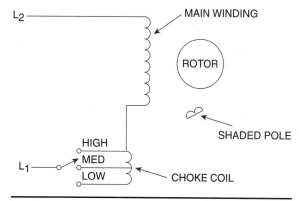

FIGURE 7-13B Speed control for shaded pole motor using a choke coil.

motor should run smoothly and quietly at a line current less than nameplate (assuming the motor is still not mechanically loaded).

Lead markings for other motors are easier to identify. For instance, capacitor motors have the one lead of the starting winding connected to the capacitor either through a starting switch or (if it is a permanent capacitor) directly to a motor lead. This lead is usually labeled T_5. The other end of this coil is marked T_8. The running winding is identified the same way as other single-phase motors previously described.

Shaded pole motor leads are also easily identified. Shaded pole motors have only one true coil connection, which is the main pole. The identification of multispeed shaded pole motors requires the use of a sensitive ohmmeter. As mentioned in Chapter 1, the speed of a shaded pole motor can be controlled by a tapped winding. This has the effect of changing the volts-per-turn applied to the winding. The higher the volts-per-turn, the higher the speed. Therefore, the higher-speed connections have fewer turns of the coil, if the motor uses full-line voltage only as the speed control. If the control circuit of the shaded pole multi-speed motor actually uses a method of changing the applied voltage to the motor, the windings are not altered, but the lowest applied voltage will give the slowest speed. (See Figure 7-13A.) Another method is to install a choke to drop the voltage applied to the motor as in Figure 7-13B.

It also is easy to identify the leads of a universal motor. (See Figure 7-14A.) The leads are either identified as field coil leads, F_1-F_2 (which correspond to the large field coils) or A_1-A_2 (which belong to the armature brush connections). Some universal motors have small embedded windings called interpoles, or compensating windings. The compensating windings are found on larger motors and are used to reduce arcing of the brushes. These are included in the field coil leads as they are wound in the stator. The compensating windings should have the same polarity as the main pole behind them in the direction of rotation.

After you have established pole continuity, the simplest way to verify polarity is to try the connections in the operating motor. If there is excessive sparking at the brushes as the motor is loaded, the compensating winding leads need to be reversed.

The brushes should also be set to the neutral plane of the motor. This refers to the

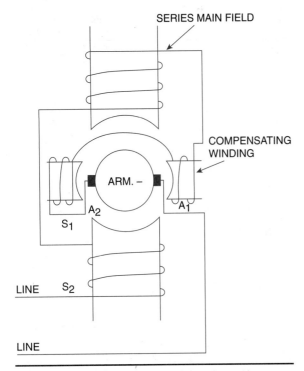

FIGURE 7-14A Markings for universal motor leads.

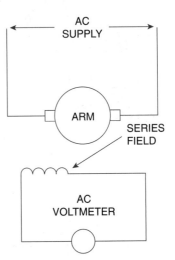

LOWEST READING INDICATES PROPER BRUSH POSITION

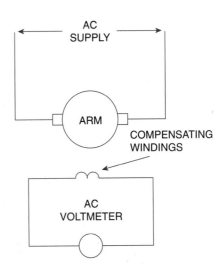

HIGHEST READING INDICATES PROPER BRUSH POSITION

FIGURE 7-14B Voltmeter used to determine the neutral plane for setting brush positions.

position on the armature where the current delivered to the rotor splits equally in the rotor. Therefore, there is a minimum voltage difference produced in the rotor halves and minimal sparking on the rotating connection point to the brushes.

To determine if the brushes are at the neutral plane, connect an AC voltmeter to the series field as shown in Figure 7-14B. Apply a low AC voltage directly to the armature through the brushes. Loosen the yoke that holds the brushes and move the brushes until the voltmeter shows the lowest reading. This is the null point and indicates that the armature will have the least amount of induced EMF at this brush location. This will give the least amount of sparking under varying loads. To reverse the direction of rotation of the motor, reverse the connection to the brushes only. (See Figure 7-15.)

Caution: When testing universal motors, be sure to have a mechanical load connected. Universal motors run at very high speed when the load is removed. This speed

Motor Maintenance and Installation

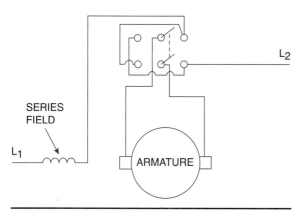

FIGURE 7-15 Reversing connections using a double-pole–double-throw switch.

Wye connection high voltage

L₁ to T₁
L₂ to T₂
L₃ to T₃
T₄ to T₇ — Connected together
T₅ to T₈ — Connected together
T₆ to T₉ — Connected together

Wye connection low voltage

L₁ to T₁ and T₇
L₂ to T₂ and T₈
L₃ to T₃ and T₉
Connect T₄, T₅, T₆ together

Delta connection high voltage

L₁ to T₁
L₂ to T₂
L₃ to T₃
T₄ to T₇ — Connected together
T₅ to T₈ — Connected together
T₆ to T₉ — Connected together

Delta connection low voltage

L₁ to T₁ and T₆ and T₇
L₂ to T₂ and T₄ and T₈
L₃ to T₃ and T₅ and T₉

FIGURE 7-16 Lead connections for high/low voltage supplied to wye- and delta-connected motors.

may be great enough to literally pull itself apart.

Three-Phase Motors

Three-phase motors were discussed in detail in Chapter 6. However, lead markings will be covered here in the section on maintenance. Three-phase motors are available in two general configurations. The motors are either delta- or wye-connected motors. If the lead markings are destroyed, the leads can be re-identified by the following method:

First determine if the motor is a wye- (sometimes called star) or delta-connected motor. For a dual-voltage motor, this can be established by checking how many leads are tied together and to what voltage the motor was connected. See Figure 7-16 for connection patterns. If the motor has nine leads open and you don't know what voltage it is connected for, you must use an ohmmeter to determine lead groups. Use the ohmmeter to determine how many groups of leads have continuity with each other. A **wye-connected motor** will have one group of three leads with continuity and three groups of leads that have only two leads with continuity. (See Figure 7-17A.) Delta-connected motors will have three groups of leads that have three leads in continuity with each other. (See Figure 7-17B.)

Star Motor Leads ID

1. Use the group of three leads as a starting point. This group is called the internal star point and the common ends of all the coils are already connected within the motor. This internal star will establish the base for the remaining lead markings.

2. Assign lead markings T₇-T₈-T₉ arbitrarily to the leads brought out from the internal star coils. *Caution:* All leads may have high voltage. Separate leads and *do not* touch open leads!

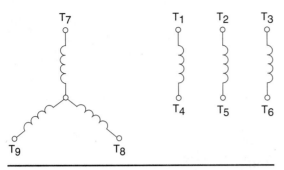

FIGURE 7-17A Groups of leads in wye-connected, dual-voltage motor.

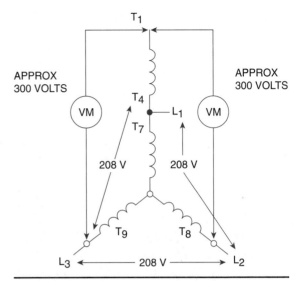

FIGURE 7-18 Lead identification on wye-connected motor.

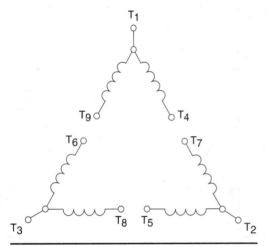

FIGURE 7-17B Groups of leads in delta-connected, dual-voltage motor.

3. Connect for the lowest rated voltage on the nameplate L_1 to T_7, L_2 to T_8, and L_3 to T_9. Be sure to mark L_1 L_2, L_3 to maintain consistency. Energize the motor and note direction of rotation.

4. Disconnect power and connect one lead of one of the two coil groups to T_7. Reconnect power as in step 3. If the coil is the correct coil group and the lead is actually T_4, the voltage between the open end and the other line leads L_2 and L_3 should be approximately 1.5 times the applied voltage and the voltage should be equal. (See Figure 7-18.) If the correct coil group was picked but the voltage on the meter is only 57 percent of the applied, then the coil is reversed and is opposing the line voltage. Change the leads around and retest so that T_4 is connected to T_7 and L_1. The open end should again have an equal reading and approximately 1.5 times the applied voltage. T_1 can now be marked as the open end of the first coil group. The wrong associated coil group will give uneven readings to the line leads T_2 and T_3. (See Figure 7-19.) If this is the case, choose another coil and retest the coils. Disconnect power.

5. Choose another unknown coil and temporarily connect one lead to T_8. Use the same test as in step 4. When the voltages measured from the open end of this coil to the line leads are equal, and 1.5 times the applied voltage, mark T_5 as the one connected to T_8 and T_2 as the other (loose) end of this coil.

Motor Maintenance and Installation

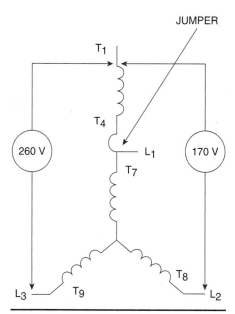

FIGURE 7-19 Uneven voltage indicates the wrong coil has been identified.

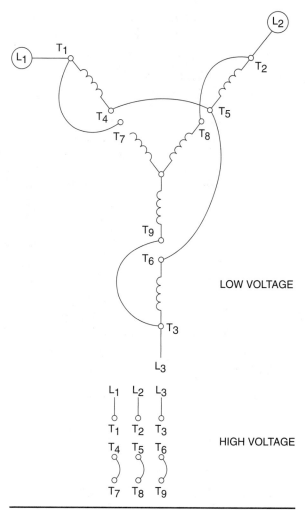

FIGURE 7-20 Connection patterns for wye-connected three-phase motor.

6. Connect the last coil to T_9. Again, apply the low voltage to the power leads L_1, L_2, L_3. Check for even and correct voltage ratings. If the voltage is correct, label the lead connected to T_9 as T_6 and the open end as T_3.

7. Connect the motor leads as if you were going to apply the higher voltage rating (L_1 to T_1, L_2 to T_2, L_3 to T_3, T_4 to T_7, T_5 to T_8, T_6 to T_9). Use the lower voltage supply as a test. The motor should operate smoothly, quietly, and in the same direction as the original direction, as in step 3.

If all the lead markings are correct you may now connect the motor for either the high- or the low-voltage connection. (See Figure 7-20.) Connect and energize the motor. Note each line current using an ammeter. Each phase L_1, L_2, L_3 should be equal and below the nameplate current if the motor has no mechanical load.

DELTA-CONNECTED MOTORS—LEAD IDENTIFICATION

Delta-connected motors are often used in larger-horsepower situations. Some motor starting schemes use a Y–delta (star–delta) starting system. If the motor is designed to run on a delta pattern, the leads must be identified in a different way than the Y (wye or star)-connected motors. As mentioned earlier, the delta motor has three sets of three leads with continuity, compared to the star motor, which has one set of three leads.

1. Select a set of three leads that have continuity. This represents one corner of the Greek letter delta (Δ). Use a low range ohmmeter to determine the center lead of the two coils. (See Figure 7-21.) Temporarily label the center lead as T_1 and the other leads as T_4 and T_9. The next coil group, mark the center T_2 and the other ends as T_5 and T_7. The last coil group mark the center T_3 and the loose ends as T_6 and T_8.

   Be careful when energizing the coils as all coils will have voltage at the open leads.

2. Connect the first coil group T_1, T_4, T_9 to the lowest voltage indicated on the nameplate (L_1 to T_1, L_2 to T_4, L_3 to T_9). Connect and run the motor. Note direction of rotation.

3. Connect the lead marked T_4 to the lead you labeled T_7. Leave the line leads connected as in step 2. Use a voltmeter to read the voltage between T_1 and T_2. If the lead markings are correct, the voltage T_1 to T_2 should be approximately twice the applied line voltage. If it reads only 1.5 times the low voltage, re-connect T_4 to the lead marked T_8.

4. Measure T_1 to T_2 again. If the voltage is twice the line voltage, the markings should be changed to indicate that T_4 is now connected to T_7. If the voltage measured T_1 to T_2 was less than line voltage then reconnect T_9 to T_7, essentially reversing both coil leads. The final connection should read twice the line voltage and T_4 should be connected to T_7.

5. Connect the third coil group to the first coil group. Use the first coil group as the reference point. Connect the coils until the lead connected to T_9 of the first group allows you to read twice the applied volt-age from T_1 to T_3. The lead connections that indicate this voltage determines that T_9 is connected to T_6 and the other loose end is T_8.

6. Mark each line lead as L_1, L_2, L_3 to check each of your lead markings. Connect L_1 to T_2, L_2 to T_5, L_3 to T_7. Energize just the second coil group from the line. The motor should run the same direction as originally connected with the first coil group.

7. The last check is to verify the third group. Connect L_1 to T_3, L_2 to T_6, L_3 to T_8. The motor should rotate in the same direction as the original group. If not, recheck your markings before connecting all the coils together.

8. If all the tests and checks are done and the lead markings are established, connect the motor for both low- and high-voltage connections as shown in Figure 7-22. The motor should run smoothly with no noise and the current in all three lines should be equal and less than the nameplate current if the motor is unloaded.

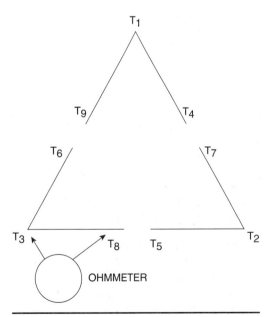

FIGURE 7-21 Use a low-range ohmmeter to determine the center of the coil group for delta-connected motor.

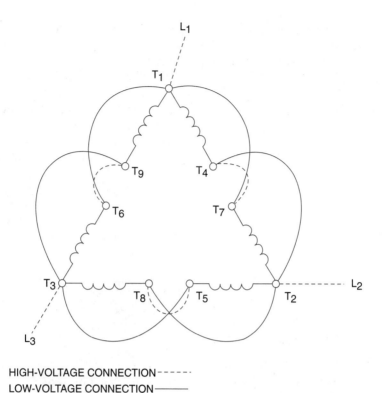

FIGURE 7-22 Connection diagram for delta-connected three-phase motor.

HORSEPOWER TESTING

Testing for output (in other words, horsepower) depends on the size of the motor and the availability of adequate test equipment. Typical ways to test a motor include the use of a dynamometer, or one could simply use a rope and pulley to give a rough approximation. For large motors, where a mechanical load test is not feasible, a calculation of output horsepower is used.

Horsepower (HP) is the measure of mechanical work done. To get an estimate of the amount of HP a particular motor should deliver, use the standard conversion factor of 746 watts equal 1 HP. This is a constant that is used to convert electrical work to mechanical work.

After you have established the approximate HP based on input, you should be able to measure the motor output. Use the formula for HP as 1 horsepower of work equals 33,000 pounds moved 1 foot in 1 minute. This formula was developed many years ago by the mining industry to determine how many horses it would take to pull carloads of coal out of the mines.

1 HP = 33,000 ft-lb/min

This equation tells us that distance traveled, weight, and time are all factors in determining HP. Motors create rotational or circular motion in a certain amount of time which is measured in RPM. The number of RPM multiplied by the torque on the motor shaft or pulley will also indicate how many pounds of weight would move how far in one minute. $2 \times \pi \times r$ is the circumference of the circular motion. Using algebra, the factors distill to

$$\frac{2 \times \pi \times \text{foot} \times \text{pounds} \times \text{RPM}}{33{,}000} = \text{HP}$$

or

$$\frac{\text{foot-pounds} \times \text{RPM}}{5252} = \text{HP}$$

or

$$\frac{F \times R \times N}{5252} = \text{HP}$$

F = Force in pounds on a scale
R = Radius of the pulley on the motor shaft
N = RPM of the shaft at the time of the test.

Now a dynamometer may be used as shown in Figure 7-23. The motor is placed in the test stand. As the motor turns, it creates a force on the scale at some specified radius from the center of the shaft. These two factors make up the force and radius. Measure the RPM with a tachometer and insert as the N in the preceding formula. The output horsepower is now calculated by that formula.

The second method that will give you approximate HP on small motors uses a rope wrapped around a pulley on the motor. The rope is wrapped one turn around the motor pulley. (See Figure 7-24.) A scale is used to measure the pull on the pulley as you force the motor to do more work. Measure the force in pounds on the scale as you pull it harder. The motor pulley will slow as it does more work against the friction of the rope. Measure the pull on the rope, the radius of

FIGURE 7-23 Motor shop dynamometer for HP testing.

FIGURE 7-24 Alternate method for estimating horsepower with rope and pulley.

the pulley, and the RPM of the motor. Apply these factors to the HP formula to determine the HP at the time of test. Be sure to monitor line current during these tests to make sure the rated current of the motor is not exceeded or that the HP is measured at rated current and voltage.

Input Minus Loss Horsepower Testing

Another test for HP is used on large motors where test loads are not practical. This method is referred to as the *input minus loss* method. It is a calculated value based on some accepted assumptions.

The first assumption for this calculation is that the internal stray losses in a motor remain constant for a specified motor. The internal motor stray losses include **eddy current** losses in the rotor and stator and **hysteresis** losses in the iron. These are referred to as iron losses and actually are watts lost in heat; they convert electrical energy to mechanical energy. A second stray loss that is assumed to stay constant is bearing friction. This energy loss is consumed by the motor in the form of actual bearing friction. It takes some energy to keep the rotor moving against the bearing surfaces. Another stray loss is windage loss. Windage loss is the energy required to have the fan mounted on the motor blow air over the motor to keep it cool. Most large motors have some type of fan that is designed to blow air over the motor housing. This fan is a drag on the motor and therefore consumes a small amount of energy.

The stray losses of iron loss, friction, and windage are assumed to stay constant from no load to full load. This is not true, but the losses do remain close enough in proportion to the remaining losses.

The other losses that occur in the motor are referred to as copper losses. These include the stator copper loss of $I^2 \times R$ of the stator winding and $I^2 \times R$ of the rotor winding. These losses will change with load.

The term "input minus loss" indicates that the input watts of the motor minus all the internal losses of the motor will yield the output of the motor in watts. Then use the watts to HP conversion 746 W = 1 HP to convert to mechanical HP.

Employ the following steps to determine HP using the input minus loss method for three-phase motors.

1. Measure the resistance of the motor coils per phase. Measure (T_1 to T_2 ohms) + (T_2 to T_3 ohms) + (T_1 to T_3 ohms) and divide by 6 to get ohms per phase.
2. Connect the motor to the line without a mechanical load. Measure the current average of the line leads. Use the current in the formula. $I^2 \times$ ohms per phase $\times$ 3 to determine stator copper loss.
3. Measure the three-phase watts consumed by the motor at no load condition. These input watts minus the stator copper loss determine stray loss.
4. Load the motor to the load you wish to drive and measure the three-phase input watts again. Using the new input current, recalculate the stator copper loss now. Use loaded input watts minus loaded $I^2 \times R$ of stator minus the stray losses constant to obtain rotor input watts.
5. Measure the speed of the rotor under load to calculate percentage of slip.

$$\frac{Ns - Nr \times 100}{Ns} = \% \text{ slip}$$

6. Rotor input watts $\times$ % slip = rotor copper loss.
7. Input watts minus all losses (stray + stator + rotor) = output watts.
8. Output watts divided by 746 = output horsepower.

Refer to Figure 7-25 for a sample.

1. Three-phase 5-HP, 230-V motor draw 5.1 amps when running unloaded. The input watts unloaded are 270 W. The ohms per phase are .337 Ω.

No-load low calculation:

2. I^2 × ohms per phase × 3 = stator copper loss

   $(5.1)^2$ × (.337) × (3) = 26 watts stator copper loss

3. 270 watts input − 26 watts = 244 watts of stray loss

   Loaded motor data
   5.8-amp line current, 950-watt input
   1780 RPM

4. I^2 × R loss of loaded stator
   $(5.8)^2$ × (.337) × 3 = 34 watts of stator copper loss

   input watts − ($I^2 R$ of stator) − stray loss = rotor input watts

   950 watts − (34 watts) − (244 watts) = 672 rotor input watts

5. % slip of loaded motor =

   $\dfrac{1800 - 1780}{1800}$ × 100 = 1.1% slip

6. rotor input watts × % slip/100 = rotor I^2 × R loss

   672 × .011 = 7.4 watts of rotor copper loss

7. loaded input watts − (stray loss + stator copper loss + rotor copper loss) = output watts

   950 watts − (244 watts + 34 watts + 7.4 watts) = 664.6 watts

8. output watts ÷ 746 = output horsepower

   664.6 ÷ 746 = .89 HP at test load

FIGURE 7-25 Sample of input minus loss calculation for HP.

MOTOR EFFICIENCY

Efficiency of a motor is expressed as a percentage of the amount of energy that is converted from electrical energy to mechanical energy. The formula for finding the percentage is watts out or 746 × (HP output) divided by watts input in electrical energy × 100 = % efficiency.

$$\dfrac{\text{watts out} \times 100}{\text{watts in}} = \text{\% efficiency}$$

Efficiency of a motor depends on the design of the motor and the operating conditions. Motors tend to operate at maximum efficiency for the particular motor when operating at full load. This condition allows the maximum amount of transfer of energy from the stator to the rotor and losses in the motor make up a very small portion of the total input energy. As described previously, stray loss, friction, and windage loss are a relatively large portion of no load energy. In fact, at no-load the motor is 0 percent efficient because it does no work. However, as the motor is loaded, these losses remain fairly constant and are therefore a smaller part of the full-load energy.

Motor design also helps determine losses and efficiency. If the tolerance for the stator air gap (distance between rotor and stator) is very small, there is good magnetic transfer from rotor to stator and the motor is more efficient. If higher quality bearings are used, friction is reduced and the motor becomes more efficient. If a better grade of silicon steel is used and laminations are thinner, the hysteresis and eddy current losses in the steel is reduced. Of course, if the motor is more efficient it takes less energy to supply it for the same HP rating output of a less efficient motor. As with most advantages, there is a cost involved. High-efficiency motors cost more to purchase but cost less to operate.

Watts input to the motor may be measured on a wattmeter. Remember that on AC motors you cannot simply multiply volts × amps to get watts. Power factor is another component with AC motors. To obtain true input watts (or true power) you must use $E \times I \times \%PF$ = watts. Also to get three-phase input watts, check to be sure all the motor leads draw the same current and use the formula: $E \times I \times 1.73 \times \%PF$ = watts.

The other option for three-phase motors is use of a three-phase wattmeter or two single-phase wattmeters. They should be added according to Blondell's theorem (discussed in Chapter 9).

NEC® AND SIZE OF FEEDERS, DISCONNECTS, AND CONTROLLERS

Motor installations, whether single- or three-phase, must meet some basic guidelines to operate safely and provide the degree of protection required. The protections include safeguards against dangerous overheating of motor, controls, and supply conductors.

As described in Chapter 1, Article 430 of the NEC® governs motor installations. The installation begins with the proper motor feeder and feeder protection. Feeders can be considered as the source of electricity for multiple motors or motors and other loads. Article 430.24 to 430.26 determines the ampacity (current carrying rating) of the feeder conductors. Conductors supplying several motors shall have enough ampacity to carry the sum of all the full load currents, plus 25 percent of the largest rated motor. There are many exceptions to this general rule so you must consult the NEC® article to see if any exceptions will apply.

The motor feeders need protection against excessive current due to short circuits or grounds. See Article 430.61 to 430.63. The basic rule for protecting the feeder is to use the largest rating of the individual motor branch circuit protection, then add the sum of the other motor full load currents. If two ratings

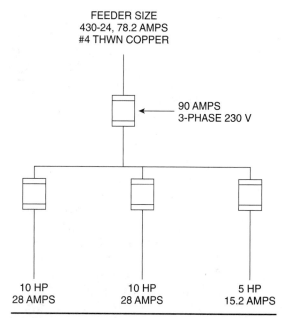

FIGURE 7-26 Sample branch and feeder circuit calculation.

are the same for large motors, one of the ratings will be considered the larger of the two. For example, if two 10-HP motors and one 5-HP loads were connected to a 230-V three-phase feeder, each of the motors would have an individual overcurrent based on Article 430.52 and Table 430.250. This tells us that the 10-HP motors use 175 percent × 28 amps as individual dual-element fuse protection and the 5 HP needs 175 percent of 15.2 amps. The feeder protection would use (175% × 28) + 28 + 15.2 amps as a feeder fuse rating or 49 amps + 28 + 15.2 = 92.2A. A 90-A fuse is required. (See Figure 7-26.)

DISCONNECTS, CONTROLLERS, FUSES, BRANCH CIRCUIT CONDUCTORS, OVERLOADS

After the feeder, each motor will have a branch circuit means for disconnecting. This is determined by Article 430.101 to 430.113.

The disconnect must be within sight of the motor controller and must disconnect the controller from the source of supply. The disconnect also must be located within sight of the motor and the driven machinery. This means you must be able to see the disconnect and it cannot be more than 50 feet from the machinery, unless the disconnect can be locked in the open (disconnect) position. Article 430.102(B) exception allows the electrician to install the disconnect *not* within sight of the motor *if* the sight specified by the main rule is impracticable or introduces increased hazards to personnel, or *if* there are written procedures in industrial installations. The disconnect still needs to disconnect the motor and the starter and be able to be locked in the disconnected position. This is also an OSHA requirement for lockout and tag-out procedures. Types of disconnecting means are listed in Article 430.109. In general, the disconnect for motors 600 V or less shall have an ampere rating of at least 115 percent of the full-load (code book) current of the motor (Article 430.110). Generally, the disconnect is rated in amperes, or HP, or both. Common sizes are 30-A, 60-A, and 100-A disconnects. Using the 10-HP motor just listed, 115 percent × 28 A = 32.2 A so a 30-A disconnect will not suffice. Also, the disconnect usually holds the branch circuit over the current device, which normally is a fuse. (See Figure 7-27 for a disconnect sample.)

Motor Branch Circuit Protection

Now as the circuit branches off of the feeder, the branch circuit also needs protection. In fact, this protection needs to be calculated in order to determine feeder protection. The branch circuit protection is sized according to Article 430.51 to 430.58. For the 10-HP motor in the installation, the rules in Article 430.52 direct us to Table 430.52. Use the 28 amps as the full-load current even though the actual nameplate current may be something less. For this example, assume full-voltage starting. Use a dual-element fuse. The table directs you to use a fuse not over 175 percent of 28 amps. Article 430.52 tells us not to use a fuse exceeding this value 1.75 × 28 = 49 A. Therefore, first try to use a fuse size as listed in Article 240.6 (45 amps). See Exception No. 1 of Article 430.52(C) to determine if the next larger size may be used.

Motor Circuit Conductors

Motor conductors extend from the branch circuit protection to the **motor controller** and then to the motor. Article 430.21 to 430.29 covers these conductors. For general-use motors, Article 430.22 stipulates that the conductors must be sized to carry 125 percent of the full-load (code book) value of the motor. The 5-HP motor in our installation requires 125 percent × 15.2 amps or ampacity of 19 amps.

The size of the wire needed to carry this load without overheating is determined by Table 310.16 of NEC®. THWN copper wire #14 AWG is rated to carry 20 amps. Care must be taken when selecting the wire from these tables. When terminating the wire, the temperature rating of the terminations must be considered. Even though the wire may be rated for 90 °C, the terminations are often only 75 °C; therefore the 75-degree column of Table 310.16 must be used even if THWN or other 90-degree wire is used.

See Article 240.4(D) and (G).

Motor Controllers

Motor controllers act as the on–off or speed-control method to control motor functions in normal operations. These controllers are covered in Article 430.81 to 430.91. The controller must be capable of stopping and starting the motor and interrupting the locked rotor

Motor Maintenance and Installation

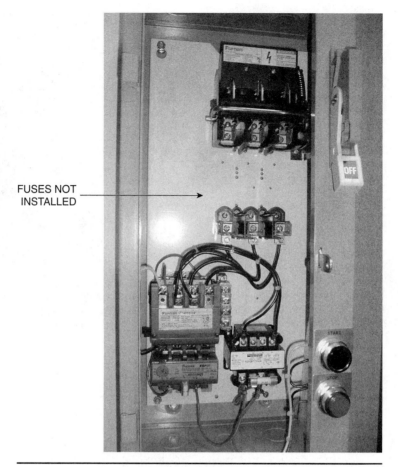

FIGURE 7-27 Disconnect switch with branch circuit protection.

current. It must have a HP rating for the motor it controls at the operating voltage of the motor. (See Figure 7-28.) For the 5-HP motor in the example, the controller must be rated at least 5-HP at 230 V. If using a NEMA-rated starter, use a size 1. (See Chapter 3.)

Running Overcurrent Protection

The last portion of the motor installation at this point is the selection of the running overload device. This is the system designed to protect the motor against overheating due to overloads while running or failure to start. Article 430.32 refers to it as motor and branch circuit overload protection. It is not intended to protect against short circuits or ground faults. For the motors in this text's examples, the motors are more than 1 HP and do not have a service factor or a temperature rise rating. Therefore, the overload device shall trip at not more than 115 percent of nameplate current (actual nameplate). If this device continually trips on starting, Article 430.32(C) allows you to increase the overload device one size, but do not exceed

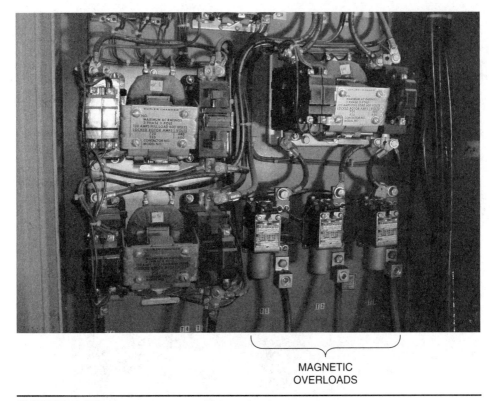

FIGURE 7-28 Large starter with overloads.

130 percent for the aforementioned motors. Remember to check actual trip current of the overload devices (see Chapter 3 for overload style selection). (See Figure 7-29 for an installation example.)

SUMMARY

Motor maintenance procedures were described in this chapter. You studied different types of bearings that are used for different loads. Basic cleaning and lubrication of the bearings and bearing replacement techniques were presented. Other maintenance and repair checks included testing of coils for proper insulation resistance, shorts, and opens.

Methods of lead identification, including single- and three-phase motors, were described to enable you to remark damaged or mismarked leads.

Once the motor was cleaned, lubricated, and the leads reidentified for connection, horsepower testing could take place. Several methods of horsepower testing were described.

A basic motor installation was presented in accordance with the NEC®. This included feeders, disconnecting means, branch circuits, motor controllers, and overload devices.

QUESTIONS

1. Explain what NEMA stands for and how different manufacturers' motors may fit the application.

Motor Maintenance and Installation

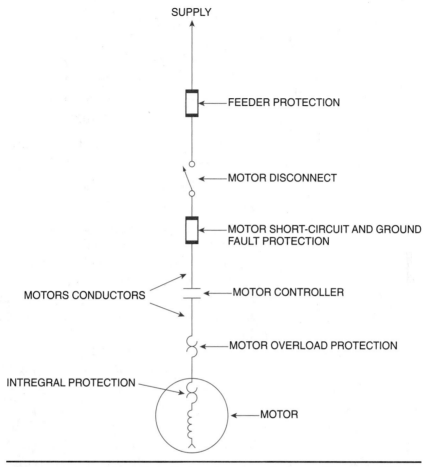

FIGURE 7-29 Motor installation diagram, similar to Diagram 430.1 in the National Electrical Code.

2. What is the difference between thrust bearings and radial bearings?

3. Give several indications of bad bearings on a motor.

4. What is the purpose of a megohmmeter?

5. When would you use an internal growler?

6. Explain how to determine the starting and running windings of a split-phase motor.

7. How would you determine the null point or neutral plane on a universal motor?

8. Explain briefly how you can identify whether a motor is wye- or delta-connected.

9. Connect the following wye-connected motor for high-voltage connections:

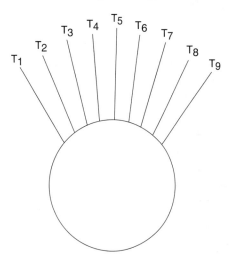

10. Show schematically how to connect a dual-voltage delta motor for low voltage.

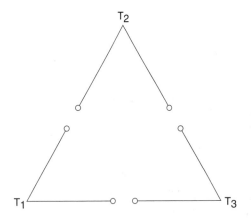

11. Write the formula for determining HP based on measurable values.

12. What are some basic assumptions in input minus loss HP calculations?

13. Explain why motors are more efficient at full load.

14. How is the motor feeder fuse protection calculated?

15. Explain what is meant by "in sight from."

16. Which NEC® article determines branch circuit protection for motors?

17. Fill in the following chart for the unknown values.
    Calculate branch circuit short-circuit protection size, disconnect size, branch circuit conductor ampacity and size of THWN conductor, rating of controller, and overload trip current rating for the following motors. Calculate the feeder protection and fuse protection if all the motors are fed by the same feeder.

    1. 10-HP, 230-V, 3-phase, 26.5-A nameplate, Code K, 40 °C temperature rise
    2. 7.5-HP, 230-V, 3-phase, 20.5-A nameplate, Code K, 40 °C temperature rise
    3. 5-HP, 230-V, 3-phase, 15-A nameplate, Code H, 50 °C ambient temperature

Branch Circuit Calculations

Motor	10 HP	7.5 HP	5 HP	NEC
Motor disconnect ampere rating	1.15 × 28 A = 32.2 A	___ A	1.15 × 15.2 = 17.5 A	430.110
Branch circuit short-circuit ground fault protection	1.75 × 28 A = 49 A 50 A FRN fuse	1.75 × 22 A = 38.5 A ___ FRN fuse	1.75 × 15.2 = 26.6 A 30 A FRN fuse	430.___
Motor circuit conductor	1.25 × 28 = ___ A = #___ AWG	1.25 × 22 = 27.5 = #10	1.25 × 15.2 = 19 A = #14	430.22 240.4 G 310.16
Controller HP rating	10 HP @ 230 V	7.5 HP @ 230 V	5 HP @ 230 V	430.83
Motor overload trip current	1.25 × 26.5 = 33.125-A trip	1.25 × 20.5 = 25.625-A trip	= ___ A trip	430.___

CHAPTER 8

SPECIAL MOTORS

OBJECTIVES

After completing this chapter and the chapter questions, you should be able to

- Identify and connect wound rotor motors
- Understand the principles and identify types of repulsion motors
- Determine the types of synchronous motors used in various applications
- Recognize stepper motors and calculate step angle and resolution
- Identify torque motor uses

KEY TERMS IN THIS CHAPTER

Permanent Magnet Motor: Motors that use permanent magnetic material as one of the magnetic fields. This magnetic field reacts with the applied magnetic field.

Repulsion–Induction Motors: A style of motor that is started as with the repulsion principle but run as an induction motor.

Repulsion Motor: A style of motor that does not use a rotating magnetic field. The rotor has a commutator and carbon brushes that are used to create a current path from the induced EMF. The current in the rotor creates poles that are repulsed by the stator poles.

Stepper Motor: A motor that does not rotate freely, but in incremented steps. Each step is controlled by a stepper controller as pulses of power are applied.

Torque Motors: A variation of conventional induction and DC design, used for slow-speed motion and tensioning applications.

Wound Rotor Motor: A type of three-phase motor that uses a standard stator but uses a rotor with coil windings rather than a squirrel cage. The rotor has slip rings that permit the rotor windings to be connected to external resistors for control purposes.

INTRODUCTION

Many of the standard motors in use are AC squirrel cage induction motors that use a variety of starting methods. Other induction motors have different windings or additions such as capacitors for particular applications. Still other standard motors such as DC motors are used in business and industry. (See Chapter 10 for discussion of DC motors.) This chapter discusses less common motors used for special applications.

WOUND ROTOR MOTOR

A **wound rotor motor** is a variation of the induction motor but does not use a squirrel cage winding. The stator of the wound rotor motor is the same as the standard three-phase induction motor, in that it produces a three-phase rotating magnetic field. The rotor is not a squirrel cage winding. Squirrel cage windings have cast conducting bars shorted together in end rings and installed in the laminated iron. The rotor of the wound rotor motor actually consists of conductors (magnet wire) wound into coils on the rotor. (See Figure 8-1.)

In Figure 8-1, notice there are three slip rings mounted at the end of the shaft. These slip rings are for connections to the rotor windings through carbon brushes. The brushes are used for connection to secondary resistors, not to line power. These resistors are referred to as secondary resistors because the rotor is called the secondary circuit of the motor. The rotor will have voltage induced into it by the stator or primary circuit.

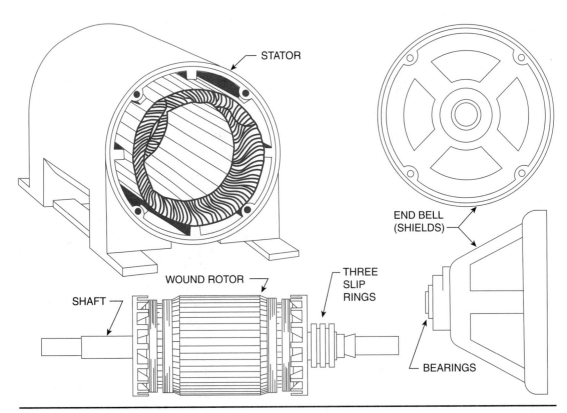

FIGURE 8-1 Parts of a wound rotor motor.

Principle of Operation

Three-phase AC is applied to the stator through leads marked T_1, T_2, T_3. (See Figure 8-2.) The three-phase rotating magnetic field is established in the stator with the desired direction of rotation as in any other three-phase induction motor. The secondary windings (rotor) have voltage induced into them in a manner similar to the squirrel cage windings. The big difference in the wound rotor motor is that the rotor circuit can have resistors connected into (or removed from) the rotor circuit to control the amount of current flow in the rotor. (See Figure 8-3.) The secondary resistors are connected by slip ring and brushes to terminal M_1, M_2, M_3. M_1 connection is the outside ring closest to the bearing, M_2 is the middle ring, and M_3 is closest to the rotor windings.

When starting the motor, remember that high starting current results in the rotor and the stator (locked rotor current). By inserting resistance into the rotor windings during starting,

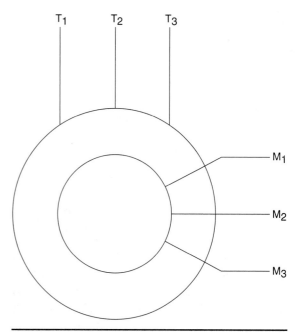

FIGURE 8-2 Wound rotor motor schematic symbol shows three primary and three secondary leads.

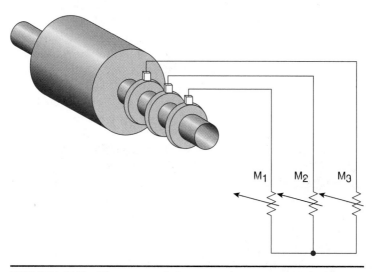

FIGURE 8-3 The rotor of the wound rotor motor has resistors that are connected in and out to control the rotor current.

the secondary current flow will be reduced; therefore, the primary line current will be reduced. As the motor speed increases, the induced EMF in the rotor decreases and so does the primary starting current. Now the secondary resistors are stepped out, or taken out in increments, and the motor increases acceleration. By removing resistance from the secondary circuit, more rotor current is allowed to flow and more torque is produced in the motor. If more torque is produced, but the mechanical load is constant, the speed will increase. The secondary resistors may be used for speed control as well as starting current reduction. By putting the resistors back into the secondary circuit, the rotor speed will be reduced.

The wound rotor motor has higher starting torque per line amps than do most AC squirrel cage motors. The reason that the starting torque per amp is higher is that resistance is added to the rotor. By adding resistance, the power factor of the rotor is improved (phase angle is reduced). When the rotor and stator flux have a closer phase angle, there is better magnetic interaction and the torque of the motor is increased. By taking resistance out of the circuit, the phase angle is increased until all the resistance is out and the motor characteristics are similar to comparable-sized squirrel cage motors.

Most motor starters used with wound rotor motors have a provision that will not allow you to start the motor if all secondary resistance is shorted out of the secondary. This condition would be a shorted secondary and would have the same characteristics as a squirrel cage winding. Also, if you attempted to start the motor with the secondary circuit open, there would be no secondary current flow, thus no rotor torque.

The wound rotor motor nameplate lists the secondary current and voltage. (See Figure 8-4.) NEC® Article 430.23 lists the ratings of the resistors and requirements for the conductors that connect the controller to the resistors and the controller to the motor. If the resistors are

PRI VOLTS: 220/440	FREQ: 60 HZ
PRI AMPS: 15/7.5	PHASE: 3
FULL SPEED: 1700 RPM	
SEC. VOLTS: 77	SF: 1.15
SEC. AMPS: 33.1	TEMP RISE: 40°
H.P.: 5	

FIGURE 8-4 Example nameplate of wound rotor motor.

used continuously for speed control, the conductors must be rated at 125 percent of the full-load secondary current on the nameplate. If the resistors are less than continuous duty, use the ratings and percentages in Table 430.22(c). If the resistors are separate from the controller enclosure and you provide the connecting conductors, the ampacity is determined by the application and NEC® Table 430.23(c).

Controllers for the wound rotor motor may be drum-type or automatic sequence-type starters. Figure 8-5 shows a diagram of a secondary resistor starter used on a wound rotor motor. As the start button is pressed, the M coil is energized and line power is applied to the stator. The control relays on lines 3 and 5 are deenergized and all the resistance is inserted into the secondary (rotor) circuit. As TR_1 times out, TR_1 contact on line 3 closes and CR_1 and TR_2 are both energized. CR_1 contacts in the secondary power circuit close, shunting out the lower bank of secondary resistors. As TR_2 times out, CR_2 is energized and the second bank of resistors is shorted out; now the rotor runs as a normal rotor with no extra resistance. In this case, the resistance is used only for starting duty. Pressing the stop button resets

Special Motors

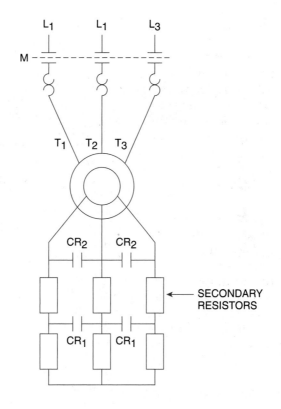

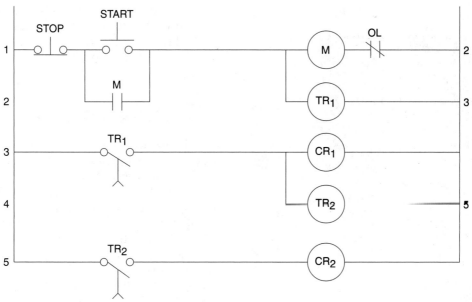

FIGURE 8-5 Automatic two-step acceleration using secondary resistors and a wound rotor motor.

all relays and resistance is reinserted into the rotor circuit.

If the motor is also to be used for speed control, a drum controller is often used. This would consist of a manual controller where you provide the mechanical movement to change taps on resistors or alter the secondary resistor circuit. Speed control also can be accomplished with magnetic control schemes.

With the increased use of solid-state soft-start and variable-frequency drives, the wound rotor motor is becoming less desirable. The wound rotor motor is a higher maintenance motor because of the windings that are fitted into the rotor. Also, the slip rings and brushes require more maintenance. The secondary resistor requires more controller equipment as the resistors consume power. Generally, the controllers and resistors are expensive. Few new wound rotor motors are being installed, but there are wound rotor motors that require maintenance, so you should familiarize yourself with their operation.

See Figure 8-6 for an electronic wound rotor drive.

SMRSM6: Synchronous (Bypass)
- RSM6B with integrated/ programmable synchronous excitation controls
- Automatic power factor control
- Also available as stand-alone synchronous excitation package

WRSM6: Wound Rotor
- RSM6 with integrated/ programmable wound rotor controls
- High torque starting or speed control

FIGURE 8-6 Electronic drive for wound rotor motor. *(Courtesy of Benshaw)*

Wound Rotor and NEC®

The wound rotor motor is addressed in the NEC® in several places. By viewing Chart 430.1, you will see that secondary controllers, conductors, and resistors are at the bottom of the chart and refer you to Part II of Article 430 and also Article 470.

Article 430.23 was presented previously. It is used for sizing the conductors for the wound rotor secondary. Article 430.32(D) refers to continuous duty motors with wound rotor secondaries. The secondary circuits are considered to be protected by the primary overloads. Article 430.52 again refers you to Table 430.52 for short-circuit protection for wound rotor motors. Notice that the protection is typically quite close to the full-load amps. This is because the inrush current level is small when using secondary resistance starters.

If a special wound rotor motor is to be kept in operation, the control methods are updated to take advantage of new technology. To update a wound rotor motor installation, the electrician should tie all the secondary leads together to form a shorted secondary. Then, new electronic speed controls can be applied to the primary to control starting current and torque and speed control where needed.

REPULSION MOTORS

Repulsion motors are special single-phase motors that are installed infrequently. Repulsion motors are designed in three styles: repulsion motor, repulsion start-induction run motor, and **repulsion–induction motor**. All of these motors work off the induction principle, where voltage is induced into the rotor. However, they differ greatly from other AC motors in that they do not rely on a rotating magnetic field of the stator. These motors use a wound rotor, not a squirrel cage, but they are not considered wound rotor motors. The rotor current produces magnetic poles in relation to the stator poles. The two poles are repelled or repulsed; thus the name "repulsion motor."

The basic operation of the repulsion motor utilizes a set of brushes connected to the rotor winding via a commutator assembly. The commutator is a segmented sliding contact system where each commutator segment is connected to a rotor coil end. (See Figure 8-7.) The stator has salient poles that produce a certain number of magnetic poles. Salient poles mean that the pole pieces are separate iron poles that are recognizable from the surrounding iron. There are two salient poles on the stator in Figure 8-7. Brushes are part of the commutator assembly but are not used to conduct line power to the rotor. The brushes are short-circuited together to provide a connecting path between commutator segments, which allows a complete electrical circuit to be established in the rotor. There are several positions that the brushes may be moved to on the movable brush yoke.

Brushes may be moved and then locked into a specific location or changed to meet the varying needs of the load. The first brush positions mentioned are neutral; this means that no torque will result when the brushes are in this location. Figure 8-7 shows the brushes located in the *soft neutral* position. The soft neutral is located 90 degrees away from the stator poles. In this position, no current flows in the rotor and no magnetic poles are established in the rotor. The motor will establish no torque when the brushes are in the soft neutral position. The motor is an AC motor and the magnetic polarities of the fields change two times per cycle of applied frequency. The diagrams reflect only one half-wave instantaneous value.

The other position of the brushes is directly beneath the stator salient poles. This position is called *hard neutral,* as there is maximum amount of current flow in the rotor and magnetic rotor poles are established directly beneath the stator poles. Because there is no angular displacement (in other words, the poles are directly aligned with each other), the rotor will not be repulsed in the clockwise (CW) or counterclockwise (CCW) direction of rotation. Figure 8-8 shows brushes moved to hard neutral. These two neutral positions can be determined not only by visual observation, but also by monitoring line current and rotor movement. If the rotor is stationary with power applied to the stator, then the motor is in neutral. If there is high line current in the primary winding and the rotor does not move, then the motor is in hard neutral. If the line current is low and the rotor does not move, then the motor is in soft neutral. To get the motor to develop torque and begin to turn, the brushes are shifted away from neutral.

If the brushes are shifted away from hard neutral as shown in Figure 8-9, the rotor poles will be shifted so they are above the brushes.

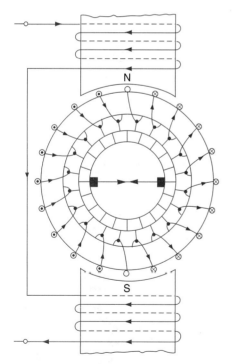

NO TOURQUE CREATED, EQUAL VOLTAGE VALUES OPPOSE EACH OTHER (SOFT NEUTRAL)

FIGURE 8-7 Repulsion motor diagram shows "soft neutral" position of brushes.

NO TORQUE EVEN THOUGH CURRENT VALUE IN ARMATURE IS HIGH (HARD NEUTRAL)

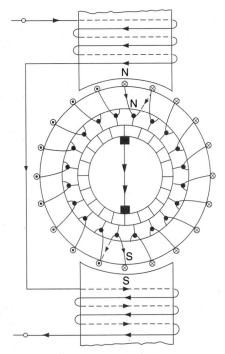

FIGURE 8-8 Repulsion motor with shorted brushes in "hard neutral."

COUNTERCLOCKWISE ROTATION: BRUSHES IN CORRECT POSITION

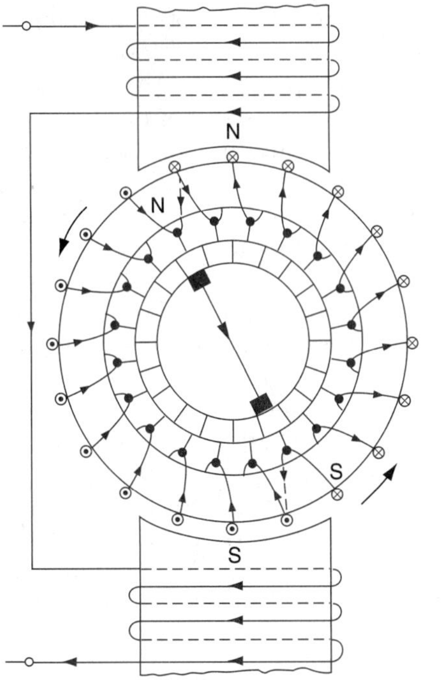

FIGURE 8-9 Brushes shifted to correct operating position away from hard neutral.

Notice in Figure 8-9 that the north rotor pole is moved CCW from the north stator pole. The two poles repel (repulse) each other and the rotor begins to spin CCW.

The normal movement that gives the best torque is approximately 15 degrees off of hard neutral. This position gives the best reaction between the two magnetic fields.

This single-phase motor has the highest starting torque of any single-phase motor. To change the direction of rotation, shift the brushes to the opposite direction from hard neutral, or 15 degrees clockwise off of hard neutral. Speed can be adjusted by moving the brushes away from hard neutral toward soft neutral. As the brushes move toward soft neutral, the flux relationship between rotor and stator weakens and the torque decreases. For repulsion motors you may also control supply voltage to change speed. By reducing line voltage you will reduce primary and secondary current and thereby reduce torque and resultant speed.

Applications for the repulsion motor are where high-torque, low-current starting is required for single-phase systems. The speed is not constant as it will rapidly decrease with increased mechanical load. Conversely, the speed can increase to very high levels without a load attached. The motors can be used for constant torque applications where load is constant, but high start torque is required.

Repulsion Start Induction Run

There are two styles of the repulsion start-induction run (RSIR) motors. The brush-lifting type actually has a mechanism that lifts the brushes off the commutator while the motor is in the running mode. See Figure 8-10 for the centrifugal weight mechanism used to lift the brushes and reduce wear. The brush-lifting mechanism is used on radial-type commutators where the brush moves parallel to the rotor shaft to connect the commutator segments that radiate from the center of the shaft. These segments are pie-shaped pieces and the brushes are designed to meet the radial commutator.

The brush-riding-type RSIR motor does not lift the brushes as the motor operates as an induction motor. This type of motor uses an axial commutator as in Figure 8-11. This motor has a brush assembly that is perpendicular to the rotor shaft.

Both styles of RSIR motors use a short-circuiting necklace that is attached to the commutator. During starting, the necklace is not in contact with the commutator and the motor is allowed to start as a repulsion motor as described earlier. At about 75 percent of full speed a centrifugal weight system forces the necklace (a flexible conductor) against the commutator segment and shorts all the segments together.

Now the motor operates as an induction motor with a shorted squirrel cage winding. Starting windings on the stator are still not necessary as it starts as a repulsion motor with the high-torque, low-current advantage of a repulsion motor. The running characteristics are more like a squirrel cage with good speed regulation. If the brush-lifting mechanism is used, there is less drag on the rotor due to brushes, less wear of the carbon brushes, and less noise during operation.

To reverse the direction of rotation of the RSIR motor, move the brushes to the opposite side of hard neutral just as in the repulsion motor. The connection diagrams for the repulsion motor and the schematic symbol are shown in Figure 8-12.

Repulsion–Induction Motor

The **repulsion–induction motor** operates similarly to the brush-riding style of the RSIR motor. The major difference is that there is no centrifugal short-circuiting necklace to short out the commutator. To accomplish the induction running characteristics, a squirrel cage rotor is added to the rotor. (See Figure 8-13.) As the motor is started, it acts as a repulsion motor with high torque and low starting current. This is due to the shorted brushes and the impedance of the repulsion coils. As the rotor speed increases and the rotor frequency drops, the deeper squirrel cage winding has more effect on operations. Now the motor runs with the better speed regulation of a squirrel cage rotor. The disadvantage of this motor is that it still has brushes and a commutator that require maintenance and the cost of two windings on the rotor is much higher than other motors.

SYNCHRONOUS MOTORS

Three-phase synchronous motors were discussed in Chapter 6. Those motors were the larger three-phase synchronous motors that required special controllers to apply DC to the rotor to keep the rotor poles energized. There are also small AC single-phase motors that can run at synchronous speed. These small, fractional horsepower motors provide constant speed for electric clocks, timers, and other speed-sensitive equipment. The motors run through use of a rotating magnetic field produced by the shaded pole effect (see Chapter 1's discussion of shaded pole motors). Because the rotor portion of the motor uses a high hysteresis magnetic rotor, they are sometimes referred to as hysteresis motors.

ELECTRIC MOTORS AND MOTOR CONTROLS

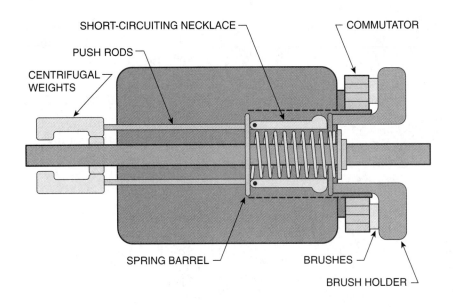

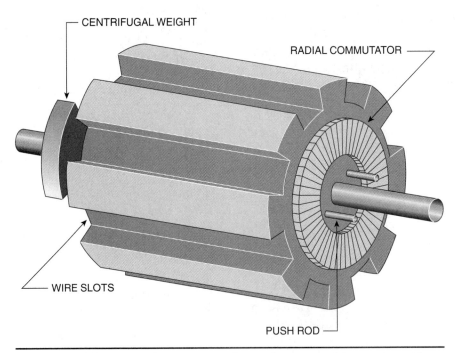

FIGURE 8-10 A radial commutator is used on a repulsion motor with a brush-lifting mechanism.

Special Motors

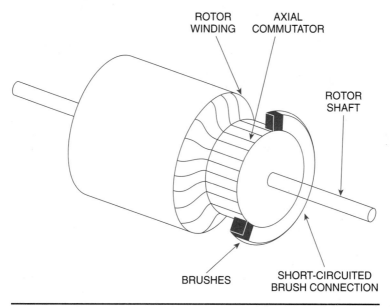

FIGURE 8-11 Axial commutator on a repulsion motor has short-circuited brushes and no brush-lifting mechanism.

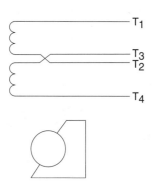

	L_1	L_2	TIE TOGETHER
LOW VOLTAGE	T_1 T_3	T_2 T_4	
HIGH VOLTAGE	T_1	T_4	T_2 T_3

FIGURE 8-12 Connection diagram and schematic symbol for a repulsion motor.

One variation of the hysteresis motor is the Holtz motor, with a shaded-pole-style stator. (See Figure 8-14.) The two-pole stator produces a rotating (moving) magnetic field at 3600 RPM when 60 Hz is applied (synchronous speed formula). The rotor is a combination of the squirrel cage winding embedded deep in the rotor slots and high hysteresis steel salient poles assembly. As the motor is started, the rotating magnetic field of the stator cuts the squirrel cage winding and starts the rotor spinning. As the rotor approaches synchronous speed, the core material creates a low opposition magnetic path between the stator poles. The six salient (or protruding) poles of the rotor line up with the two stator poles during each half-cycle of the rotating magnetic field. The six poles of the rotor determine the synchronous speed of the rotor. If the stator field rotates at 3600 RPM, the rotor will follow the same direction of rotation at 1200 RPM.

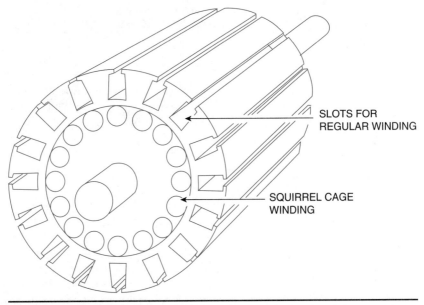

FIGURE 8-13 A squirrel cage rotor is added to a repulsion motor rotor to create a repulsion–induction motor.

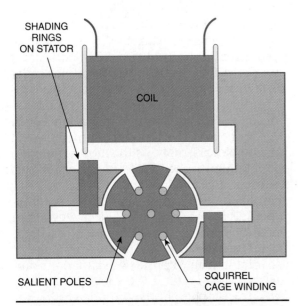

FIGURE 8-14 A Holtz motor uses a squirrel cage and a salient pole made of high-hysteresis steel.

Two stator poles and six rotor poles produce a speed of 2/6 or 1/3 of stator speed. Remember, these are fractional horsepower motors so they do not have a great deal of torque.

Warren Motor

The other version of a hysteresis motor is the Warren or General Electric motor. This hysteresis motor design does not use a squirrel cage winding in the rotor. It uses the principle of the same shaded-pole-style stator to produce a synchronous rotating magnetic field. The difference is in the rotor construction. The rotor is constructed of high permeability iron. The iron forms magnetic paths through the rotor as seen in Figure 8-15. As the stator is energized, the induced voltage into the iron causes the rotor poles to follow the stator poles. As the motor approaches synchronous speed, the stator poles align with the rotor's

Special Motors

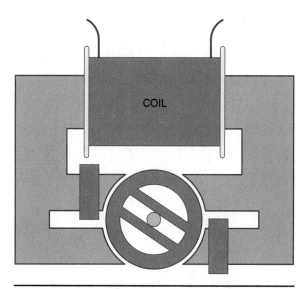

FIGURE 8-15 A Warren motor has a high-permeability rotor and a shaded pole stator.

high-permeability iron. With high permeability (magnetic lines of force pass through easily), the magnetic field of the stator pulls the rotor at synchronous speed.

PERMANENT MAGNET MOTORS

Still another type of small synchronous motor uses a **permanent magnet** as the rotor core. If the rotor has a permanent magnetic field established, there is no need for induction to take place. As the stator is energized, the rotor will attempt to follow the magnetic poles of the stator. There is some initial slip as the rotor comes up to speed, but the rotor will operate at the synchronous speed of the stator.

NEC® Requirements

Because these motors are small and generally have high-impedance stator windings, they may be included in NEC® Article 430.32(B)(4). If the impedance of the windings is high enough to prevent overheating if the rotor does not turn, you may protect them using the branch circuit, short-circuit, and ground fault device if sized according to NEC® Article 430, Part III. The fine-print note (FPN) under Article 430.32(B)(4) helps explain the proper circumstances.

STEPPER MOTORS

Stepper motors are specialized motors that also create incremented steps of motion rather than a smooth unbroken rotation. The basic stepper concept is explained using a permanent magnet on the rotor with two sets of poles as shown in Figure 8-16. As the stator is energized with pulses of DC, the permanent magnet rotor will be repelled or attracted to line up with the stator magnetic poles. The pulses are provided by a stepper controller as illustrated in Figure 8-17.

The electronic controller provides timing and sequencing of the motor, but it operates electronically to provide circuit closure as shown in Figure 8-16. For example, by moving switch 1 and 2 to position A or B, the top poles or the side poles can be reversed. By following the first switch sequence where both switches are set to "A," the top and right poles become north magnetic polarity and the rotor will align between the poles. Step 2 changes switch 1 to B. If power is left on the top poles, the rotor will align top to bottom. When switch 1 connects to point B, the rotor will again move, the south rotor pole aligning between the top and left poles. This will result in CCW rotation. To reverse the direction of rotation, use the second set of steps. Notice that by changing the sequence and length of time the coils are energized, the direction and speed of the steps are controlled.

The rotor could be one of three different styles: the variable reluctance rotor, the permanent magnet rotor, or a combination of the two—a hybrid rotor. Rather than being simply

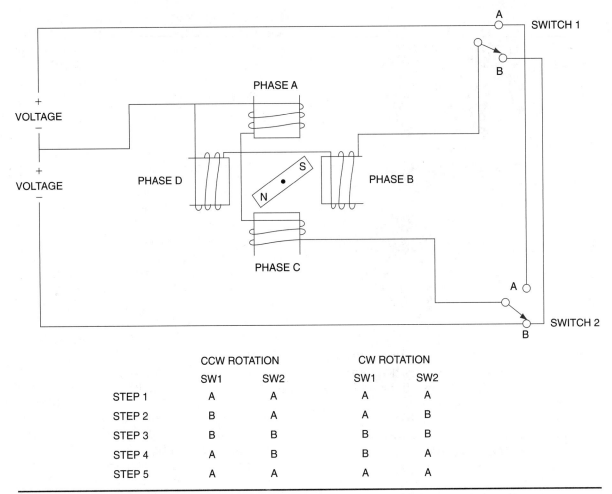

FIGURE 8-16 Stepper motor showing switch position and DC voltage source.

two magnetic poles, the rotor is many magnetic poles, lined up with the teeth on the rotor. The rotor's teeth are spaced so that only one set of teeth remains in perfect alignment with the stator poles at any one time. (See Figure 8-18.)

By taking the number of times the power must be applied to the stator poles to move one tooth through 360 degrees of mechanical rotation, you can compute the step angle. For example, if the stator needs 200 pulses of power to move one tooth 360 degrees, then divide 360 by 200 to get 1.8 degrees of motion per step. This is the step angle of the motor. The motor will move 1.8 degrees per pulse of stator power. Steppers are available in 90, 45, 15, 7.5, 1.8, and .9 degrees of step angle. The resolution of the motor is a measure of how fine the steps are divided. Resolution is determined by dividing 360 degrees by the step angle. The 1.8-degree step angle motor takes 200 steps to move around a full revolution, so the resolution is 200. Step angle and resolution are inversely proportional to each other.

FIGURE 8-17 Stepper motor and associated circuitry.

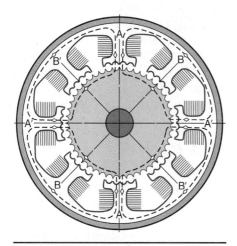

FIGURE 8-18 Stepper motor construction shows teeth on rotor and the stator.

The permanent magnet rotor was used for the example on step motion control. These permanent magnet rotors are used with the four pole pieces to provide either 90 or 45 degrees of step angle. This allows the motors to turn at higher speed, but with less resolution. The poles can be physically larger because there are only four of them; thus the stator windings can carry higher current. Higher current capability means that more torque is available.

Servo Motors

Synchronous servo motors typically use permanent magnet rotors that are allowed to follow the stator field more closely than would an induction-type squirrel cage rotor. These motors can then be used for more precise speed and motion control. To monitor speed control and rotational position, a monitoring system is used. Typically an encoder is used to feed position information back to the servo controller. Asynchronous servo motors do use a squirrel cage–type rotor and rely on induction to create current in the rotor to develop magnetic poles. Again, a feedback system is needed to monitor the speed, direction, and position of the rotor.

The speed of a servo motor is controlled through an electronic drive that produces a Pulse Width Modulation (PWM) output that supplies the stator field. (See Chapter 5 for PWM information.) The servo controller must set speed and rotational direction, as well as total number of revolutions, to direct the motor to an exact position at the proper speed and then hold in place or continue to a next position. To track the motion either an encoder or a resolver can be used to provide feedback as to the actual number of rotations that have occurred, compared to the drive command. Servos drive a component in a smoother and usually in more precise increments than a comparable stepper motor.

Variable Reluctance Rotor

The variable reluctance rotor has the same number of teeth as there are stator poles. The rotor bars (or teeth) are made of high-permeability iron. The iron will be magnetized and align with the stator field in small steps. When the rotor is in operation, the rotor retains the magnetism for a short time and will continue to follow the stepped stator field. When the motor is deenergized, the variable reluctance rotor loses its magnetism and does not lock into place, as does the permanent magnet rotor.

HYBRID MOTORS

The hybrid motor uses some of the characteristics of the variable reluctance rotor and some of the permanent magnet rotor characteristics. The stator is wound with two sets of windings on each pole. This method is called the *bifilar method*. The windings are smaller gauge and have a center tap point brought out for connection. Figure 8-19 shows how a bifilar winding is controlled.

The advantage of this motor is that it provides a very small step angle or high resolution with fewer stator windings. Also, because of

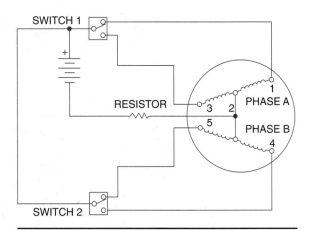

FIGURE 8-19 Bifilar winding on a stepper motor.

the hybrid rotor, it has both strong operating and holding torque for stopped motion. Acceleration and deceleration torques are good as they allow this motor to handle heavier loads at higher stepping speeds but still maintain accuracy and precision motion control. Most stepping motors are used for precision and incremental motion control. Usually the signals are provided by microprocessor circuitry and fed to a controller board that processes the signal and applies power to the motor.

TORQUE MOTORS

Torque motors are special motors that will deliver maximum torque even if stalled in a locked rotor position. This is useful in applications such as spooling machines or tape drives. They will provide constant tension even if stalled for long periods.

The torque motor can be used where there is no rotation required, but only tension applied (similar to a spring). They also can be used to open or close valves that require only a few turns of the motor shaft to be held in position. Torque motors can be used for normal or very slow speed operations. AC torque motors are normally a polyphase or permanent split capacitor (PSC) motor design.

SUMMARY

This chapter discussed special motors that are not as frequently used. It is necessary to be familiar with these different types of motors even though you may not work on them regularly.

Wound rotor motors were first presented as an older method of controlling speed and starting characteristics of three-phase motors. Repulsion motors were used to obtain higher starting torque with lower inrush currents than other motors provided. Several styles of repulsion motors were explained so that you will recognize this type of motor and be able to diagnose some of the problems that occur.

Small, synchronous motor styles frequently are used in time-synchronized equipment. A basic understanding of how they work will help you analyze problems with the system. Stepper motors are used frequently in motion control systems to move devices a specific distance. Their incremented steps allow precise control and positioning in both forward and reverse directions.

QUESTIONS

1. How is the rotor of a wound rotor motor different from a squirrel cage rotor?
2. What is connected to M_1, M_2, and M_3 leads on a wound rotor motor?
3. Where does the NEC® address the protection of secondary conductors on wound rotor motors?
4. Give a brief description of a repulsion motor principle.
5. What is meant by *hard neutral* in a repulsion motor?
6. How is a repulsion motor reversed?
7. Explain how to compute the step angle of a stepper motor.
8. What is an application for a torque motor?

CHAPTER 9

POWER DISTRIBUTION AND MONITORING SYSTEMS

OBJECTIVES

After completing this chapter and the chapter questions, you should be able to

- Understand the basics of power transmission
- Explain electrical transformer fundamentals
- Identify types of transformers
- Perform transformer calculations
- Connect and test transformers in different three-phase patterns
- Size overcurrent protection for transformers
- Perform voltage drop calculations on feeder conductors
- Understand the principles used in demand monitoring

KEY TERMS IN THIS CHAPTER

Ampere Turns: The number of turns in a coiled conductor multiplied by the current through the conductor. This is an indication of the magnetic strength of the coil.

Copper Loss: The watts lost in heat due to the resistance of the coiled conductor and the current flow, calculated by $I^2 \times R$.

Eddy Current: Currents that are produced in the iron of a magnetic circuit. The currents flow within the iron as it is cut by a magnetic field.

Faraday's Law: An electrical law that states that the amount of voltage produced is dependent on the rate that a conductor cuts through a magnetic field. The greater the number of flux lines cut in a period of time, the greater the induced voltage.

Hysteresis Loss: The watt loss in the form of heat that occurs in the iron of a magnetic circuit as the magnetic fields are constantly reversed.

Open Delta: One method of supplying three-phase power to a system using only two single-phase transformers.

Percent Impedance: A percentage calculated for a transformer that indicates the percent of the rated primary voltage that is needed to supply full current to a shorted secondary winding. It is an indicator of the amount of the transformer's internal coil impedance.

Primary Winding: The coiled conductor in the transformer that receives energy from the source.

Secondary Winding: The coiled conductor in the transformer that delivers power to the load.

Transformation Ratio: The calculated ratio of primary to secondary line voltages in three various transformer configurations.

Wild Leg: One of the terms given to the conductor with the higher voltage to ground found on a three-phase, four-wire, delta-connected system. It is the B-phase conductor.

INTRODUCTION

Use of electricity and power-generating facilities has grown through the years. Generating plants have been placed in locations that have the proper natural resources. In some cases, power is generated at hydroelectric facilities and the electrical energy is transported hundreds of miles to consumers. In other instances, generating plants have been located close to coal supplies although intended consumers live many states away. Nuclear plants usually are placed away from large centers of population, but the output energy is used for the large population centers. The United States is tied together through power lines in a national electrical grid. Regional areas of the country are designated to help utility companies on the power grid buy and sell power from each other. This helps them take advantage of another utility's excess generating capacity, or supply power for another region's peak needs.

Most of this power is transported by AC transmission lines and some by DC transmission lines. To make the transmission efficient, and to keep electrical losses to a minimum, voltage on AC lines is raised to very high levels. Many cross-country lines are 345,000 V (345 KV) or higher between conducting phases. The higher the voltage, the lower the current for the same amount of electrical load.

To make the system work, transformers are used to increase or decrease the voltages to meet the customers' needs. Other methods also are used to help control, monitor, and correct power systems.

TRANSFORMERS

Single-phase transformers are in use constantly and serve as the basis for understanding three-phase transformers. Transformers are designed to provide specific functions. Transformers used to increase (step up) or decrease (step down) the voltage without electrical connection between the two voltages are called *isolation transformers*. Autotransformers are used to change voltage levels but provide a common electrical connection from the higher voltage to the lower voltage. Instrument transformers are specifically designed to give accurate transformation of electrical quantities for use with measuring instruments.

To understand how a transformer works, you must understand the principles of magnetic induction. The principles you have used for motors will also be used for transformer action. The same three essential elements for electrical generation or magnetic induction are employed: (1) a magnetic field, (2) a conductor, and (3) motion. The *primary* of the transformer is the coil of wire, or winding, that is connected to a source of current flow. (See Figure 9-1.) The *secondary* of the transformer is the load side of the transformer. It is the secondary coil of wire or winding that will supply voltage to an electrical load. All transformers work off the same principle. Magnetic lines of force created by current flow in the primary winding will expand as sine wave current increases. These magnetic lines of force also will induce voltage in the secondary winding.

Figure 9-2 shows a cross section of a primary and secondary conductor. As current flow increases in the primary circuit, following the increasing value of the sine wave, the

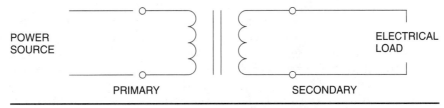

FIGURE 9-1 Primary and secondary windings shown separated in a schematic.

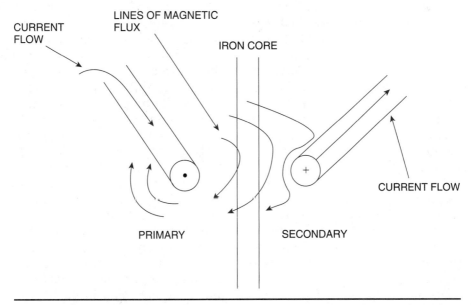

FIGURE 9-2 Expanding magnetic flux from the primary induces voltage into the secondary.

magnetic field around the primary conductor also expands. Use the left-hand rule for a conductor to determine the direction of flux. Grasp the conductor with your left hand, thumb in the direction of electron flow, fingers in the direction of magnetic flux (north to south). As the flux expands through the highly permeable iron core, the flux lines will pass through or cut the secondary conductor. As the flux cuts through the secondary conductor, a voltage is induced into the conductor. If the secondary circuit is complete, a resultant current will flow in the direction indicated. Use the left-hand rule for the secondary conductor. Grasp the conductor with the left hand, fingers in the direction of flux (CCW for this example) and the thumb in the direction of resultant current flow in the secondary winding. The amount of secondary voltage depends on the number of conductors that are cut by the primary flux.

All the factors needed to produce voltage are now present. The magnetic field is provided by the primary current flow. The conductor is provided by the secondary coiled conductors and the motion is provided by the expanding and contracting magnetic field.

As the magnetic field expands, it cuts the secondary conductor as shown in Figure 9-2. The primary current wave form will increase and then begin to decrease from its peak value. As the primary current falls, the primary flux lines begin to contract. This collapsing magnetic field cuts through the secondary conductors from the opposite direction and the induced voltage causes the secondary current to flow in the opposite direction. This will continue in the same direction until the primary flux has expanded out in the opposite direction from the original, due to the opposite peak of the primary current. The AC primary current has produced an AC secondary voltage and resultant current.

Transformers are AC devices. The movement of the flux is due to the AC waveform on the primary. Transformers do not transform DC because once the magnetic field is first expanded, there is no more movement.

VOLTAGE TRANSFORMATION

The basic equation for voltage relationships is provided by the transformer theory: Volts per turn of the **primary winding** are equal to the volts per turn of the **secondary winding**. This equation is based on the assumption that there is 100 percent flux linkage between the primary and secondary. If transformers were 100 percent efficient, there would be no magnetic field lost and "volts per turn in" would precisely equal "volts per turn out." See Figure 9-3. Transformers are not 100 percent efficient in the transformation of power from one voltage level to another. They are between 90 percent to 99 percent efficient when operating at the full-load rating of the transformer. For most calculations, the transformer is assumed to be 100 percent efficient. If transformers are assumed to be 100 percent efficient, the amount of input power (measured in (volt × amps)) will equal the output power measured in volts × amps (VA). Now, transformer formulas can be established. If you count the number of turns of wire in the primary and multiply by the amperage of the supply, that product will equal the product of the secondary turns of wire × the output current. The formulas are presented in Figure 9-3.

The values of turns, voltages, and currents are usually presented as ratios. For instance, the first ratio compares N_p or primary turns to secondary turns N_s. It also compares E_p or primary voltage to E_s or secondary voltage. These two ratios are directly proportional. If the primary has twice as many turns, it also will have twice as much voltage. Notice that the current ratios are inverted. This means that the volts × the amps input must equal the volts × the amps output. A transformer does not create or consume energy (if 100 percent efficient). If the primary voltage is twice as much as the

VOLTS PER TURN IN = VOLTS PER TURN OUT

INPUT $\dfrac{\text{VOLTS}}{\text{TURNS}} = \dfrac{\text{VOLTS}}{\text{TURNS}}$ OUTPUT

INPUT VOLTS × AMPS = VOLTS × AMPS OUTPUT FORMULAS ASSUME 100% EFFICIENCY

$$\dfrac{E_p}{E_s} = \dfrac{N_p}{N_s} = \dfrac{I_s}{I_p}$$

E_p = PRIMARY VOLTAGE
E_s = SECONDARY VOLTAGE
N_p = PRIMARY TURNS
N_s = SECONDARY TURNS
I_p = PRIMARY CURRENT
I_s = SECONDARY CURRENT

FIGURE 9-3 *Transformer formulae assuming 100 percent efficiency.*

secondary voltage, the primary current will be one-half of the secondary current. When using ratios, remember to cross multiply to find the missing value. For example, if the primary voltage is 230 V and the secondary voltage is 115 V and the secondary load draws 10 amps at full load, what is the primary current at full load?

$$\dfrac{E_p}{E_s} = \dfrac{I_s}{I_p}$$

To solve: $E_p \times I_p = E_s \times I_s$
$230 \times I_p = 115 \times 10$
$I_p = 1150/230$
$I_p = 5$ amps

Transformers are generally used to transfer power from one voltage level to another. They do not always have to change voltage levels. If the level is kept the same (in other words, the transformer has the same number of turns on the primary and secondary), it has a 1:1 ratio. These transformers provide electrical isolation. Because there is no electrical connection between primary and secondary, the two systems (primary and secondary) are isolated from each other. Isolation may be used to reduce primary line spikes as these are not transferred completely through the magnetic core. Isolation may be used to remove references to ground or grounded conductors.

If the transformer is used to increase the voltage level in the secondary compared to the primary, it is considered *a step-up transformer*. If the secondary voltage is lower than the primary voltage, it is *a step-down transformer*. Step up or step down is always in reference to the voltage change. The current capability is the inverse of the voltage change. Voltage changes are also expressed as ratios. A transformer that is used to step down the voltage when the primary voltage is 460 V and the secondary voltage is 115 V is said to be a 4:1 transformer.

TRANSFORMER DESIGN

Transformers are designed to transfer energy from one AC system to another AC system by electromagnetic means, usually at different voltage levels. Although this transfer of energy is the purpose of all transformers, they are designed for different specific purposes.

FIGURE 9-4 Power transformer 500 KVA or larger. *(Courtesy of Cooper Power Systems)*

FIGURE 9-5 Cutaway view of a distribution transformer.

There are three main classes of transformers.

1. Power transformer over 500,000 VA (500 KVA) used to step up or step down large amounts of power at generating stations or substations. (See Figure 9-4.)
2. Distribution transformers range from 1.5 KVA to 500 KVA. These are used to deliver power to the customer at desired levels. (See Figure 9-5.)
3. Instrument and control transformers range in size from VA to several KVA. They are used to operate control circuits, or provide voltage or current to instruments. (See Figure 9-6.)

There are also several core types of transformers, which are usually selected to provide the proper level of transfer with the least amount of loss. The transformer core is usually shown schematically as in Figure 9-7, which does not indicate the actual core type.

One style of core design is the *core-type* transformer shown in Figure 9-8. The core is made of laminated iron pieces. The iron usually is a high-silicon steel to reduce hysteresis losses. It is cut into thin wafers called *laminations*. The laminations actually become electrically insulated from each other because a thin layer of iron oxide, an electrical insulator, is formed on the pieces before assembly. Laminations are used to keep **eddy currents** induced into the iron to a minimum. In the core-type transformer the coils are wound and then the

FIGURE 9-6 A current transformer (CT) type of instrument transformer.

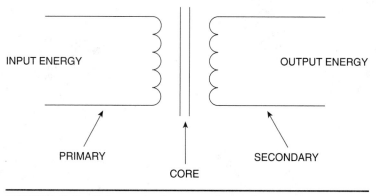

FIGURE 9-7 Schematic diagram of transformer.

iron E and I (or C and I) pieces are assembled. The coils are usually wound so that the lower voltage winding is placed next to the iron core and the higher voltage winding is wound on top. This method keeps flux leakage to a minimum. Also, this method helps prevent breakdown of electrical insulation of the lower voltage winding, which could short out to the core.

To improve efficiency the *shell type* of transformer was developed. Figure 9-9 shows a shell type. This transformer has both coils wound and then E and I pieces are assembled. Again, the lower voltage winding is wound closer to the iron core. There is very little flux leakage in this style. Of course, if the transformer is more efficient in operation, it costs more to build.

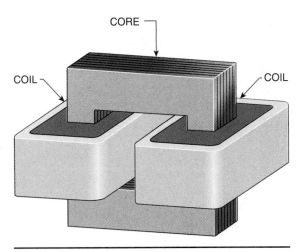

FIGURE 9-8 Core-type transformer.

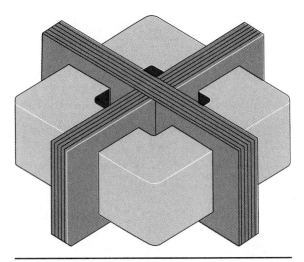

FIGURE 9-10 An H-core-type transformer.

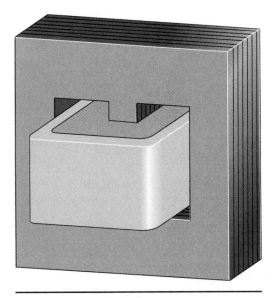

FIGURE 9-9 A shell-type transformer.

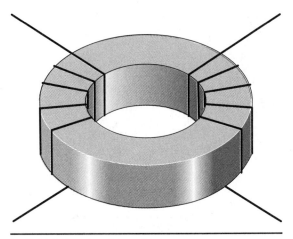

FIGURE 9-11 A toroidal-core-type transformer.

Another style of transformer design is called the *H-core-type*. (See Figure 9-10.) This core further encloses the central primary and secondary coils to provide even better flux linkage. There are four paths around the periphery of the coil rather than the two paths found in the shell type. Better efficiency means higher initial cost.

One of the most efficient core designs is the *toroid core*. Figure 9-11 illustrates a *toroidal-core-type*. This core is different than the others in that it is not laminated pieces of iron stuck together. Instead, it is one long piece of silicon-steel ribbon wound into a circular core. The idea is to provide even better flux paths to reduce flux leakage to the outside air. Flux leakage is flux that is not used in linking the primary to secondary coils.

TRANSFORMER OPERATION

To understand the relationship between voltage, current, and polarity on transformers and the efficiency of a particular transformer, you should understand the waveforms represented in Figure 9-12. As voltage is applied to the primary winding, a primary current will flow. This current lags the primary voltage by 90 electrical degrees because of the inductance of the coil of wire. The primary current with no load connected to the secondary is called the *exciting* or *magnetizing current*. This is the current that creates the magnetic flux around the primary conductors. Primary flux expands and contracts in phase with the primary current. Remember, as discussed earlier in this chapter, primary flux cuts through the secondary conductors. If you notice the primary flux at the 180-degree line, it has stopped expanding and there is no cutting action taking place. At this moment, the secondary voltage is at zero potential. As the primary magnetic flux collapses, it produces a secondary voltage opposite to the primary voltage. Note also that the maximum secondary EMF is induced as the primary flux moves through the zero point at the 270-degree line. This is due to the fact that this is the greatest rate of change in the flux. **Faraday's law** is intact. It states that the higher the rate of cutting lines of flux, the higher the voltage induced.

Now, if you connect an electrical load to the secondary circuit, you will provide a current path for secondary current. Because the secondary winding is also a coil and has inductance, the secondary current will lag the secondary voltage by 90 electrical degrees. When current flows in the secondary circuit, it also produces a magnetic flux. The secondary

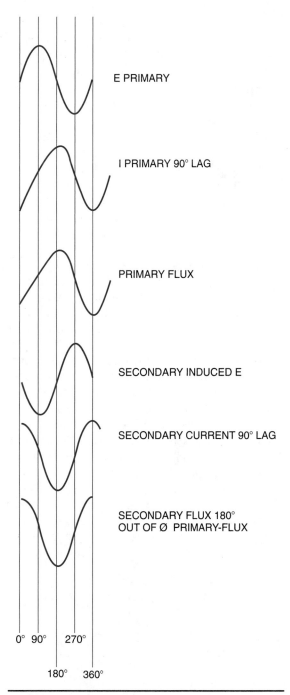

FIGURE 9-12 Voltage and current waveforms in transformer primary and secondary windings.

flux is 180 degrees out of phase with the primary flux just as the secondary voltage is 180 degrees out of phase with the primary voltage. However, the higher the current drawn by the secondary circuit, the more secondary flux is produced. The secondary flux cancels some of the effect of the primary flux and the impedance of the primary decreases. In effect, the flux makes the core less permeable and the primary impedance decreases. This causes primary current to increase due to less primary impedance. The result is that the more current drawn from the secondary, the more current enters the primary.

TRANSFORMER POLARITY

The description on transformer operations mentions that there is a 180-degree phase shift from primary to secondary. If transformers are to be connected in series or parallel, the instantaneous values of voltage must be known to prevent one transformer lead that is reaching peak positive from being connected to a transformer lead reaching peak negative.

As transformer coils are wound, the leads are brought out through the case. H numbered leads are assigned to the higher voltage windings with H_1 always the top-left terminal on the transformer diagrams. (See Figure 9-13.) The lower voltage windings are marked with an X and an associated number. Transformers are marked according to American National Standards Institute (ANSI) standards that determine the instantaneous polarity of the secondary lead with relation to the primary lead. If a transformer is subtractive polarity, it means that the secondary leads are brought out and that H_1 has current entering the transformer as in Figure 9-14. This is a load to the power source and creates a voltage drop as represented by the battery symbols in Figure 9-15. At that exact instant in time, the induced voltage creates a source of voltage for the transformer load with polarities as indicated by the battery symbols in Figure 9-15. This source of voltage will cause current to flow away from X_1 to the load and back to X_2. The term *subtractive polarity* originates from the way that you can identify these leads using a voltmeter.

In Figure 9-15 an AC voltmeter is connected from H_2 to one of the secondary leads and a temporary jumper wire is connected from H_1 to the other secondary lead. Apply AC power to the primary circuit. The instantaneous voltages are represented by the battery symbols and polarity marked in Figure 9-15. The voltmeter will display the difference of these two opposing voltages or the subtractive value of the voltages. In this example, the 120-V secondary opposes or subtracts from the primary 240 V and the voltmeter reads 120 V. Be careful when checking polarities. The meter must be able to read the additive value of the voltages as well. If the meter reads subtractive quantities, then the lead jumpered to H_1 is

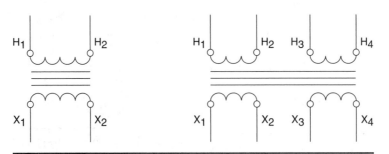

FIGURE 9-13 H leads are higher voltage leads; X leads are lower voltage leads.

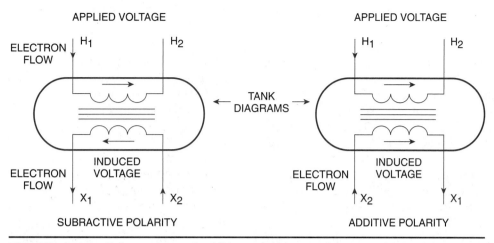

FIGURE 9-14 Subtractive and additive polarity transformers with lead markings and electron flow.

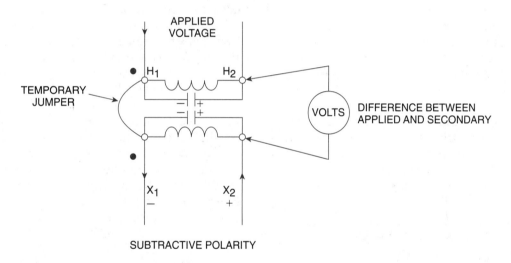

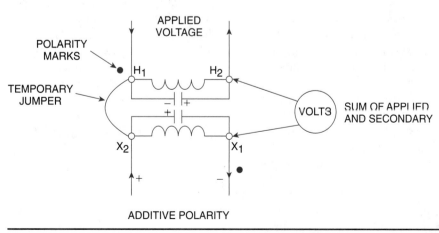

FIGURE 9-15 Polarity tests for transformer to determine lead markings.

marked X_1 and the other secondary lead is marked X_2.

Additive polarity results when the same test is performed, but the two voltages add together. This means that the leads are brought out in such a manner that the two voltage waveforms are in phase with each other and the instantaneous voltages add. H_1 still has current entering and X_1 still has current leaving the transformer, so the lead markings are consistent. However, note that the X_1 and X_2 leads are reversed in their physical location from the subtractive polarity transformer. Refer back to Figure 9-14.

The ANSI standards are that additive polarity is standard for all single-phase transformers up to and including 200 KVA and with any voltage rating 9000 V or less. Subtractive polarity is standard for all single-phase transformers over 200-KVA rating regardless of voltage. Subtractive polarity is used regardless of KVA if any voltage rating exceeds 9000 V. Subtractive polarity is also standard for instrument transformers such as *potential transformers* (PTs) or *current transformers* (CTs). Obviously, additive polarity fits most conditions.

TRANSFORMER LOSSES AND REGULATION

Until now the assumption has been that the transformer is 100 percent efficient. This is almost true as the losses are usually negligible compared to the power delivered. For most of the formulas, 100 percent efficiency is still used. There actually are losses in the transformer. These losses are watt losses that occur in the transfer of electrical energy from one system to another. If careful measurements are made, you will notice that the volt amps input does not equal the volt amps output. The difference between the two is the energy lost inside the transformer. These losses are more apparent at light loads and are less obvious at the transformer's full-rated load.

Losses in the transformer are made up of two parts. First, there is **copper loss.** This is the watt loss due to the resistance of the wire. It can be calculated by the formula:

$(Ip)^2 \times Rp$ = primary copper loss in watts

with the primary current and the primary resistance. Added to this loss is the secondary copper loss found by using $(Is)^2 \times Rs$. The higher the primary and secondary current, the higher the copper losses.

The second loss consists of *core losses*. Part of the core losses are caused by eddy current losses in the iron. As already discussed, the eddy currents are set up in the iron core and circulate within the iron. Lamination of the iron helps to reduce eddy currents. The rest of the core loss is due to hysteresis of the iron.

Hysteresis is the opposition to changing the magnetic field. If iron has high hysteresis, the magnetic domains within the iron do not move easily. The domains change with each one-half cycle of the applied flux and the friction of moving causes heat. The use of high-silicon or low-hysteresis steel helps prevent the friction of the moving domains and therefore reduces the losses in the transformer. The higher silicon content of the steel also increases resistance to further reduce the loss due to eddy currents in the steel.

Transformer regulation refers to the ability of a transformer to maintain a constant output voltage from no load to full load. No load and full load refer to the current load on the secondary. Transformer regulation is expressed as a percent. Use the same formula as that for other power supplies.

$$\text{Percent regulation} = \frac{Es_{NL} - Es_{FL}}{Es_{FL}} \times 100$$

Es_{NL} = E secondary at no load
Es_{FL} = E secondary at full load

Most transformers maintain voltage levels very well from no load to full load, so the percent regulation is small.

MULTIVOLTAGE TRANSFORMERS

Many transformers come with multiple voltage connections for the primary, secondary, or both. Figure 9-16A shows a typical application of a dual-voltage primary and a dual-voltage secondary. The leads often are brought out to terminal strips and are crossed to make connection easy. This transformer is a 2:1 transformer ratio. By connecting the primary coils in series with each other, 460 V can be applied to the primary. If you desire the secondary voltage to be 115 V, connect even-numbered secondary leads and odd-numbered secondary leads. The rating of the transformer is still based on the output line current × the output line voltage. Note that by connecting coils in parallel the current supply is double that of one coil.

TRANSFORMER CONNECTIONS FOR VOLTAGE

If one transformer is not large enough to carry the load, a second transformer may be added in parallel to the first. If the transformers are to be paralleled, parameters must match to get the transformers to share the appropriate amount of KVA load. The electrical characteristics that must match are as follows.

1. Same polarity or connecting the same marked leads of the primary and connecting the same marked leads of the secondary.
2. Voltage ratios must match.
3. The percent impedance of the transformer must match with the same components of resistance and reactance. (See section on impedance in this chapter.)

(See Figure 9-16B for sample diagram of two transformers connected in parallel.) Both transformers are 10 KVA and 2400//120/240 voltage rating. The secondary windings are connected in series to produce 240 volts at the output. The KVA ratings of each transformer add together to yield 20-KVA capacity to the secondary load.

Transformers used in various locations for returning voltage to standard values have many primary taps. An example of a transformer used to bring a distribution voltage back to usable levels is shown in Figure 9-17. In this example the primary windings are tapped at different points. This example will correct the number of turns per volt so that it will give the proper secondary voltage to the customer. This transformer can

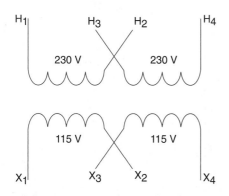

460-V PRIMARY CONNECT H_2 TO H_3

230-V PRIMARY CONNECT H_1 TO H_3, H_2 TO H_4

230-V SECONDARY CONNECT X_2 TO X_3

115-V SECONDARY CONNECT X_1 TO X_3, X_2 TO X_4

FIGURE 9-16A Dual-voltage primary and secondary for single-phase transformers.

ELECTRIC MOTORS AND MOTOR CONTROLS

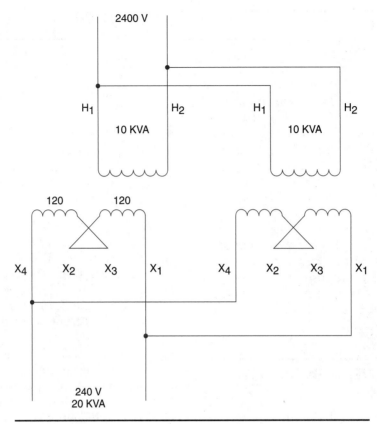

FIGURE 9-16B Two single-phase transformers connected in parallel to a single load.

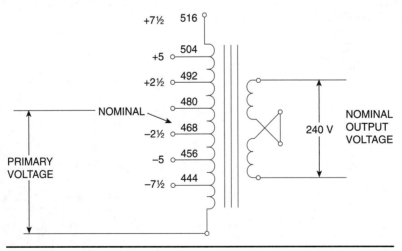

FIGURE 9-17 Distribution transformer with multiple taps to adjust for voltage.

accommodate a voltage from 7.5 percent above normal 480 V to 7.5 percent below normal delivery voltage. More turns are added at the 7.5 percent above tap and turns are taken out of the primary at the 7.5 percent below tap.

Power companies can use automatic tap changing transformers to compensate for heavy line drop on long runs. If there is heavy current use on the distribution network, the system voltage tends to drop. With voltage sensors, the transformer can automatically move to a new tap and try to hold customer secondary voltage constant.

AUTOTRANSFORMERS

An autotransformer diagram is shown in Figure 9-18. This transformer is different from other transformers described in that the autotransformer has only one winding. Voltage can be tapped off the secondary and used as a step-down transformer, or the lower voltage side may be used as a primary and used as a step-up transformer. The same transformer formulas that apply to isolation-type transformers also apply to the autotransformer.

The advantage of the autotransformer is that only one winding is used. If the voltage levels are not too far apart, an autotransformer may be a logical choice because it is cheaper than a full two-winding transformer.

The disadvantage of the autotransformer is that it does not provide electrical isolation between primary and secondary. If the lower voltage secondary winding were to open under load, the full primary voltage would appear across the secondary leads. Autotransformers are limited to certain situations by the NEC® as listed in Article 210.9. The overcurrent protection requirements for autotransformers are listed in Article 450.4.

Article 450.4B refers to transformers that are field connected to create an autotransformer. For example, there may be the case where a 208-V system is used to supply power to the building. A new piece of equipment that you install will only work satisfactorily on 240 V. The supply voltage must be raised 32 V. By using a 208- to 32-V isolation transformer, the primary and secondary can be tied together as in the additive transformer test circuit. This connection shown in Figure 9-19 is

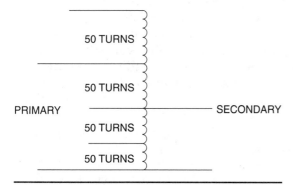

FIGURE 9-18 Autotransformer with multiple taps.

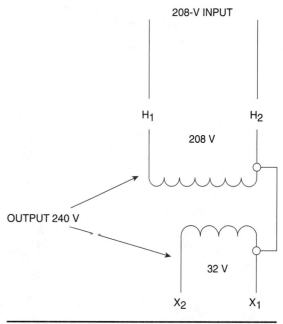

FIGURE 9-19 Boost transformer uses both windings of a transformer to create an autotransformer.

called a *boost* transformer connection. In effect, the isolation transformer windings are tied together to create an autotransformer winding. If the voltage needs to be lowered, the secondary winding may be connected to buck or oppose the primary voltage and therefore lower the voltage of the system. Grounding of autotransformers is covered in Article 450.5.

CONTROL TRANSFORMERS

Control transformers are designed to provide step-down voltage for use in motor control functions. They are able to maintain a steady secondary voltage under varying load conditions. In other words, they have good regulation. The distinguishing feature of control transformers is the ability to deliver inrush current to coils associated with motor control systems. NEC® Article 430.72(C) deals with transformers used in motor control circuits. The conditions of Article 430.72(C) are important. The exceptions and conditions of Article 430.72(C) determine how to protect the control circuit transformer. See Figure 9-20 for an example of a control transformer with mounted fuse assembly for protection.

INSTRUMENT TRANSFORMERS

When measuring large voltages or currents on a feeder or distribution system, it is not practical to have meters directly connected to the large power sources. The meters would have to have wire large enough to carry high currents or windings insulated with high-voltage insulation. Also, if meters were mounted in panels at monitoring stations, the large voltages or currents would be a danger to personnel.

To reduce these risks and to use smaller, less-expensive metering and switching controls, the *instrument transformer* was developed. Instrument transformers are specifically designed to provide accurate transformation of specific quantities. They are not designed to transform power and are rated only in the VA range rather than KVA. There are two types of instrument transformers. One is specifically designed to step-down high voltages to a standard secondary voltage. As mentioned earlier, this is the potential transformer (PT). The other transformer used to transform higher currents to a standard secondary current is a current transformer (CT).

Potential Transformers

Potential transformers look like small power transformers. They are rated from 100 VA to 500 VA and the low-voltage side is typically 115 V AC. The primary winding is connected to the power line side and is rated for the line voltage that is available. Other PTs used at voltages below 600 V on the primary have standard terminals. If a PT is to be used for measuring the voltage of a high-voltage line, the primary to secondary turns ratio is the same as the voltage ratio. For example, if the primary line is 4600 V (as shown in Figure 9-21), the secondary voltage is 115 V and the transformer is 4600/115 or 40:1 transformer.

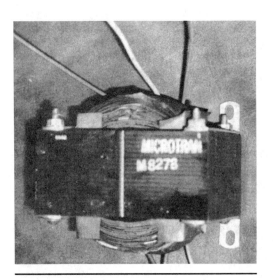

FIGURE 9-20 Motor control circuit transformer.

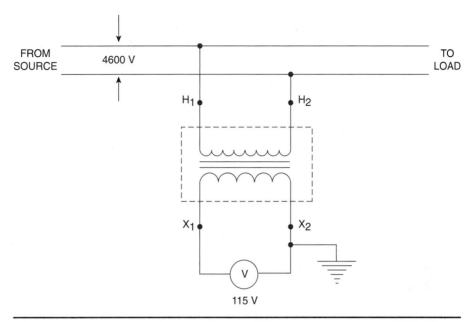

FIGURE 9-21 A potential transformer (PT) used to measure 4600-V line with a 115-V secondary.

The meter is a standard panel meter with a 115- or 120-voltmeter movement. The scale on the meter is calibrated in the primary line voltage increments. Considering the meter in Figure 9-21, if the primary line voltage dipped to 4510 V, the actual secondary voltage is 112.75 V. This lower secondary voltage is read on the meter face as 4510 V.

The PT has subtractive polarity. This is important if you wish to monitor three-phase lines. Three meters and two PTs can be connected as in Figure 9-22. Be sure to observe the lead markings. In some cases, the H_1 and X_1 leads are merely identified by a polarity mark. The polarity mark is usually a dot (.) or a plus/minus sign (±). These marks are placed by the H_1 and X_1 leads.

PTs are required to be grounded according to Part L of NEC® Article 250. Secondary circuits of CT and PT shall be grounded where there is more than 300 V on the primary, or if the secondary circuits are connected to a switchboard. Grounding usually is accomplished on the X_2 lead. (See exception to Article 250.121.) Article 250.122 and the exception refers to the cases of instrument transformers. Potential transformers need to be fused according to NEC® Article 450. (See "NEC® Requirements" section later in this chapter.)

Current Transformers

Current transformers are used to insulate low-current (5-A) meters and instruments from high-current sources. There are several styles of CTs. *A wound primary type* of CT has separate primary and secondary windings and is assembled in an iron core. The primary has a rating for the primary line current and the secondary full-load rating is always 5 amps AC. Five amps is the secondary rating of all CTs. Another style of CT is the *bar type*. As shown in Figure 9-23, the primary is actually a bus bar mount. The

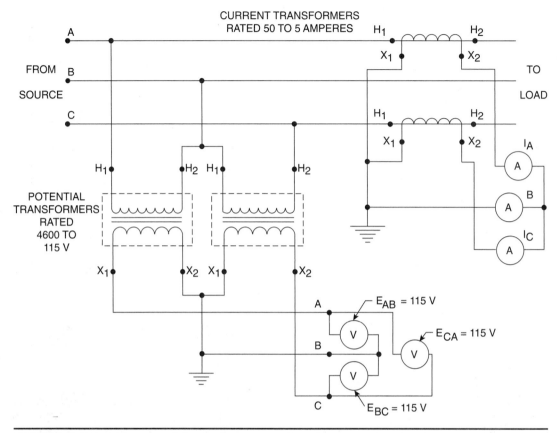

FIGURE 9-22 Two potential transformers and two current transformers can be used to measure three-phase power.

FIGURE 9-23 Bar-type current transformer.

bus bar passes through the core and provides the primary flux. The secondary is wound around the bar.

Window-type CTs are very useful. This CT has only an open window through which the primary conductor will pass. Figure 9-24 shows a window-type CT. The primary magnetic flux is provided by the conductor that passes through the window. Usually the transformer will have ratios printed on the nameplate. For example, if one conductor passes through the window, the current transformer is 800:5 ratio. If you circle the conductor through the window twice, the ratio is 400:5. In other words, if you pass a 400-amp conductor through the window twice, the CT will still deliver 5-amps

Power Distribution and Monitoring Systems

FIGURE 9-24 Window-type current transformer.

shorted, you may remove the burden. CTs need to be grounded according to NEC® Article 250, Part IX. To see how subtractive polarity marks are used to connect CTs, see Figure 9-22. Using Blondell's theorem, as discussed in Chapter 2, two CTs can be used to monitor three phases. Note the polarity marks and the grounding of the CT pattern.

Clamp-On Ammeters

Clamp-on AC ammeters work by the same principle as a split-core CT with a burden always connected. Many clamp-on ammeters are digital readout and can be automatic ranging. Figure 9-25 shows a digital readout on a

output. You can use multiple turns through the window: Divide the highest turns ratio for the window by the number of loops through the window.

Split-core CTs are like window CTs, but the window is hinged so the core can be split and clamped around the conductor primary. The same concept is used for a clamp-on amperage monitoring.

Current transformers are not fused because of the high voltage that is produced if the secondary circuit is opened.

Caution: The secondary of a CT must always have an ammeter connected to it or have the shorting bar closed if there is current in the primary. The CT is designed to transform the primary current down to 5 A from some larger primary current. Remember, if you step down the current, you step up the voltage at the inverse ratio. The secondary winding is made to have a very low ohm load on the winding. If this load (or burden as it is called for instrument transformers) is disconnected or open circuited, the very high voltage available at the CT terminals may be dangerous or destroy the CT.

Caution: Never open the secondary circuit of a CT with a current in the primary.

If you need to disconnect the secondary burden, first close the shorting switch on the CT secondary terminals. After the secondary is

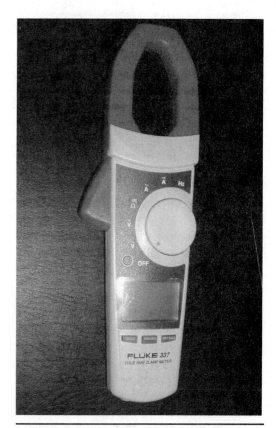

FIGURE 9-25 Clamp-on amperage sensors use current transformer principle.

clamp-on ammeter. These meters are designed to read the AC current in a conductor by using the flux produced by the AC current.

In addition to reading AC current, most clamp-on meters can also read voltage by use of leads; many also read ohms of a circuit. For purposes of measuring the peak current or inrush current to devices, a peak hold function is used. This lets the meter register the peak value and hold it for you to read while the actual line current goes back down.

Some clamp-on meters can measure AC or DC current. The DC current is registered through the use of a Hall effect generator, a device that produces voltage when acted upon by a magnetic field. See Figure 9-26 for an AC/DC clamp-on meter.

In addition to the clamp-on meters used to measure AC or DC current, digital multimeters (DMMs) (see Figure 9-27) can also be used to measure current directly. They can usually measure milliamps by connecting the leads as a series ammeter into a circuit. Many manufacturers also provide current transducers that act as clamp-on devices. These clamp-on sensors are plugged into the meter jacks and convert the amperage to a value that the meter registers as amps.

FIGURE 9-26 AC/DC clamp-on meter.

FIGURE 9-27 Digital meters for current, voltage, and so on.

THREE-PHASE TRANSFORMERS

To this point, only transformation of a single phase of voltage at a time has been presented. Now transformers used to transfer all phases from one three-phase AC system to another three-phase system will be presented. Generally there are two ways to provide three-phase transformation. One method is to use single-phase transformers and connect them into the desired pattern or to use a manufactured three-phase transformer.

Using single-phase transformers gives you many options. The same single-phase transformers can be connected in a variety of patterns and if one of the transformers fails, only one leg of the transformer pattern needs to be changed. Two of the most common patterns for three-phase transformers are the wye and delta patterns. The delta can be a closed delta using three transformers to form a polyphase bank or an open delta or V pattern that uses only two single-phase transformers to provide a reduced three-phase capacity. Patterns can be different from the primary to the secondary. For instance, the primary could be a connected Δ and the secondary a connected wye. Depending on the application, the patterns are selected to give the proper transformation of voltages.

Wye–Wye-Connected Transformers

The wye or star connection is made by connecting the H_1 leads of each of the primaries of the single-phase transformers to the high-voltage lines. (Fuses and other equipment will be presented later in this chapter.) The H_2 leads are connected into a wye point. Figure 9-28 shows a wye primary and secondary (Y–Y).

To understand the transformer voltages, calculate each transformer ratio independently. Each single-phase leg of the wye would have the same transformer ratio. If the primary line voltage is 480 volts line-to-line, the coil voltage applied to each transformer primary is

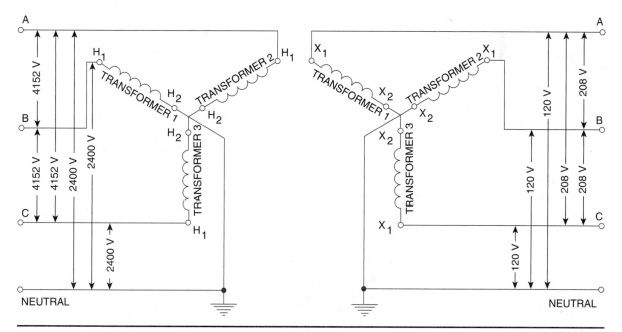

FIGURE 9-28 A wye–wye-connected three-phase transformer bank.

only 277 volts. Remember, in a wye pattern, line E divided by 1.73 equals phase or coil voltage. Now the transformer windings have 277 volts applied to the primary. Assume that each transformer is rated 277// 120 for this example. If 277 volts are applied to the primary, 120 volts will be available at the secondary. By supplying each phase of the secondary with 120 volts, the phase-to-phase or line-to-line voltage will be 208 volts. Again, use the 1.73 factor applied to three-phase wye systems to determine the line-to-line voltages.

Each transformer is capable of delivering the full KVA rating marked on the nameplate. In most instances for Y–Y systems, all KVAs will be the same. This is not essential, so if there is an emergency, another KVA-size transformer can be substituted. It is essential that the voltage ratings be the same. To determine the KVA capacity of the three-phase bank, add the three individual capacities: 10 KVA in this example; the sum is a 30-KVA capacity.

The line currents for primary and secondary can be calculated by using the single-phase values of each transformer. In this example, each transformer is 10 KVA and operated at 277 V on the primary. The primary current is 10,000 VA divided by 277 V or 36.1 A from the line. The secondary current is established by using the same 10,000 VA divided by 120 V or 83.3 A. In a wye connection, the coil or phase current equals the line amps. Therefore, the primary line current is 36.1 A at 480 and the secondary line current is 83.3 A at 208 V. To confirm these figures, use the three-phase capacity formula for three-phase VA. Line E × line I × 1.73 = three-phase VA. Primary KVA = 480 V × 36.1 × 1.73, or approximately 30 KVA. Secondary KVA = 208 V × 83.3 × 1.73, or approximately 30 KVA. These figures do not match exactly because of rounding and using the nonexact 1.73 constant. However, the figures do verify that the three-phase capacity does match the three single-phase capacities and the voltages and currents do equate.

Polarity checks should be made on each transformer individually to verify lead markings. If you connect all the H_1 leads to the line and the H_2 leads to the star point, you should get secondary voltage on each coil.

A double-check of the secondary windings and lead markings should be completed before the leads are tied together or connected to a load.

Caution: Energize primaries to check voltages; then deenergize all windings before making secondary connections. If each secondary winding puts out the rated voltage (120 V for this example), then connect the X_2 leads of two of the secondary windings. The voltage across the open leads should be 208 V. (See Figure 9-29.) If the lead markings are wrong, the voltage will read only 120 V. One of the coils needs to be reversed. If the first two coils are correctly marked, the X_2 leads are tied together. Now connect the third leg of the transformer bank. All three line voltages from each X_1 lead to another X_1 lead should be 208 V or 1.73 times the phase voltage. If it is not balanced, check for reversal of the last coil.

The Y–Y connection is used extensively to bring voltages down to the 120/208 three-phase, four-wire wye connection. The load on the system should be as evenly balanced as possible. Figure 9-30 shows how a three-phase, four-wire wye primary is connected to three single-phase transformers to supply a three-phase, four-wire wye secondary at 120/208 V for lighting and light motor loads.

When calculating the neutral current returning to a wye-connected neutral point, the following formula is useful:

Neutral current = $\sqrt{a^2 + b^2 + c^2 - ab - bc - ac}$.

If all the currents in the three line leads are equal, the neutral current is zero. If two lines are equal but the third line is zero current, as when using two lines to drive a single-phase 208-V load, the neutral current will be the same as the two hot conductors. If all three

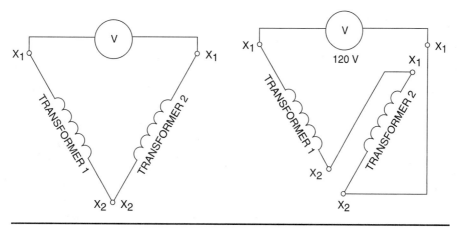

FIGURE 9-29 Testing lead markings on transformer secondary leads.

lines are unequal, using the formula will calculate how much neutral current to expect.

Delta–Delta-Connected Transformer

The closed delta system (Δ) is also used for delivery of power. Three-phase transformers are not often wound Δ–Δ, but three single-phase transformers are often connected in a Δ–Δ bank. The transformer primary winding must be capable of having full primary line voltage applied, not just 58 percent of line voltage as in the Y-connected primary. Therefore, for a given voltage (480 V, in the last example) there are more primary turns in a delta-connected transformer. There are 1.73 times as many windings, but the windings do not carry as much current as the Y-connected windings. Remember that delta coil current is only 58 percent of the line current. This means that for the same KVA, the windings can be 58 percent smaller wire. An advantage to the delta secondary is that secondaries are often center tapped. If the transformer is rated 480//240/120, it means that the secondary has 240 V line-to-line, but 120 V is available at midpoint to line. This pattern is called a three-phase, four-wire delta system. See Figure 9-31 to show how the middle transformer is connected to provide 120/240 V three-wire, single-phase power. (See "High Leg or Wild Leg" section in this chapter.)

A standard Δ–Δ connection without secondary taps is shown in Figure 9-32. Connect the primary H_1 lead of one single-phase transformer to H_2 of the next transformer. Do the same for each transformer but *do not* connect the secondaries yet.

In this example, assume the primary line voltage is 2400 volts and the transformers are 2400//240 or 10:1. Each of the individual transformers should be checked for proper voltage output and correct polarity marking of the leads.

Connect the secondary leads with the primary windings deenergized. Connect X_1 of one coil to X_2 of the next coil. With the primary energized again, the secondary voltage between the open leads should read 240 V. If it reads 415 V, or 1.73 × 240 V, then one of the coils is mismarked and needs to be checked. Deenergize the primary and connect the third coil's X_1 to form a delta pattern, but *do not* connect the last leads yet. Use a voltmeter between X_2 and X_1 at the last connection point. (See Figure 9-33.) The voltage between these points should be 0 volts. If it is not, do not connect the last points to close the delta. The last coil is

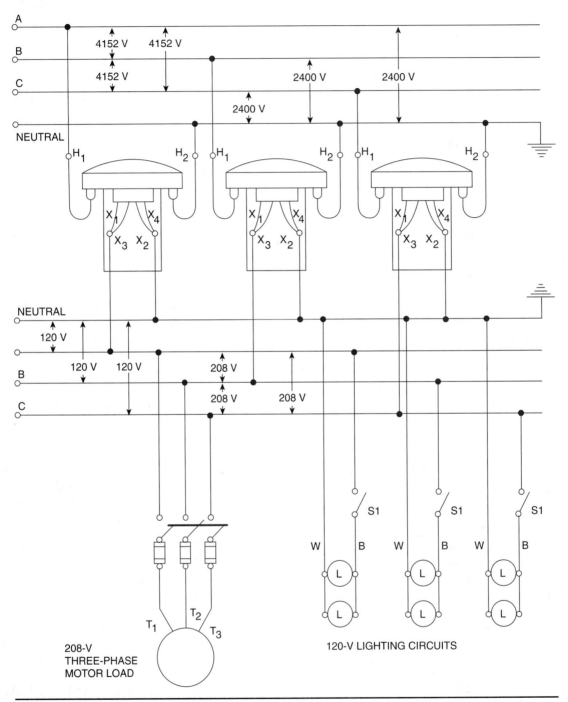

FIGURE 9-30 Y–Y transformer used for three-phase and single-phase secondary loads.

Power Distribution and Monitoring Systems

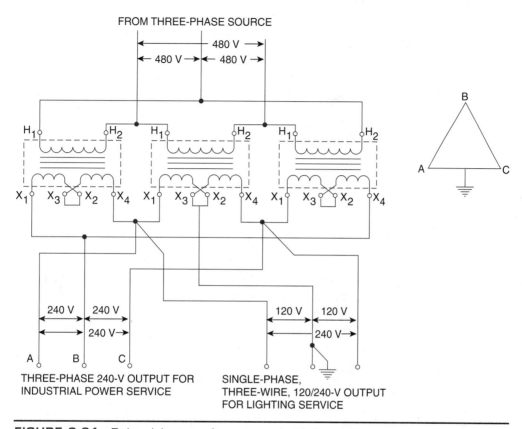

FIGURE 9-31 Delta–delta transformer used with four-wire secondary to supply three-phase and single-phase loads.

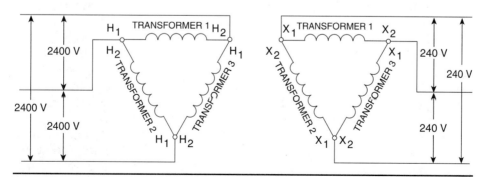

FIGURE 9-32 Elementary diagram of delta–delta transformer connection.

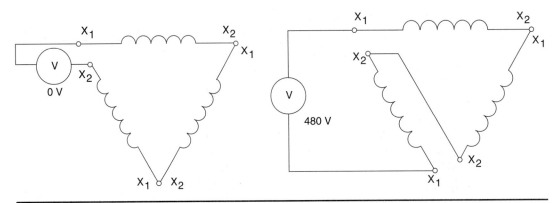

FIGURE 9-33 Use a voltmeter to check delta transformer lead markings on the secondary.

reversed and must be checked before final connections can be made. If all the voltages are correct, you may close the Δ secondary and connect load to the Δ–Δ transformer bank.

Three-Phase Capacity Calculations

As in the wye-connected pattern, the three-phase KVA is three times the single-phase KVA of each transformer, if they are all the same size. For example, consider that each is the same 10-KVA single-phase transformer for this first example. Analyze the capacity, voltages, and currents one phase at a time. If 2400 V are applied to the single-phase 2400//240-V transformer in Figure 9-32, then the phase or individual primary coil current is 10,000 VA divided by 2400 V or 4.17 A. The line current, however, is 1.73 × 4.17 A, or 7.2 A of primary line current. The three-phase capacity is three times 10 KVA or 30 KVA. Three-phase KVA can also be calculated by line E × line I × 1.73 = three-phase VA. In this example, 2400 V × 7.2 A × 1.73 is approximately 30 KVA. The secondary voltage is 240 V line to line. Ten KVA of the single transformer divided by the 240 V equals 41.7 A. The secondary line current in a delta pattern is 1.73 times the coil current so the line current is 72 A. Three-phase capacity is verified by line E × line I × 1.73, or 240 V × 72 A × 1.73 = approximately 30 KVA, which is the three-phase capacity.

Delta–Delta Transformers of Different Size

The Δ–Δ banks are not always the same KVA ratings. If one transformer is larger, it is used to supply a single-phase load as well as three-phase loads. When a three-phase bank also supplies a single-phase, three-wire system (such as in Figure 9-31), then the middle transformer used for three-phase and single-phase loads is often a larger KVA. Many of the single-phase transformers used will have center-tapped 120/240-V secondaries. Connect the two secondary windings in series to create 240-V output. On the larger transformer secondary, bring out the center tap to create a 120/240-V single-phase, three-wire circuit. If you ground the center tap as is standard practice, do not ground the center taps of the other windings, as short circuits will result.

To calculate capacity assume that two transformers are 10 KVA each and the third one is 25 KVA. In this example, you will use 10 KVA of the 25-KVA transformer to supply a three-phase load and the remainder 15 KVA will supply a single-phase load at 120/240 V. If 10 KVA is dedicated to the three-phase load, then the total three-phase capacity will be 30 KVA and the 120/240-V single-phase load will have 15 KVA supplied. The fuses and conductors are based on the individual transformer's single-phase capacity. The 25-KVA transformer use 25,000 VA divided by 2400 V or 10.4 A coil current. The secondary coil is 25,000 VA divided by 240 V or 104 A.

HIGH LEG OR WILD LEG

When connecting the delta pattern with one transformer center-tap-grounded, the phase conductor opposite the ground will have a higher voltage to ground than the other two-phase conductors. This system is referred to as a three-phase, four-wire delta-connected system. Phases A and C are required to be the two-line wires that are 120 volts to ground; B phase will have a higher voltage to ground. The B phase is 1.73 times the center-tap voltage. In this example, the B phase will be 208 volts, or 1.73 × 120 volts.

Article 384.3(e) indicates that the phase that has the higher voltage to ground must be permanently marked orange in color. Article 384.3f states that the **wild leg** or high leg must be the B phase and indicates the arrangement of bus bars. Utilities require the high leg to be marked and may have requirements as to where this phase is to be placed in metering systems.

Be careful! You must exercise caution when dealing with a three-phase 4-W Δ system. Two phases to ground will measure 120 V and the third phase measures 208 V. When connecting loads, be sure you measure the values of the lines you are to connect. If the voltage to ground exceeds 250 volts on any conductor in a metal raceway, the NEC® requires special bonding. In this example, the high leg was only 208 V, but you must exercise caution.

Open Delta- or V-Connected Transformers

If necessary, three-phase power can be transformed using only two transformers. In this situation, two transformers are connected to provide three-phase power using the **open delta** or V pattern.

This method is used frequently if one of the three single-phase transformers fails and an immediate replacement is not available. The defective transformer can be removed and still supply three-phase power to the customer although at a reduced level. Figure 9-34 shows the schematic diagram of an open delta to

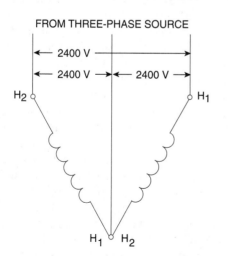
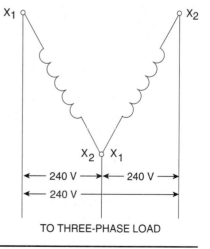

FIGURE 9-34 Open delta pattern for transformers.

open delta transformer pattern. If three 10 KVA transformers were used in a Δ–Δ transformer pattern, the three-phase KVA would be 30 KVA. (See previous explanation.) However, if one of the 10-KVA transformers fails, the other two can continue to supply three-phase power. The KVA rating of the two transformers is only 58 percent of the rating of all three. In this example, the two open delta transformers supply .58 × 30 or approximately 17.3 KVA. This is the result of one transformer (KVA × 1.73). 10 KVA × 1.73 equals 17.3-KVA capacity for the two-transformer bank.

The calculations for an open delta are different because there is no third coil producing power. On the primary side of Figure 9-34, there are 2400 V applied to each primary coil. If each transformer is 10 KVA, then each transformer primary current is 10 KVA ÷ 2400 V = 4.16 A. This coil current is the same as the primary line current as the current does not split between two coils as it does in a full delta. Likewise, the secondary coils produce 10 KVA ÷ 240 V = 41.6 A per coil. This is the current available to the load. By using the consistent three-phase KVA formula, three-phase VA = line E × line I × 1.73, you would calculate the new KVA of the three-phase bank of two transformers. Secondary values are 240 V × 41.6 A × 1.73 = approximately 17.3 KVA. One other method of calculating the three-phase KVA of a two-transformer bank is to add the two KVA ratings together and multiply the sum by 86.5 percent. In this example, two 10-KVA transformers are used. The sum is 20 KVA. Next, multiply by .865, which creates a three-phase capacity of 17.3 KVA. It is important to remember the actual line voltages and currents when determining if an open delta pattern will satisfy the three-phase load requirements.

Three-Phase Transformation Ratios

Transformers can be connected with different patterns on the primary and secondary to provide needed step-up or step-down ratios. A delta–wye pattern is used frequently to step up the voltage from a generating station so that the secondary voltage is increased to provide high-voltage transmission voltages. In Figure 9-35, the 13,800-V generated voltage is connected to the delta-connected primary. If each transformer is 50 KVA, then the primary coil current is 50 KVA divided by 13.8 KV or 3.6 A, and the line current on the primary is 6.27 A. To verify the three-phase KVA of the bank, take 13.8 KVA × 6.27 A × 1.73 = approximately 150 KVA.

The secondary is connected into a wye pattern. Analyze the bank of transformers, one transformer at a time. If one 50-KVA transformer voltage ratio is 13.8 KV:69 KVA step-up, the secondary single-coil voltage would be 69,000 volts. However, this secondary is connected in a wye or star pattern. By doing this, the line voltage output is increased by a factor of 1.73. Secondary line voltage is 1.73 × 69 KV or 119.37 KV. This type of pattern gives a **transformation ratio** higher than the transformer ratio.

The transformer ratio is the primary coil to secondary coil ratio. In this instance, it is 13.8 KV to 69 KV, or a step-up of 1:5. The transformation ratio refers to the effects on the primary and secondary line voltages. If the primary and secondary patterns are different, the transformation ratio and the transformer ratio will be different. In this example, the transformation ratio is 13.8 KV:119.37 KV or 1:8.65 step-up. This Y-pattern secondary is used to boost the secondary coil voltage by a factor of 1.73 just by making the Y connection. The idea is to get the voltage as high as possible to reduce the current in the line wires and still deliver the same VA rating to a load.

If you use a Δ–Y to step up voltage, a Y–Δ transformer pattern will give the greatest transformation ratio to step down the voltage. Referring to Figure 9-36, the schematic layout of a Y–Δ transformer is presented. In this example, assume the primary line voltage is 4160 V. If the primary is connected in a Y pattern, the

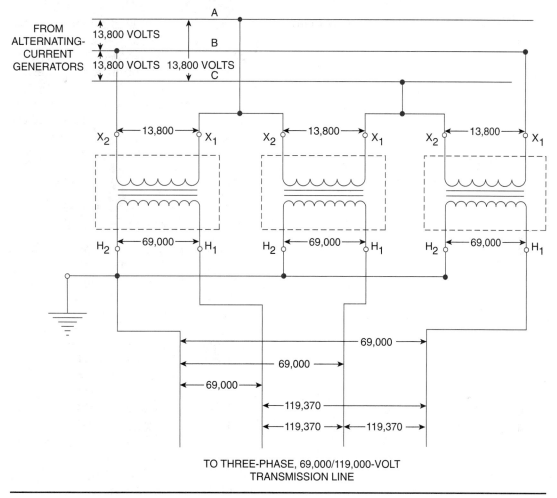

FIGURE 9-35 A delta–wye pattern used to step up voltage.

coil voltage of the transformer need only be 4160 ÷ 1.73 or 2400 V. Using a 10:1 transformer ratio, the secondary Δ voltage is 240 V on the coils and secondary lines. The transformer ratio is 10:1, but the transformation ratio is 4160:240 or 17.3 divided by 1.

Three-Phase Transformer Units

Transformers may be purchased as a three-phase unit. This means that all the coils have been connected internally and the cores are mounted in one unit. The transformer is normally in a single enclosure and may have supplemental cooling to help cool the transformer cores. With three-phase transformers, the lead markings are slightly different. The primary leads are marked as H_1, H_2, and H_3. These leads are either connected directly to the supply voltage leads or to tap points within the transformer. Figure 9-37 shows how internal primary taps may be connected to compensate for actual field conditions where the exact primary

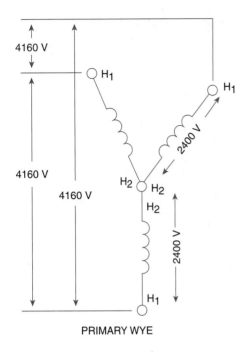

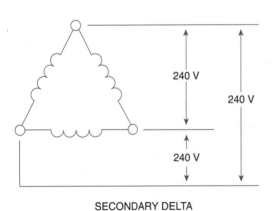

FIGURE 9-36 A wye–delta bank of transformers is used to step down the voltage.

phase rotation should be ABC on X_1-X_2-X_3, respectively.

The three-phase transformer is wound on a common core, so it uses less space than three single-phase units. Because there is a common core, the three-phase transformer is also more efficient.

The biggest disadvantage to a three-phase transformer compared to a three-phase bank of single-phase transformers is that if one coil or phase winding fails, the entire transformer must be replaced.

TRANSFORMER NAMEPLATE

Transformer nameplates must have basic information. Figure 9-38 shows a transformer nameplate. Information should include manufacturer's name, transformer type, serial number, VA ratings, nominal voltages, transformer impedance, and other details as needed (other details may include coolant type and capacity, tap changer if available, and so on). NEC® Article 450.11 determines other markings required.

All leads should be identified on a symbol diagram. The leads should include terminal numbers, tap percentages, any other identifications, as well as polarity marks where needed. On three-phase transformers, primary and secondary voltage vectors are drawn to indicate the phase angle between primary and secondary.

Impedance is a necessary element when determining if transformers can be connected in parallel to share the load according to the rated KVA. If the **percent impedance** is not shown, or if there is any doubt, perform an impedance check. Refer to Figure 9-39 for the connections to determine transformer impedance. First, disconnect all power to the transformer. Next, connect an ammeter across the secondary leads with the desired output voltage connection pattern completed. You also can short-circuit the secondary coil and place a

rated value is not available. The phase rotation is a factor in three-phase transformer banks. If the phase rotation is A-B-C on the primary connected to H_1-H_2-H_3, respectively, the secondary

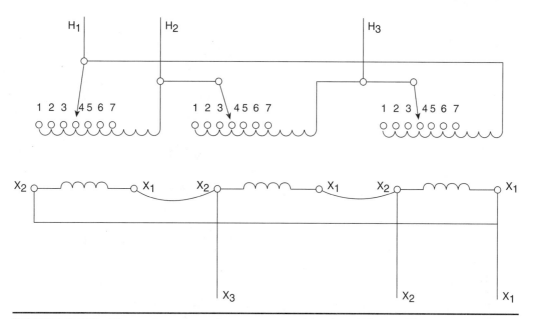

FIGURE 9-37 Three-phase transformer external lead markings with a tapped primary winding.

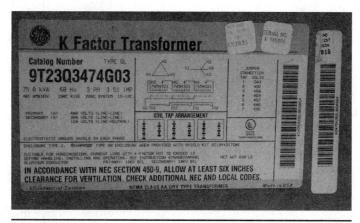

FIGURE 9-38 Typical three-phase transformer nameplate.

clamp-on meter around the shorting conductor to measure current. Connect a variable voltage supply to the primary winding. Also, measure the voltage that will be applied to the primary winding. Energize the primary winding with a very low value of AC voltage. Increase the AC applied to the winding until the secondary ammeter reaches full-rated secondary current. Read the voltage that you applied to the primary and disconnect all power. Divide the applied voltage by the rated primary voltage. For example, if the primary is rated at 277 V and the voltage needed for rated secondary current is 13.85 V, then the percent

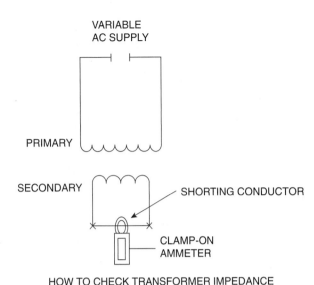

FIGURE 9-39 Method used to check transformer impedance.

FIGURE 9-40 Transformer with radiator cooling design.

impedance is 13.85 divided by 277 × 100 = 5 percent.

Methods of cooling also may be abbreviated on the nameplate. As you know, transformers are not really 100 percent efficient. There are copper and core losses that produce heat. To increase the life of the transformer, the heat must be dissipated. The following abbreviations are used to help determine the method of cooling.

1. AA is a dry transformer that is self-cooled by convection.
2. AA/FA means that the transformer is a dry type with convection cooling, but a thermostat will turn on forced air fans as the transformer temperature rises.

Other transformers are immersed in cooling oil. The oil is an insulating mineral oil that has good heat transfer capability. The oil collects the heat from the transformer core. As it is heated, it rises to the top of the tank and is cooled. As it cools, the oil lowers in the tank. A natural convection of moving oil within the tank helps dissipate the heat. Some transformers have fins that help radiate heat. Other transformers (as shown in Figure 9-40) have a radiator design where the hot oil rises to the top, then cools as it moves into the oil-filled tubes and reenters the bottom of the transformer as cooled oil. Still other transformers have fans that force air over the oil radiator and occasionally oil pumps to force the oil through the cooling fins.

The following abbreviations are indicated on oil-immersed transformers.

1. OA—oil-immersed self-cooling with convection
2. OA/FA—oil-immersed self-cooling, but fan-cooled if necessary
3. OA/FA/FA—same as No. 2, but with two stages of fan cooling
4. FOA—oil-immersed with forced oil (oil pumps) and forced air (fan) cooling

These are the most common, but other cooling schemes also are used.

Insulation Classes

Insulation class ratings are an indicator of how much heat a transformer winding can withstand without insulation breakdown. Class A insulating materials can withstand a temperature rise of 55 °C over the 40 °C ambient temperature, plus a hot-spot temperature of 10 °C above the 55 ° rise. This means that the average conductor temperature at full load should not exceed 95 °C when operating at 40 °C. Class A can be used in oil-filled transformers because of lower temperature rise in operating temperature. Class B insulation can withstand an 80 °C rise over 40 °C ambient plus a 30 °C higher hot-spot temperature. The full-load temperature should not exceed 120 °C with a 150 °C hot spot. This used to be a standard insulation for dry transformers.

Class F insulation has a 115 °C temperature rise rating over the 40 °C ambient. The hot-spot temperature can be 30 °C above that. Full-load winding temperature should not average over 155 °C and a hot-spot temperature of 185 °C. This class of insulation is not used as much in favor of the next class—H insulation.

Class H insulation has a 150 °C rise and another 30 degrees additional hot-spot temperature. The winding insulation can withstand 190 °C and a hot spot of 220 °C without breaking down. Most dry-type distribution transformers with air self-cooling are insulated with class H insulation.

NEC® REQUIREMENTS

Transformers are covered by the NEC® Code, primarily in Article 450. At the beginning of Article 450 (specifically 450.1), transformers that are used for special applications are listed with the appropriate code article. The definition of a transformer includes any single transformer or polyphase transformer identified by a single nameplate.

Transformer overcurrent protection is determined by following the applicable paragraphs of NEC® Article 450.3. When applying this article, "transformer" means a single transformer, or two or more transformers operating as a polyphase unit.

If a transformer or transformer bank has voltages present over 600 volts nominal, the transformer shall have primary and secondary protection set to open based on NEC® Table 450.3(A). Note the following exceptions to this table.

- If fuses or circuit breakers are used and Table 450.3(A) does not correspond to a standard rating, the next higher size is permitted.
- If the transformer is installed where only qualified persons will have access, then Table 450.3(A) states transformers can be protected on the primary side provided fuses used do not exceed 250 percent of the rated primary current.
- If circuit breakers or electronically actuated fuses are used, they must not exceed 400–600 percent of the primary current. (See Figure 9-41.)

Transformers that are rated either primary or secondary at 600 V nominal or less comply with NEC® Article 450.3(B). If the overcurrent protection is in the primary leads, the fuse rating should not exceed 125 percent of the primary current. Exceptions include using the next higher size fuse if 125 percent does not correspond to a standard fuse. Standard fuse ratings are found in Article 240.6. If the primary current is less than 9A, 167 percent of the rated current may be used for protection calculations. If the rated current is less than 2A, protection shall not exceed 300 percent of rated current.

Article 450.3(B) provides for protection in the primary and the secondary. This may be used to reduce costs where multiple transformers are fed by the same feeder. If secondary fuses do not exceed 125 percent of rated secondary current, then the primary need only be protected by feeder protection set at a maximum of 250 percent of primary transformer current.

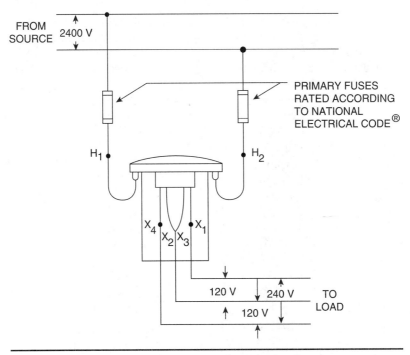

FIGURE 9-41 Primary fuse protection of transformer.

Potential transformers (PTs) used for instruments and metering shall be protected with primary fuses according to Article 408.52, which states the primary fuses will be rated at 15 amps or less. Remember, PTs are used in very low current situations.

Autotransformer protection is based on Article 450.4. The overcurrent protection is placed in each ungrounded input conductor. This protection shall be set to open at no more than 125 percent of the full-load primary current. (See exceptions in the NEC® Code.) Do not install a fuse in the winding that is common to the input and output. (See Figure 9-42.) Article 450.5 refers to grounding connections for autotransformers.

Article 450, Part II refers to the installation of different types of transformers. It is best to refer to the code that applies to your specific installation. It is not the intent of this book to reproduce all applicable requirements of the NEC®.

SINGLE- AND THREE-PHASE LINE DROP

When power is distributed either on a distribution network to the customer or within a building, line drop or voltage drop in the lines must be considered. In Article 210.19(A)(1) (F.P.N.4), the NEC® specifies that conductors for branch circuits should be sized to prevent a voltage drop of over 3 percent. The maximum voltage drop on the feeder and branch circuit should not exceed 5 percent.

To calculate the voltage drop on single-phase conductors, use the volt drop formula. The formula is a version of the Ohm's law formula: $V = I \times R$. To calculate the single-phase

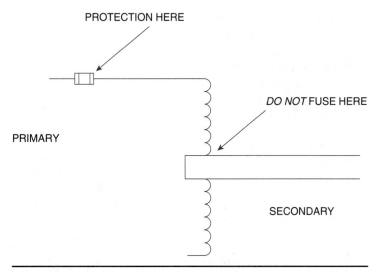

FIGURE 9-42 Autotransformer with proper fuse protection.

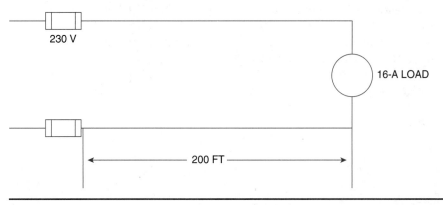

FIGURE 9-43 Voltage drop calculation requires distance to and from the load.

volt drop, find the expected or rated current of the conductor and multiply by resistance of the total length of the circuit. Figure 9-43 indicates that the distance to and from the source of current is needed. You can calculate the R of a particular conductor by using the formula

$$R \text{ in ohms} = \frac{K \times L}{CM}$$

where K is the resistance of the conducting material, L is the length in feet of the circuit, and CM is the circular mills. Use 10.4 as the K or constant of copper. Table 8 in Chapter 9 of the NEC® book gives the CM values for different wire gauges. For example, if the motor is 200 feet from the branch circuit panel, the total distance in feet will be 400 feet. Using No. 12 copper wire and 16 amps, the volt drop will be found by

$R = 10.4 \times 400 \text{ ft} \div 6530 CM = .637$ ohms

$16 \text{ A} \times .637 = 10.2 \text{ V}.$

If the circuit is a 120-V circuit, then 10.2 V represent an 8.5 percent drop. This exceeds the code requirements and a larger size wire is required. K is a constant and is usually stipulated when making code calculations. However, K changes for wire when it is stranded or solid, or if the circuits are heavily loaded and cause heating.

Another method of calculating volt drop is to calculate the overall resistance of the wire by using Table 8 in NEC® Chapter 9. Figure 9-44 illustrates the circuit used for volt drop calculations. Use a No. 12 AWG wire solid copper uncoated. The resistance, as listed in the table, is 2.01 ohms per 1000 feet. You have a 400-foot circuit, so the resistance is 40 percent of 2.01 ohms or .804 ohms. If the ampacity is 16 amps, the volt drop in the wires equals .804 × 16 or 12.86 volts. Again, this exceeds the 3 percent value recommended in the Code.

To reverse the process, use 3 percent of 120 V as a maximum value. This is a 3.6-volt allowable drop. If the current is 16 amps, the maximum ohms of the wires will be .225 ohms for the 400 feet. This translates into .56 ohms per 1000 feet. Using NEC® Chapter 9, Table 8, the conductor will have to be a No. 6 conductor.

For three-phase circuits, use the three current-carrying conductors and the formula

$$\frac{K \times 2 \times L \times I \times .866}{CM \text{ diameter}} = \text{volt drop.}$$

This is the same basic formula as for single-phase circuits, except the volt drop is reduced to .866 times the single-phase drop. For example, a three-phase 30-amp circuit operating at 230 V is 50 feet from the feeder panel, using a No. 10 stranded conductor of coated copper. The original formula is

$$\frac{12 \times 2 \times 50 \text{ ft} \times 30A \times .866}{10380} = 3.0 \text{ volts.}$$

The ohms per 1000 feet of No. 10 wire according to NEC® Chapter 9, Table 8 is 1.29 ohms. This translates to .129 ohms for 100 feet. The equivalent formula is

.129 ohms for 100 ft of No. 12 copper × .866 × 30 amps = 3.35 volts.

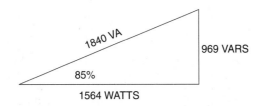

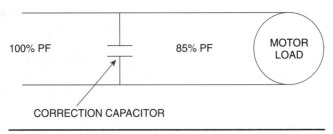

FIGURE 9-44 Power factor correction for motor loads.

There are slight discrepancies in these answers depending on the K factor and the table resistance.

POWER FACTOR CORRECTION

Power factor is a *factor* applied to the *apparent power* to yield *true power*. It is the ratio of true power (watts) to apparent power (VA) expressed in the formula

watts/voltamps × 100 = % power factor

The formula to correct power factor is based on the number of VARS represented in the power triangle. See Figure 9-44 for the power triangle of a typical single-phase motor. If the power factor is to be corrected to 100 percent, 969 VARS of reactive power will have to be supplied by capacitive action.

By adding a capacitor in parallel to the load or motor (across the line), the power factor of the circuit can be corrected to 100 percent without affecting the actual current to the motor. The capacitors can be purchased in VAR or KVAR ratings. Choose a value as close to 969 VARS without exceeding this value. You should not overcorrect the power factor. In fact, correcting past 95 percent PF usually is not required.

The process is exactly the same for three-phase power factor correction. Be sure to use the three-phase formulas for VA and watts. (See Figure 9-45.) The capacitor is calculated in KVAR of correction. Charts (as shown in Figure 9-46) help determine the actual three-phase correction capacitor needed to correct for a beginning power factor to a desired ending power factor. Use the left-hand column of the chart as the circuit's original power factor. The top row of the chart is the desired ending power factor. By crossing the two rows, a multiplier is found that can be applied to the circuit KW. By using the multiplier and the KW, the desired KVAR capacitor size is determined.

DEMAND METERS

To help determine generating capacity requirements, power companies monitor peak demands on the electrical systems. For commercial and industrial customers, metering includes the KVAR meters used in determining power factor. The KWH or kilowatt hour meter measures the actual amount of kilowatts used over time. One thousand watts left on for one hour equals 1000 watt hours or 1 KWH. This is the basis for the energy charges on a power bill. However, the power company also needs to know how many kilowatts were used over short periods of time.

Demand is the kilowatt load that is held for a short period. The best way to explain the need for a demand meter is to use an example of two customers. If each customer uses 10,000 KWH over a one-month period, they would be charged the same energy charge. If customer no. 1 uses the electrical power spread out over an entire month of operation, and customer no. 2 uses all the power in 10 days, then stops, the demands on the power system would be quite different. Customer no. 1 uses only 1/30 of the energy in any one day while customer no. 2 uses 1/10 of the energy in any one day. The actual energy used to create the electricity is the same for both customers, but the cost of transmission equipment is much higher for the second customer. All equipment must be sized to carry much higher power loads for the second customer. The transmission line, transformers, and generation equipment must have higher capacities for customer no. 2. The cost of distribution equipment, breakers, substation equipment, and transtormers are higher for customer no. 2 because of a higher peak demand. Customer no. 2 has a higher demand for electricity over a shorter period of time.

Usually the demand will be registered on the demand meters if the KW load is maintained over a 15-minute period. This is because the distribution equipment can carry temporary peak values for less than 15 minutes

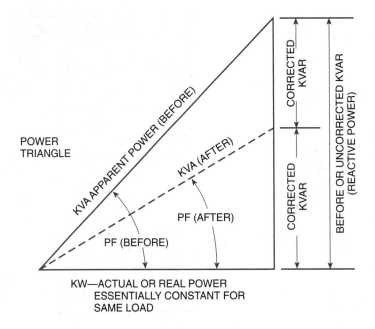

FIGURE 9-45 Representative power factor correction diagram.

without overheating. If the load is maintained over 15 minutes, then overheating occurs and the power company will need to increase the size of their equipment. The 15-minute average prevents the surge that occurs by starting a large motor from establishing a new demand.

ENERGY MANAGEMENT

There are many new methods used for controlling demand. The demand charge for commercial and industrial customers may make up to one-half of the total billing for the power company. Through the uses of demand monitoring systems, steps can be taken to reduce system demand as a facility begins to reach a new, higher demand. Through the use of light harvesting (in other words, using sensors to turn off electrical lighting when natural light is available), demand can be reduced. With variable air-volume controls and variable-frequency electronic drives, less power is consumed when the air handling systems require less air flow. Motion sensors are becoming

Power Distribution and Monitoring Systems

KW MULTIPLIERS FOR DETERMINING CAPACITOR KILOVARS

Desired Power Factor in Percentage

ORIGINAL POWER FACTOR PERCENTAGE	80	81	82	83	84	85	86	87	88	89	90	91	92	93	94	95	96	97
50	0.982	1.008	1.034	1.060	1.086	1.112	1.139	1.165	1.192	1.220	1.248	1.276	1.306	1.337	1.369	1.403	1.440	1.481
51	0.937	0.962	0.989	1.015	1.041	1.067	1.094	1.120	1.147	1.175	1.203	1.231	1.261	1.292	1.324	1.358	1.395	1.436
52	0.893	0.919	0.945	0.971	0.997	1.023	1.050	1.076	1.103	1.131	1.159	1.187	1.217	1.248	1.280	1.314	1.351	1.392
53	0.850	0.876	0.902	0.928	0.954	0.980	1.007	1.033	1.060	1.088	1.116	1.144	1.174	1.205	1.237	1.271	1.308	1.349
54	0.809	0.835	0.861	0.887	0.913	0.939	0.966	0.992	1.019	1.047	1.075	1.103	1.133	1.164	1.196	1.230	1.267	1.308
55	0.769	0.795	0.821	0.847	0.873	0.899	0.926	0.952	0.979	1.007	1.035	1.063	1.093	1.124	1.156	1.190	1.227	1.268
56	0.730	0.756	0.782	0.808	0.834	0.860	0.887	0.913	0.940	0.968	0.996	1.024	1.054	1.085	1.117	1.151	1.188	1.229
57	0.692	0.718	0.744	0.770	0.796	0.822	0.849	0.875	0.902	0.930	0.958	0.986	1.016	1.047	1.079	1.113	1.150	1.191
58	0.655	0.681	0.707	0.733	0.759	0.785	0.812	0.838	0.865	0.893	0.921	0.949	0.979	1.010	1.042	1.076	1.113	1.154
59	0.619	0.645	0.671	0.697	0.723	0.749	0.776	0.802	0.829	0.857	0.885	0.913	0.943	0.974	1.006	1.040	1.077	1.118
60	0.583	0.609	0.635	0.661	0.687	0.713	0.740	0.766	0.793	0.821	0.849	0.877	0.907	0.938	0.970	1.004	1.041	1.082
61	0.549	0.575	0.601	0.627	0.653	0.679	0.706	0.732	0.759	0.787	0.815	0.843	0.873	0.904	0.936	0.970	1.007	1.048
62	0.516	0.542	0.568	0.594	0.620	0.646	0.673	0.699	0.725	0.754	0.782	0.810	0.840	0.871	0.903	0.937	0.974	1.015
63	0.483	0.509	0.535	0.561	0.587	0.613	0.640	0.666	0.693	0.721	0.749	0.777	0.807	0.838	0.870	0.904	0.941	0.982
64	0.451	0.474	0.503	0.529	0.555	0.581	0.608	0.634	0.661	0.689	0.717	0.745	0.775	0.806	0.838	0.872	0.909	0.950
65	0.419	0.445	0.471	0.497	0.523	0.549	0.576	0.602	0.629	0.657	0.685	0.713	0.743	0.774	0.806	0.840	0.877	0.918
66	0.388	0.414	0.440	0.466	0.492	0.518	0.545	0.571	0.598	0.626	0.654	0.682	0.712	0.743	0.775	0.809	0.846	0.887
67	0.358	0.384	0.410	0.436	0.462	0.488	0.515	0.541	0.568	0.596	0.624	0.652	0.682	0.713	0.745	0.779	0.816	0.857
68	0.328	0.354	0.380	0.406	0.432	0.458	0.485	0.511	0.538	0.566	0.594	0.622	0.652	0.683	0.715	0.749	0.786	0.827
69	0.299	0.325	0.351	0.377	0.403	0.429	0.456	0.482	0.509	0.537	0.565	0.593	0.623	0.654	0.686	0.720	0.757	0.798
70	0.270	0.296	0.322	0.348	0.374	0.400	0.427	0.453	0.480	0.508	0.536	0.564	0.594	0.625	0.657	0.691	0.728	0.769
71	0.242	0.268	0.294	0.320	0.346	0.372	0.399	0.425	0.452	0.480	0.508	0.536	0.566	0.597	0.629	0.663	0.700	0.741
72	0.214	0.240	0.266	0.292	0.318	0.344	0.371	0.397	0.424	0.452	0.480	0.508	0.538	0.569	0.601	0.635	0.672	0.713
73	0.186	0.212	0.238	0.264	0.290	0.316	0.343	0.369	0.396	0.424	0.452	0.480	0.510	0.541	0.573	0.607	0.644	0.685
74	0.159	0.185	0.211	0.237	0.263	0.289	0.316	0.342	0.369	0.397	0.425	0.453	0.483	0.514	0.546	0.580	0.617	0.658
75	0.132	0.158	0.184	0.210	0.236	0.262	0.289	0.315	0.342	0.370	0.398	0.426	0.456	0.487	0.519	0.553	0.590	0.631
76	0.105	0.131	0.157	0.183	0.209	0.235	0.262	0.288	0.315	0.343	0.371	0.399	0.429	0.460	0.492	0.526	0.563	0.604
77	0.079	0.105	0.131	0.157	0.183	0.209	0.236	0.262	0.289	0.317	0.345	0.373	0.403	0.434	0.466	0.500	0.537	0.578
78	0.052	0.078	0.104	0.130	0.156	0.182	0.209	0.235	0.262	0.290	0.318	0.346	0.376	0.407	0.439	0.473	0.510	0.551
79	0.026	0.052	0.078	0.104	0.130	0.156	0.183	0.209	0.236	0.264	0.292	0.320	0.350	0.381	0.413	0.447	0.484	0.525
80	0.000	0.026	0.052	0.078	0.104	0.130	0.157	0.183	0.210	0.238	0.266	0.294	0.324	0.355	0.387	0.421	0.458	0.499
81		0.000	0.026	0.052	0.078	0.104	0.131	0.157	0.184	0.212	0.240	0.268	0.298	0.329	0.361	0.395	0.432	0.473
82			0.000	0.026	0.052	0.078	0.105	0.131	0.158	0.186	0.214	0.242	0.272	0.303	0.335	0.369	0.406	0.447
83				0.000	0.026	0.052	0.079	0.105	0.132	0.160	0.188	0.216	0.246	0.277	0.309	0.343	0.380	0.421
84					0.000	0.026	0.053	0.079	0.106	0.134	0.162	0.190	0.220	0.251	0.283	0.317	0.354	0.395
85						0.000	0.027	0.053	0.080	0.108	0.136	0.164	0.194	0.225	0.257	0.291	0.328	0.369
86							0.000	0.026	0.053	0.081	0.109	0.137	0.167	0.198	0.230	0.264	0.301	0.342
87								0.000	0.027	0.055	0.083	0.111	0.141	0.172	0.204	0.238	0.275	0.316
88									0.000	0.028	0.056	0.084	0.114	0.145	0.177	0.211	0.248	0.289
89										0.000	0.028	0.056	0.086	0.117	0.149	0.183	0.220	0.261
90											0.000	0.028	0.058	0.089	0.121	0.155	0.192	0.233

Example Total KW input of load from wattmeter reading 100 KW at a power factor of 70%. The leading reactive KVAR necessary to raise the power factor to 95% is found by multiplying the 100 KW by the factor found in the table, which is .691. Then 100 × .691 equals 69.1 KVAR. Use 70 KVAR.

FIGURE 9-46 Chart used for capacitor multiplier for power factor correction. *(Courtesy of Arco Electric Products Corp.)*

more popular to turn off lights in rooms that are not being used.

More sophisticated energy management systems monitor the demand at the demand meter installed by the power company. See Figure 9-47 for a sample of a demand meter. The energy management systems will have computer programs that will begin turning off electrical loads as the demand meter approaches a predetermined load.

Computer programs are designed to begin trimming nonessential loads first. If the load continues to grow, the programs will initiate temporary run cycles. This system will turn off selected air conditioning or air handlers for short periods, then turn these back on and turn others off, all in an attempt to keep the demand on the electrical system to a predetermined load.

Power companies encourage customers to control their own loads. Incentives are often offered to commercial customers to curtail their maximum usage during peak use days. This allows the power companies to avoid running expensive peaking plants for power generation. In the United States these peaks usually occur in the summer on hot and humid days when cooling demands are greatest.

Also, there is a large change in technology that is providing more energy savings. Through the use of electronic ballasts and more efficient fluorescent lighting, light illumination levels are being maintained while lowering the demand and the total KW usage. The idea is not to waste valuable resources and to conserve and reduce energy waste.

SUMMARY

This chapter introduced the concepts used in transformation and distribution of electric power. The basics of transformers were explained to help you determine the proper selection and uses of various types of transformers. Many different types and styles of transformers were presented, each with a specific function or application.

Transformer calculations were presented in addition to the methods of connecting transformers into different patterns to attain the desired levels of voltage and current. Methods of testing transformers for proper operation and checks to verify safe installations were also explained. NEC® requirements as related to transformer installations were reviewed.

Power distribution and line drop calculations were introduced as a portion of the electrical distribution sizing considerations. Other considerations of the distribution system (such as power factor correction methods and demand control) also were presented.

FIGURE 9-47 Demand meter used to monitor peak energy needs.

QUESTIONS

1. What does a *step-up* transformer refer to?
2. Explain how the three factors for induced voltage are obtained in a transformer.

3. Give the voltage and current relationships between the primary and secondary of a transformer.

4. Why are transformer cores laminated?

5. Why is a shell-type transformer more efficient than a core-type transformer?

6. What is meant by *magnetizing current*?

7. Explain how to determine if a transformer has subtractive polarity.

8. List some of the losses that occur in a transformer.

9. What is meant by a transformer percent regulation?

10. Show how to connect two single-phase transformers in parallel.

11. What conditions must be met before paralleling transformers?

12. How do autotransformers differ from isolation transformers?

13. What are two types of instrument transformers?

14. Why would you never open a CT secondary while there is a primary current?

15. When is transformer ratio different from transformation ratio?

16. Explain how to check the secondary connection of a delta–delta bank before final connections.

17. What is a name for the B phase on a three-phase, four-wire, delta secondary?

18. An open delta pattern reduces the output capacity to _____ percent of the original three-phase capacity.

19. What does AA/FA mean on a transformer's nameplate?

20. Which class of insulation has the highest temperature rating?

21. Write the formula for calculating three-phase voltage drop.

22. In what units is electrical demand measured on a demand meter?

CHAPTER 10

DC MOTORS, GENERATORS, AND CONTROLS

OBJECTIVES

After completing this chapter and the chapter questions, you should be able to

- Understand the theory involved in DC generation
- Be able to connect, test, and operate series, shunt, and compound generators
- Determine if a self-excited generator is building the proper output voltage, or if the residual magnetism needs to be restored
- Select the proper connections for the application involved (such as cumulative or differential compounding and degree of compounding)
- Understand DC motor theory and make selections of motors according to their operating characteristics
- Determine the requirements needed in DC electronic speed control

KEY TERMS IN THIS CHAPTER

Armature Reaction: The distortion of the plane of the main magnetic field caused by the reaction to current flow through the armature.

CEMF: The counter electromotive force or back EMF that is developed by a spinning rotor within a magnetic field. In DC machines, the CEMF is produced in the rotor and opposes the applied voltage.

Commutator: The segmented connection point to the rotating armature coils. A carbon brush makes the electrical connection to the commutator.

Compound Fields: Connections that are made in motor or generator fields that connect both the series and shunt field into the circuit.

Cumulative Compound: The connection that allows the shunt field flux and the series field flux to aid each other.

Differential Compound: The connection that allows the series field flux to oppose the shunt field flux.

Differential Voltage: Voltage difference between the applied EMF and the CEMF that is produced in the DC machine. Differential voltage or effective voltage changes as the motor speed changes.

Self-Excited Generator: A condition where the generator does not need an external source of electrical current to establish the magnetic field in the stator windings. The generator produces its own field excitation by using residual magnetism.

Separately Excited: A generator that needs an external current source to develop the main magnetic field.

INTRODUCTION

As with AC motors and generators, DC motors and generators are energy converters. The DC generator converts mechanical energy (HP) to electrical energy (watts). Likewise, the DC motor converts watts of electrical energy to horsepower of mechanical energy. DC motors have some advantages over AC motors and are used in applications that require the DC motor characteristics. Until the 1960s, DC motors were used extensively in speed control situations. Even later many electronic drives were developed to control DC motors.

DC GENERATION

DC generators were used to produce the DC needed for DC motors. Often these DC generators were driven by AC motors. This system of an AC motor driving a DC generator was called a *Ward-Leonard system*. DC generators are also used on railroad locomotives. Large diesel engines drive a DC generator to provide DC power to the DC traction motors used on the drive wheels. A brief discussion of DC generation should help in the explanation of DC motors.

Generation is based on the same three factors that are involved in AC generation as described in Chapter 5. These factors are

1. A magnetic field
2. A conductor
3. Relative motion

Contrary to most AC generators (alternators), the DC generator uses a rotating armature as it moves past stationary fields. If you use the electron flow theory, electrons will flow from negative to positive potential. Referring to Figure 10-1, and using the left-hand rule for generators, as the conductor moves past the poles a voltage is induced that forces electrons in the direction indicated. The symbol ⊙ means electrons flow toward you. The symbol ⊕ means electrons flow away from you. Similar to the alternator, the spinning conductor produces a sine wave with peak values occurring as the conductor rotates directly under the strong magnetic poles; the zero points as it travels parallel to the magnetic flux. The difference in a DC generator is the method of connection made to the revolving armature.

Figure 10-2 illustrates how the carbon brushes are connected to a split-ring sliding connection. Notice that at this point, the

DC Motors, Generators, and Controls

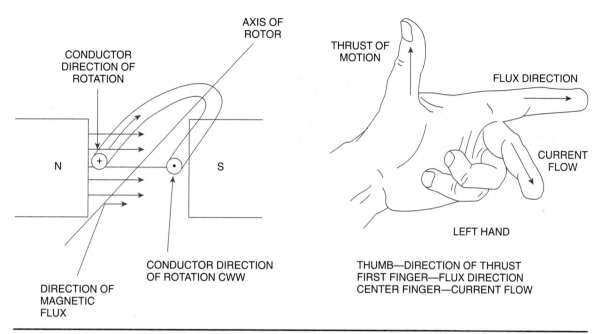

FIGURE 10-1 Induced voltage and direction of current flow.

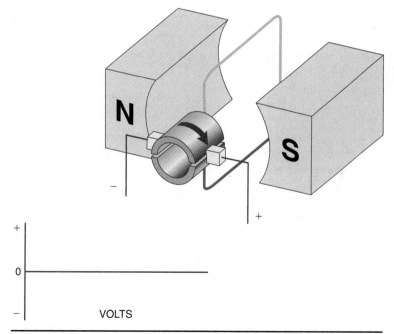

FIGURE 10-2 Carbon brushes are used for connection to commutator.

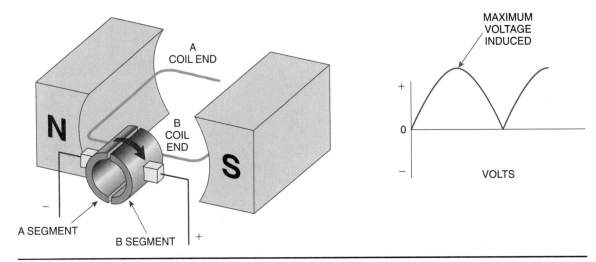

FIGURE 10-3 Brushes connected to commutator when coil is directly under the magnetic pole.

conductors are parallel to the lines of flux and no voltage is induced because the rotor conductors are not moving through the magnetic lines of force. At this instant there is no voltage induced and there is no voltage at the brushes.

As the rotor is moved by mechanical energy, it produces a varying amount of voltage. Figure 10-3 shows how the brushes are connected to the **commutator** as the rotor is at the maximum voltage point. The commutator is the connection point for the coil ends. The brushes make contact with the commutator to collect the voltage and complete the electrical circuit to the load. In this example the "A" commutator segment is always connected to the "A" side of the spinning coil as it moves across the north pole face, and the "B" segment to the "B" coil as it moves across the south pole face. The brushes and commutator are mechanical rectifiers changing the generated AC within the coils to DC output. The negative carbon brush is always connected to whichever coil side crosses beneath the north pole. This means that the brush will always be connected to the same polarity of induced voltage. The output will be a pulsating DC that moves from maximum to zero but never reverses. (See Figure 10-4.)

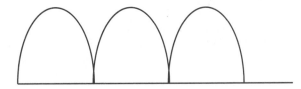

FIGURE 10-4 Output voltage has peak and zero points but does not cross zero or reverse direction.

ARMATURES

Armatures are wound in different ways to give different applications. A number of turns of wire are wound so the voltage in each turn is added to the next to obtain a higher voltage. The larger the number of conductors spinning through the magnetic fields, the higher the voltages. In addition to more coil windings, there are usually more than two magnetic poles (as shown in Figure 10-5). By increasing the number of poles, the DC output will be smoother. The brush riding on the commutator will make contact with the coil traveling directly under the pole pieces and will connect to the peak generation points making the DC appear smoother.

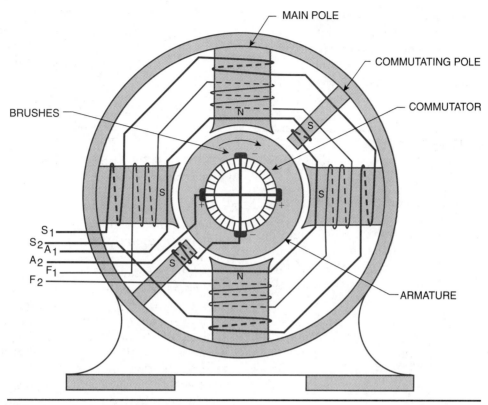

FIGURE 10-5 DC generator with four poles.

Armature Windings

Armatures such as those used in the starter motor (motor application) are wound with lap windings. These *lap-wound armatures* produce low voltage, but they produce a higher current because there are many coils in parallel.

Wave-wound armatures produce a high voltage but have low current output. These are characteristically two-brush machines. Their windings are connected in series so that all the winding-induced voltages add up, but the current capacity is low.

Frog-leg-wound armatures use a combination of series and parallel connections of the coils to produce a compromise between the first two styles. Figure 10-6 shows a schematic view of a frog-leg-wound generator armature.

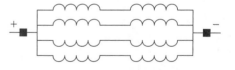

FIGURE 10-6 Frog-leg winding produces a combination of wave-wound and lap-wound characteristics.

These armatures are used for intermediate voltage and current.

Magnetic Field Windings

The magnetic field on the stator poles is produced by several different methods. The field is an electromagnetic field produced by coils wound to produce north or south poles. Depending on the type of generator, the poles can have a series- or shunt-winding coil or a combination of both (called a *compound winding*).

SERIES GENERATOR

In a series generator, the pole piece is wound with large-diameter magnet wire to form an electromagnet. See Figure 10-7A. If the large, heavy current conductor is the only winding on the generator pole, the winding would be connected in series with the electrical load. A series field refers to the way the stator magnetic field is connected in the generator

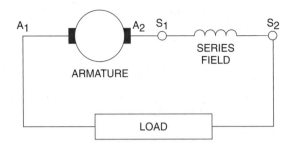

FIGURE 10-7B Schematic diagram of load connected to a series generator.

circuit. Figure 10-7B shows the schematic relationship of the armature and the series field winding.

In the generator in Figure 10-7B, the series DC generator is **self-excited.** This means that the generator can begin producing voltage without needing an outside source of electricity to establish the stator poles. The process of self-excitation relies on the fact that there will be some residual magnetism left in the field poles even after the electromagnetic field has been deenergized. As the rotor begins to spin through a weak residual magnetic field, a small voltage is generated. This small voltage will produce a small current flow to the load. As the current flows through the series field wound on the pole piece, more magnetic flux is produced by the pole. The increased flux continues to produce more output voltage until the output voltage is at rated value. If there is no residual voltage (produced by residual magnetism), the magnetic field must be reestablished. (See the "Flashing the Fields" section.)

After the series field is established, the output voltage (therefore, the current) can be controlled by changing the speed or strength of the series field. Increasing the speed results in more flux being cut per second; thus more output voltage is achieved (Faraday's law).

Maintaining the speed but strengthening the magnetic field also increases output voltage. As more electrical load is added (current

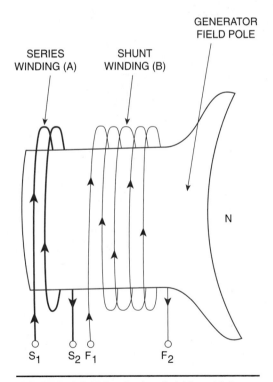

FIGURE 10-7A Series field is a high-current winding wound on the stator salient pole. Also shown is the shunt field winding.

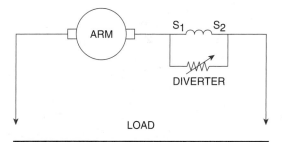

FIGURE 10-8 A series field diverter can be used to "divert" current around the series field.

SHUNT-WOUND GENERATORS

Only shunt-wound generators have the shunt windings. Figure 10-7A shows both windings, but the shunt generator does not have the series winding in place. Shunt generators use a magnetic pole winding designed for lower current, but more turns of the coil, than a series winding. The shunt generator can use the residual magnetism in the poles to create residual voltage as the armature begins to spin. However, the shunt field is connected across the armature, shunting the armature and the load. Figure 10-9 shows how the schematic diagram of a shunt generator would look.

The shunt winding is connected across the armature and uses the armature voltage as a source of DC to produce field excitation. The generator is a self-excited shunt generator if the residual magnetism is used to begin the voltage buildup. Again, the principle of self-excitation uses the generator's own output from the armature to provide a DC source for the field winding. Unlike the series generator, the output voltage (or field strength) does not depend on the load current. The field current is a relatively small portion of the output current and is controlled by a *field rheostat*. Field

draw) to the generator, the magnetic pole strength will also be increased until magnetic saturation occurs. Saturation means that the iron core of the pole piece cannot hold any more flux lines even if the field winding current increases.

To decrease the effect of the series winding current (in other words, to prevent magnetic field change with every change in load), use a *diverter* to shunt current around the series field. Figure 10-8 shows how a series field diverter shunts current around the series field winding. The diverter is a very low ohm, high-current rheostat that is designed to bypass current around the low-resistance series field.

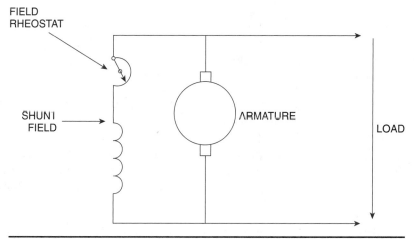

FIGURE 10-9 The shunt field is connected across the armature.

rheostat resistance is adjusted to control field current until the desired output voltage is reached based on field strength and rotor speed.

The shunt generator also can be operated as a **separately excited** shunt generator. In this case, the shunt field is maintained by a separate source of DC independent of the DC generator output. Voltage regulation tends to be more stable in a separately excited generator as it is not subject to some of the variables of a self-excited generator. If output voltage must remain constant, a sense circuit at the load can provide field current adjustment to the shunt field. (See voltage output curves in Figure 10-11.)

COMPOUND FIELD GENERATORS

A compromise between the series generator (which has a rising voltage characteristic) and the shunt generator (which has a drooping voltage characteristic) is the compound field generator that uses both characteristics. Rising voltage characteristics mean that the output voltage rises with an increase in electrical load. A drooping characteristic means the output voltage decreases with an increase in current load. Figure 10-10A shows a long shunt compound connection where the shunt field is connected across both the armature and the series field. This is the most used motor connection but could be used for generator connection. Figure 10-10B shows a short shunt connection where the shunt field is shorted across the armature. In both of these connections, the series field remains in series with the load and the armature, while the shunt field is connected across or shunts the generator voltage.

Cumulative Compounding

If the series field and the shunt field are connected so that the series field produces the

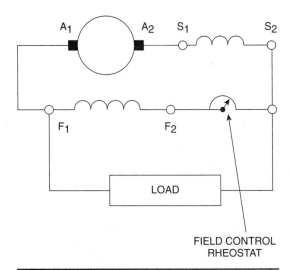

FIGURE 10-10A Long shunt connection of a DC compound generator.

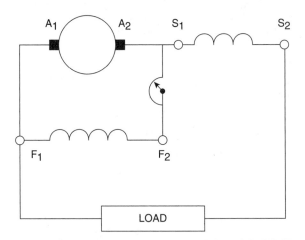

FIGURE 10-10B A short shunt connection of a DC compound generator is usual.

same magnetic polarity as the shunt field, the magnetic fields are aiding each other and the connection is called a **cumulative compound**.

Figure 10-7A shows both fields producing the same magnetic polarity, thus establishing a cumulative compound winding. If the series field adds to the flux as the load increases, the

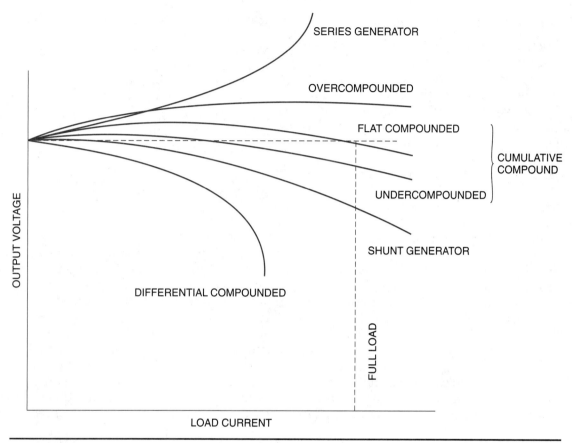

FIGURE 10-11   Voltage output curves of DC generators.

output voltage will rise as it does with a series generator. This is called *overcompounding*. (See Figure 10-11.)

If the series field has some of the load current diverted by a diverter, the voltage could be the same at full load as it was at no load. This connection is called *flat compounding*. The voltage output is not necessarily always the same value, but the no-load and full-load values are the same.

The last possibility is that the output looks more like a shunt generator. If the series field does not add enough magnetic flux to the pole, the output voltage will tend to fall with an increase in load. (See the "Generator Output Voltage" section.) In this case, the generator is undercompounded and the generator acts more like a shunt generator but does not drop as far. (See Figure 10-11.)

Differential Compounding

Another possibility is to connect the series and shunt field so the series field flux opposes the shunt field flux. This results in a **differential compounding** of the two fields. Usually the shunt field establishes the no-load voltage. As load is added, the series field cancels some of the shunt field flux and the output voltage drops rapidly.

Generator Output Voltage

Several factors affect the amount of output voltage for any DC generator. As you know, the speed of the rotating armature will affect output voltage. The higher the speed, the higher the output voltage. As the generator is electrically loaded, the motor effect is the same as that in AC generators. That is, as the current in the armature increases, a countertorque is produced and attempts to drive the armature in the opposite direction. The right-hand rule for motors proves that the output current produces a countertorque. Thus, the rotor slows with increased electrical load. As the rotor slows, less voltage is available at the brushes because it is cutting flux at a slower pace. If less voltage is available at the brushes, the voltage at the shunt field is also reduced and shunt field strength weakens. If there is less shunt field flux, output voltage decreases. These changes are small, but they do cause output voltage to drop with increased electrical load.

Another factor to consider is the resistance of the armature windings. As output current load increases, there is more line or voltage drop within the armature due to the current and resistance of the armature windings. Again, this means there is less voltage available at the brushes as the load increases.

All of these factors are apparent when examining the shunt generator curves of Figure 10-11. The compound generator uses the series field to compensate for the drooping voltage of the shunt field.

ARMATURE REACTION

The rotor has current flow in the armature conductors. This means that there is a magnetic field produced around the armature conductors. This magnetic field reacts with the main stator fields and tends to distort the main field flux lines. The result is called **armature reaction.**

Armature reaction shifts the magnetic field and causes excess sparking in the commutator. In a generator, the armature field distorts the main stator field flux as illustrated in Figure 10-12. To compensate for armature reaction, *commutating poles* or *interpoles* are

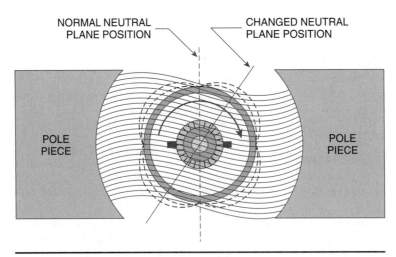

FIGURE 10-12 Effects of armature reaction show distortion of main magnetic field.

inserted in the stator fields. (See Figure 10-5.) The interpoles are connected in series with the armature so that every time there is an increase in load current, current will flow through the interpoles and these magnetic fields realign the main fields into the original shape.

Flashing the Fields

If the generator is a self-excited generator, the residual voltage depends on a small amount of residual magnetism left in the pole pieces. If the residual magnetism is lost, the fields can be reestablished by momentarily applying an external DC voltage source. This process is called *flashing the fields*. If possible, disconnect the shunt field windings marked F_1 and F_2. Note which one is the positive lead. Provide a low-voltage DC source to the field to reestablish the magnetic field in the winding.

After reconnecting the field, you can discern if you flashed the field correctly. Run the generator in the proper direction of rotation. If the residual voltage shows downscale, or reversed polarity, flash the field again but change the polarity on the DC source applied to the shunt field.

Next, connect the series field and observe the output voltage as you add a small load. If the voltage begins to drop quickly, the series field is opposing and not assisting. This is a differential connection. If you want a cumulative compound, reverse the series field winding.

PARALLEL OPERATION OF DC COMPOUND GENERATORS

Similar to AC generation, one generator may not be able to supply the total KW rating of the load. If so, DC generators can be operated in parallel so the output power adds, to deliver power to the load. If the generators are compound DC generators, an additional connection is required between the generators (in addition to the output positive and negative leads being paralleled). Figure 10-13 shows an additional parallel point called an *equalizer* connection. The equalizer parallels the two armatures and the two series fields. Without

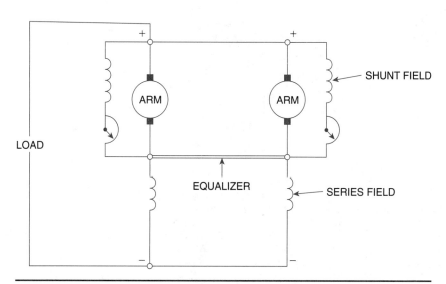

FIGURE 10-13 Equalizer connection is required for parallel DC compound generators.

this connection, one generator will grab all the load and operate into overload while the other generator will supply no load current at all. In fact, the dominant generator will attempt to drive the other generator as a motor.

Without the low-ohm parallel equalizer connection, the generator with a slightly higher DC output voltage would also supply slightly higher load current. If the load current increased, the series field would also be strengthened and result in higher output voltage, which increases series field strength, and so on.

The first generator would continue to be stronger. The second generator, which had less current output, would have a weaker series field. Thus, the second generator would yield less voltage output and so on until it could be driven as a motor. Change in loads occurs so quickly that there is no time to correct the changing loads through adjustment of the shunt field rheostats or diverters. The equalizer connection resistance must be less than 20 percent of the series field resistance. That means it is a large conductor that must have solid connections.

DC MOTORS

DC motors serve many applications. They have a variety of applications where the speed, torque characteristics, and size are specially suited. The DC motor lost popularity for a while because AC motors were cheaper and much easier to maintain. As industrialization of electronics made DC power readily available by rectifying AC, DC motors became popular where speed control was needed.

The DC motor principle is the same principle used in every motor. That is, as current flows through a conductor and the conductor is placed in a magnetic field, the two fields react to move the conductor. Figure 10-14 shows what happens as a conductor is placed into the middle of an electromagnetic field. Using the right-hand rule for motors and electron flow theory, you can establish in which direction the conductor will move. Figure 10-15 illustrates Fleming's right-hand rule. If the conductor placed in the field is attached to a rotating member, the member will move in the same direction. The electromagnetic field

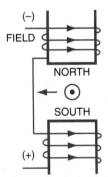

ORIGINAL CONNECTIONS GIVE FIELD POLARITY AS SHOWN. ARMATURE CURRENT AS SHOWN WOULD PRODUCE A COUNTERCLOCKWISE ROTATION.

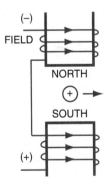

ARMATURE CONNECTION CHANGED TO GIVE OPPOSITE DIRECTION OF CURRENT IN ARMATURE WHILE MAINTAINING FIELD DIRECTION RESULTS IN CLOCKWISE ROTATION.

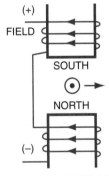

REVERSE FIELD POLARITY AND ARMATURE POLARITY FROM MIDDLE DIAGRAM WILL RESULT IN THE SAME CLOCKWISE DIRECTION OF ROTATION.

FIGURE 10-14 Resultant motion with current flow in the armature and the magnetic field established as depicted.

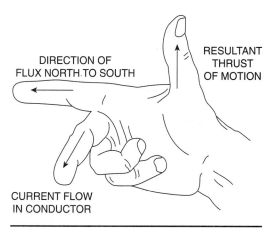

FIGURE 10-15 Right-hand rule for motors using electron flow theory.

will be made up of the series field, the shunt field, or both fields, just as the DC generator was constructed.

Actually, the conductors in the field are rotor conductors. The electron flow to the rotor is provided by a commutator and brush, sliding contacts. The brushes are connected to the power source and make connection to the proper rotor coil so that there is always the same direction of current flow under the same magnetic pole. The concept of commutation is the same as for DC generators in that the rotor conductor is connected to the same polarity voltage as it rotates under a specific pole.

There are several styles of standard DC motors and some variations. The brushless DC motor is actually an AC synchronous or permanent magnet motor. Conventional DC motors are classified like the DC generators. These motors include shunt, series, and compound motors. The conventional motors all use the principle of developing a stationary field by electromagnetic coils wound on the stationary field poles.

If the poles protrude from the surrounding iron, they are called *salient poles*. Current flow is conducted to the rotor by the commutator and brush movable contacts.

All *conduction motors* produce torque by creating a magnetic field around the rotor conductor that reacts with the strength of the stator magnetic field. Therefore, torque is controlled by how much magnetic field is produced by the rotor and stator. Rotor torque is based on current flow and number of windings. The magnetic field of the stator is also based on current flow and number of windings. How these fields and rotor currents react with each other depends on the motor style.

Starting current and the effects of CEMF are common to DC conduction motors. As a motor is first energized, there is very little opposition to the flow of DC current. Because the rotor has not begun to spin, there is a high inrush current through the rotor conductors. This large current reacts with the field winding flux to create starting torque, which starts the rotor spinning. As the rotor begins to spin, it begins generating a voltage. The rotor has the three factors needed to generate: a magnetic field, motion, and conductors. The voltage produced in the rotor is counter or opposite to the applied voltage. The countervoltage produced is referred to as *counter electromotive force* or **CEMF**, also referred to as back EMF. The faster the rotor spins, the more CEMF is produced and the effect counters, or opposes, the applied voltage to create an opposition to line current. In fact, the line current decreases as the rotor spins faster. The difference between the applied voltage and the CEMF is called **differential voltage** or *effective voltage*. This relationship is expressed in Figure 10-16A. The effective voltage will cause a no load current to flow in the motor. As mechanical load is added, the rotor will slow and less CEMF will be produced. This creates a larger differential or effective voltage. The result is a higher running current that flows through the armature conductors and creates more torque needed to spin a heavier load. See Figure 10-16B for a graph of change in differential voltage.

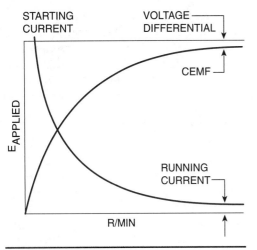

FIGURE 10-16A Voltage differential is small at no load, creating a small effective voltage.

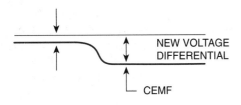

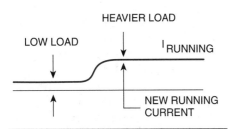

FIGURE 10-16B As load is added, the CEMF drops and this creates a higher voltage differential.

SPEED AND DIRECTION CONTROL

To change the direction of rotation of a DC conduction motor, you will need to change either the polarity of the stator fields, or the direction of the rotor conductor current flow. *Do not change both.*

Figure 10-14 shows what happens as you change only the rotor current in diagram B, or if you change both rotor and field from diagram B to C. The direction of the rotor is the same between B and C. Therefore, the usual method to change the direction of the motor output is to reverse the connections to the brushes marked as A_1 and A_2, or armature leads.

Magnetic field strengths are adjusted to change the speed of a DC conduction motor. To decrease the speed below normal operating point, weaken the field of the armature. By reducing the voltage only to the armature, there is less flux produced in the armature and less torque is produced by interaction of rotor and stator.

To operate the motor above rated speed you may weaken (but do not remove) the stator field flux. Weakening the field results in decreased CEMF and increased voltage differential. The resulting higher current flows through the armature. This armature current reacts with a weaker stator field but produces more twisting effort at a light load. Of course, weakening the field too much will cause a heavy load to stall.

Caution: Never open the shunt field circuit of a DC motor, especially with no mechanical load attached. The speed may increase so far that the centrifugal force on the rotor will cause all the rotor bars to fly off. This is a runaway condition. Some motors have a circuit that disconnects the motor if the shunt field is lost.

SHUNT MOTORS

The shunt motor connection diagram is shown in Figure 10-17. The armature leads are marked A_1 and A_2; the shunt field leads are marked F_1 and F_2. The shunt field is the same type of winding as found in the shunt generator. It has a relatively high resistance and is

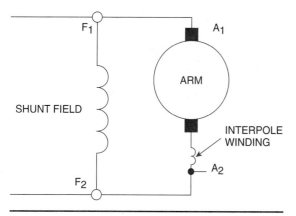

FIGURE 10-17 Lead marking and connection for a DC shunt motor.

designed to be connected across the DC power source. If speed control is needed, a field rheostat may be included. The field rheostat is not used in electronic DC motor control, but the concept of controlling field and armature current is employed.

The speed of a shunt motor is quite stable; thus speed regulation is good. It has a low percent speed regulation, so speed stays nearly constant between no- and full-load speed. This is because the shunt field is connected directly to the power source and the magnetic field does not change much. See Figure 10-18 for a comparison graph of speed regulation. If the stator field does not change, the speed and torque are controlled by the armature winding. The lower the armature resistance, the lower the percent speed regulation. Thus, any small change in voltage differential caused by a slowing rotor, due to a heavier mechanical load, causes an increase in developed torque to keep speed constant or nearly constant. Remember, no nonsynchronous motor has 0 percent speed regulation. There must be a change in speed to produce more torque. The torque curves of a shunt motor are almost linear. See Figure 10-19 for torque curve comparisons.

As mentioned, when the motor starts, there is a large DC inrush current flow. This starting current produces enough starting torque to get the rotor spinning. Normally, the inrush current is high enough that DC motor starters are designed to limit this inrush

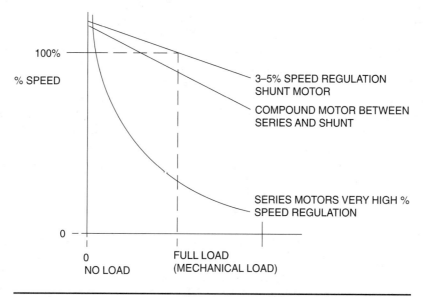

FIGURE 10-18 Comparison of speed regulation for DC motors.

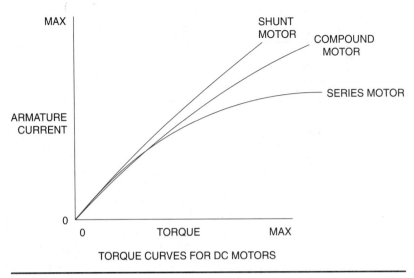

FIGURE 10-19 Comparison of torque versus armature current in DC motors.

current to about 150 percent. This corresponds to the overcurrent protection table in NEC® Article 430.52. For all but instantaneous trip breakers, the percentage used for protection is 150 percent of full-load current.

Speed Control and Direction

Shunt motor speed control is accomplished by changing the applied armature voltage. To reduce the normal speed, reduce the voltage. This will reduce the current to the armature leads A_1 and A_2. To increase the speed above nameplate speed, weaken, but do not remove, the shunt field.

Direction of rotation of shunt motors can be changed by either reversing the relative direction of the rotor flux by reversing the armature leads, or by reversing the shunt field leads.

Armature Reaction

Armature reaction occurs in DC motors exactly as it does in DC generators. If you recall, armature reaction is the distortion to the original plane of the stator magnetic field. The brushes of the DC motor could be moved in the opposite direction from the direction of rotation each time mechanical load is added and armature current increases. Moving the brushes realigns them in the neutral plane. Brushes for generators are moved in the same direction as the rotor rotation. As in the generator, other windings can be added to the motor's stator to compensate for the distorted field. *Commutating poles* or *interpoles* are added midway between the salient main poles to shift the magnetic field back to the original neutral plane. Interpoles have the same polarity as the main pole behind them in the direction of rotation. Usually, the interpole windings are connected internally directly in line with the armature and the leads are brought out at A_1 or A_2 leads. If the leads are brought out for external connections, they are labeled C_1 and C_2. They might also be labeled S_3 and S_4. Figure 10-17 shows a shunt motor with interpoles. The shunt field is represented by at least four loops in the schematic diagram.

PERMANENT MAGNET MOTORS

Permanent magnet (PM) motors are a variation of the shunt motor. Instead of wound fields in the stator, ceramic magnets (permanent magnets) have been installed. Now no power is consumed by the field and the efficiency of the motor increases. In addition, the size of the motor can be reduced. See Figure 10-20 for size comparison. Another advantage is that the PM motor is not susceptible to armature reaction. The permanent magnetic field is not distorted by the armature current. The rotor is essentially the same as the standard shunt motor. The speed is easily controlled by adjusting the applied voltage to the armature through the brushes. Overheating can result if PM motors are operated at continuous high-torque levels.

The permanent magnet field means there is less copper loss in the motor; therefore, it is more efficient and runs cooler. This feature also enables the PM motor to be used where portable battery-operated motors are required.

Additionally, the permanent magnet system provides some braking of the motor. Dynamic braking can be achieved by shorting the only two power leads (armature leads) after disconnecting power. Reversing the motor is done by reversing the DC polarities to the armature leads.

Permanent magnet motors lose some magnetic field strength at very low temperatures. *Caution:* Be careful when using PM motors for gear motors. Because PM motors have high starting torque (175 percent or more), the high torque could damage the gear motor assembly.

SERIES MOTORS

Series motors use only a series field in series with the armature. Figure 10-21 illustrates how the series field (represented by three loops) is connected in the motor circuit. The mechanical load has a large effect on series motor operations. Because the motor current flows through the armature and the series field, the motor current has a doubled effect on the torque. Referring to Figure 10-19, notice that the series motor has the most torque per amp of any of the DC motors. These motors are often used as traction motors. Motors that will produce a great deal of torque at low speed and decreased torque at higher speed are often used for drive wheels on locomotives and forklifts.

FIGURE 10-20 Permanent magnet stator field.

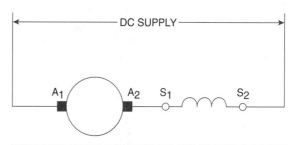

FIGURE 10-21 Series motor connection diagram.

Speed variations are extreme for the series motor. Figure 10-18 shows how the speed curves compare to the other DC motors. At light mechanical load, the speed of the motor is high. In fact, at no load the speed may be high enough to run away or tear the rotor apart by centrifugal force. Series motors should only be run with a load attached. As mechanical load is added, the CEMF of the rotor decreases and the motor current increases. The torque is produced by the reaction of the series field flux and the armature flux. Because both magnetic fields fluctuate together, there is a wide range of speed and torque that are inversely proportional.

To reverse the series motor, reverse the relationship between the fields. The preferred method is reversing the armature connection. Speed control is accomplished by adjusting the voltage to the motor, both series and armature.

Compound Motors

Compound motors are DC motors that use both series field and shunt field windings. (See Figure 10-22.) These motors are a compromise between the characteristics of the series and the shunt motors. As with the DC generators, the motor fields can be connected *short shunt* where the shunt field is shorted across the armature or *long shunt,* where the shunt field is connected across the armature and the series field. The long shunt connection is the more common motor connection because it has better speed regulation. The two fields also can be connected to aid each other in the cumulative compounded motors or connected to oppose each other in the differentially compounded motor. Cumulative compound, long shunt is the standard connection.

If a motor is mistakenly connected differentially, the shunt field will determine the direction at no load. As load is added and more current flows in the series field, the opposite series field may become stronger and suddenly reverse the rotation. To test the proper direction of each field, connect each field separately and note direction of rotation produced by each. Remember to have a mechanical load connected to the motor when testing the series field connection to prevent a runaway condition. Then reconnect the fields into the cumulative pattern.

To reverse the direction of rotation, change just the armature connections. Figure 10-18 shows how the speed/load curves

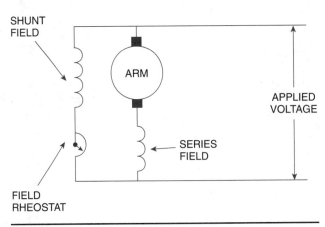

FIGURE 10-22 Schematic of long shunt compound motor.

compare to other DC motors and Figure 10-19 shows how the compound motor compares to other DC motors in torque/current curves.

BRUSHLESS DC MOTORS

The stator field of a brushless DC motor actually rotates. Commutation of the DC supply voltage is controlled electronically by switching the magnetic polarity of the stator poles to correspond to the position of the permanent magnet rotor. Feedback circuits are needed to determine the actual rotor position. The stator has windings that are energized by a digital sequencer. As the rotor north pole position is tracked, the stator south pole is energized just ahead of the rotor pole so that the rotor north follows the stator south pole. Speed of the rotation is determined by the pulsing or energizing frequency of the stator fields.

There are several advantages of brushless DC motors.

1. They do not have brushes or commutators to wear out.
2. There is no brush sparking or radio frequency interference (RFI) that accompanies brush sparking.
3. The speed torque curves are linear and there are no armature losses.
4. Higher horsepower can be obtained from physically smaller motors.

See Figure 10-23 for a brushless DC motor controller.

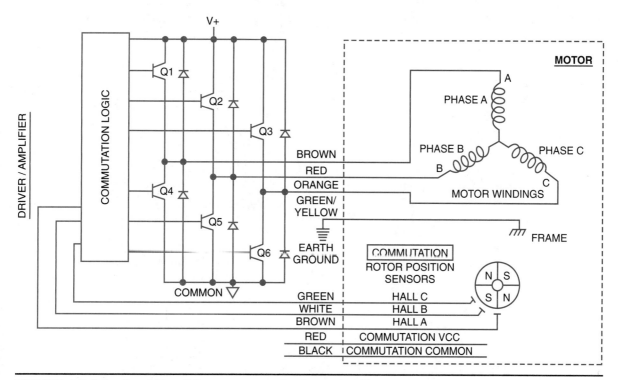

FIGURE 10-23   Brushless DC motor control schematic. *(Courtesy of Bodine Electric Company)*

ELECTRONIC SPEED CONTROL

DC motors were used extensively in applications where speed control was desired. Until the 1970s and early 1980s, DC was the choice for controlling speed of motors over a wide range. Now AC motors also have good speed control techniques using variable-frequency drives. There were several methods for starting DC motors and for controlling the speed. The older manual starters were called three- and four-point starting rheostats. The objective of these controllers (shown in Figure 10-24) was to start the motor while reducing inrush current. Notice on the three-point starter that the arm would be mechanically moved from left to right. The first position applies full voltage to the shunt field, but inserts resistance into the armature and series field circuit. As the arm moves, resistance is removed from the armature circuit and added to the shunt field circuit. The electromagnet at the far right is a keeper that holds the movable arm in the run position. If the shunt field is opened or the power source is removed, springs pull the arm back to the left (off) position.

There are not too many of these controllers left, but the concept of reducing the current and voltage to the armature during starting remains the same. Modern methods of controlling speed use electronic controllers to adjust the voltage and therefore, the current to the armature during starting.

Figure 10-25 shows a block diagram approach to an electronic controller used for a shunt motor. Notice that the shunt field in this controller has full DC applied any time the motor is on. The full DC is supplied by a full-wave bridge rectifier circuit. The armature circuit is controlled by a variable DC supply that uses silicon control rectifiers (SCRs) to adjust the amount of DC voltage and current supplied. The SCR circuit is controlled by an SCR firing circuit (either phase or amplitude control) that determines how long each rectifier will conduct. The shorter the on time, the

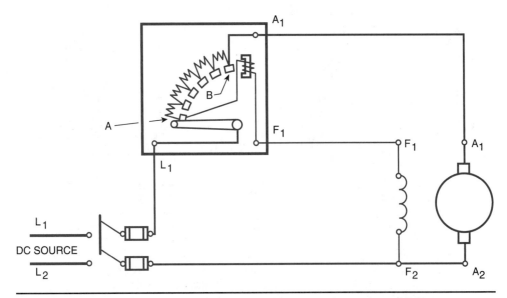

FIGURE 10-24   Three-point starters used for manual starting of DC motors.

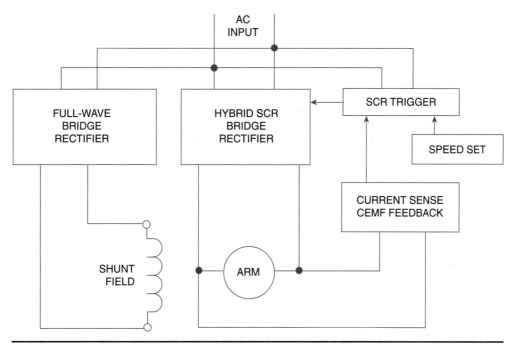

FIGURE 10-25 Block diagram of electronic controller used for a shunt motor.

lower the DC voltage applied to the armature. Usually, there are feedback circuits to sense the armature current and CEMF feedback to determine motor speed. Remember, the faster a DC motor spins, the higher its CEMF. By using resistors across the armature circuit and sense resistors in series with the armature circuit, speed and torque can be controlled by the SCR firing circuit.

MANUAL CONTROL OF DC MOTORS

Manual control can be used for small DC motors where inrush values are not too large. Figure 10-26 shows how a drum switch can be used to operate conventional DC motors.

Figure 10-27 shows a magnetic starter circuit. Note that there is an additional shunt field relay (SFR) relay added. This relay is a low-resistance, series-type relay electrically in series with the shunt field. The SFR contact is in series with the M coil. If the shunt field is disconnected or an open circuit results, the shunt field relay will open the circuit to the M coil and disconnect the power to the armature.

DC MOTORS AND NEC®

Motor conductors are covered in NEC® Article 430.22. General conditions are specified in Article 430.22(A). Exception 11 will determine ampacities for conductors that supply motors

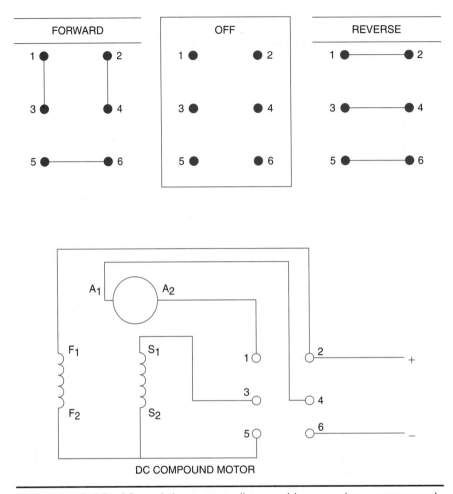

FIGURE 10-26 Manual drum controller used in reversing a compound motor.

from single-phase AC sources. Half-wave rectifiers need 190 percent of full-load ratings and full-wave rectifiers need 150 percent. Table 430.247 lists currents for DC motors with various armature voltage ratings.

Article 430.29 refers to sizing resistor conductors for constant voltage DC motors. If the resistors are used for the accelerating resistors or dynamic braking and are separate from the controllers, Table 430.29 is used to determine ampacity of the connecting conductors.

Article 430.88 refers to DC motors used for adjustable speed applications. If the speed is adjusted by weakening the shunt field, they are not to be started with a weakened field (see exception in NEC®). Also, Article 430.89 requires some DC machines to have speed-limiting devices or speed-limiting means.

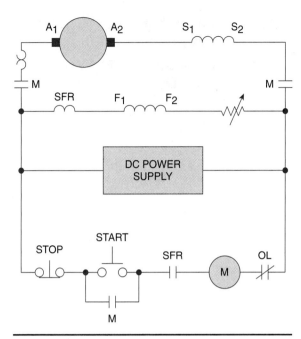

FIGURE 10-27 Magnetic controller used for compound motor control.

Remember, series motors may run away without a mechanical load permanently attached. Do not use belts or chains to connect a series motor to the load.

SUMMARY

This chapter introduced you to the various types of DC generators and motors. DC generation was discussed first to explain how a DC generator works to produce a DC output from a generated AC within the armature coils. Self-excited and separately excited generators were presented as well as series, shunt, and compound generators.

Connections for paralleling DC generators were shown. Methods to connect generators for different output voltage characteristics were explained. A description of the procedure used to flash the field to restore residual magnetism was included.

Motors of various styles were presented to acquaint you with different operating parameters. Methods used to control the speed and direction of DC motors were explained. Finally, there was a review of NEC® articles that pertain to DC motors and generators.

QUESTIONS

1. What is the name of the segmented conductors where the carbon brushes make contact?
2. Explain what is meant by "pulsating DC."
3. What is an advantage of "frog-leg-wound armatures"?
4. How does a self-excited generator begin a voltage buildup?
5. How can you control the amount of current that goes through the series field of a series generator?
6. Why is a field rheostat used?
7. What is the standard connection for a compound generator?
8. How is cumulative compounding established?
9. Explain what is meant by armature reaction.
10. Explain how to "flash the field."
11. What essential connection is required when paralleling DC compound generators?
12. Explain how to reverse DC motors.

13. How is effective voltage influenced by motor speed?

14. Explain how speed is controlled on a permanent magnet motor.

15. Which DC motor has the greatest torque? Explain why.

16. Why shouldn't you remove the shunt field from a shunt motor?

17. What is the current NEC® book value for a 10-HP DC motor with 120-V armature voltage?

APPENDIX

A

WIRING DIAGRAMS

ELECTRICAL SYMBOLS

CONTROL CIRCUIT

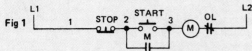

Fig 1 — Three Wire Control Giving Low Voltage Protection Using Single Two Button Station

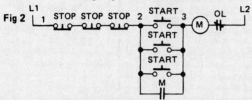

Fig 2 — Three Wire Control Giving Low Voltage Protection Using Multiple Two Button Stations

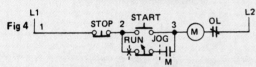

Fig 3 — Three Wire Control Giving Low Voltage Protection with Safe-Run Selector Switch

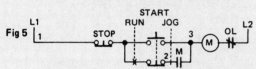

Fig 4 — Three Wire Control for Jog or Run Using Start Stop Push Buttons and Jog-Run Selector Switch

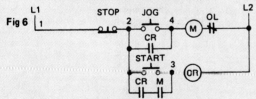

Fig 5 — Control for Jog or Run Using Stop Push Button and Jog-Run Selector Push Selector Switch. Selector Push Contacts are Shown for "Run" (Three Wire Operation). Rotate Switch Sleeve and Selector Contact Opens Between "2" and "Stop" Button (Two Wire Operation).

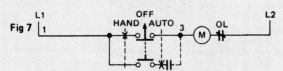

Fig 6 — Three Wire Control for Jogging, Start, Stop Using Push Buttons

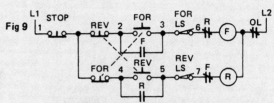

Fig 7 — Two Wire Control Giving Low Voltage Release Only Using Hand-Off-Auto Selector Switch

Courtesy of Furnas Electric Company

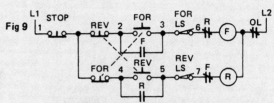

Fig 8 — Two Wire Control for Reversing Jogging Using Single Two Button Station

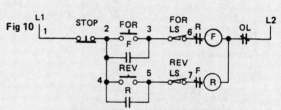

Fig 9 — Three Wire Control for Instant Reversing Applications Using Single Three Button Station

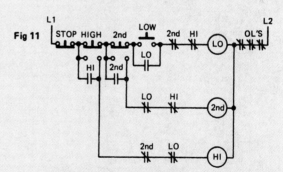

Fig 10 — Three Wire Control for Reversing After Stop Using Single Three Button Station

Fig 11 — Control for Three Speed with Selective Circuitry to Insure the Stop Button is Pressed Before Going to a Lower Speed

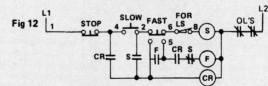

Fig 12 — Three Wire Control for Two Speed with a Compelling Relay to Insure Starting on Slow Speed

310 ELECTRIC MOTORS AND MOTOR CONTROLS

CONTROL CIRCUITS, CIRCUITS WITH TRANSFORMERS

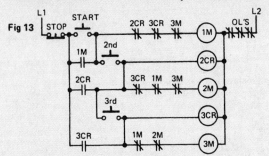

Control for Three Speed with a Compelling Relay to Insure Starting on Low Speed

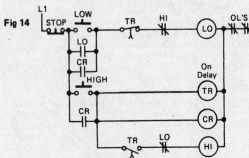

Control for Two Speed to Provide Automatic Acceleration from Low to High Speed

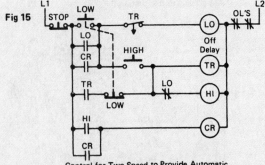

Control for Two Speed to Provide Automatic Deceleration from High to Low Speed

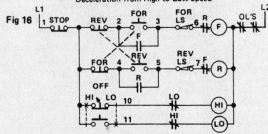

Control for Two Speed Reversing Starter Using Forward, Reverse, Stop Push Buttons and High-Low-Off Selector Switch

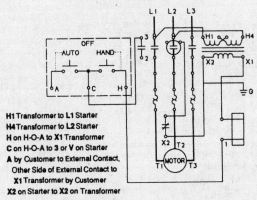

Size 0–2½ Starter with Transformer and 3 Position Selector Switch

H1 Transformer to L1 Starter
H4 Transformer to L2 Starter
H on H-O-A to X1 Transformer
C on H-O-A to 3 or V on Starter
A by Customer to External Contact, Other Side of External Contact to X1 Transformer by Customer
X2 on Starter to X2 on Transformer

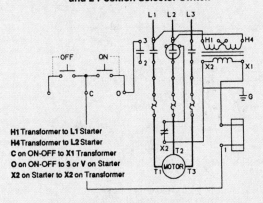

Size 0–2½ Starter with Transformer and 2 Position Selector Switch

H1 Transformer to L1 Starter
H4 Transformer to L2 Starter
C on ON-OFF to X1 Transformer
O on ON-OFF to 3 or V on Starter
X2 on Starter to X2 on Transformer

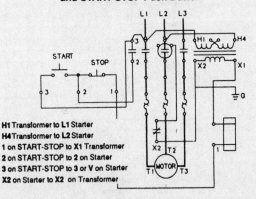

Size 0–2½ Starter with Transformer and START-STOP Push Button

H1 Transformer to L1 Starter
H4 Transformer to L2 Starter
1 on START-STOP to X1 Transformer
2 on START-STOP to 2 on Starter
3 on START-STOP to 3 or V on Starter
X2 on Starter to X2 on Transformer

Courtesy of Furnas Electric Company

Wiring Diagrams

COIL CONNECTIONS, PILOT CONTROL

AC Coil
DC Coil

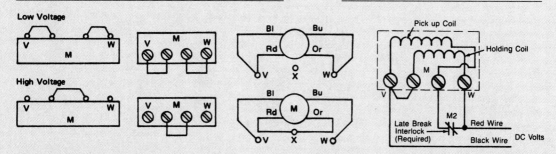

Non Reversing Pilot Control

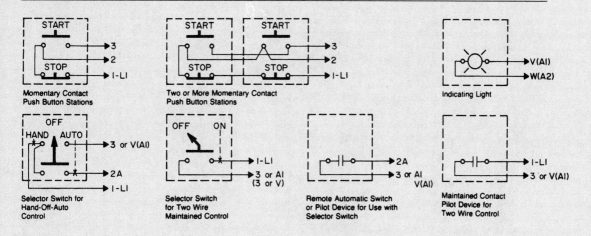

Terminal Markings shown in ( ) indicate IEC Style.
For separate control voltage source remove Jumper A shown in individual wiring diagrams. Connect separate voltage source to terminal 1 on the pilot device as shown and to the terminal X2 on the overload relay, or W(A2) on the coil if there is no overload.

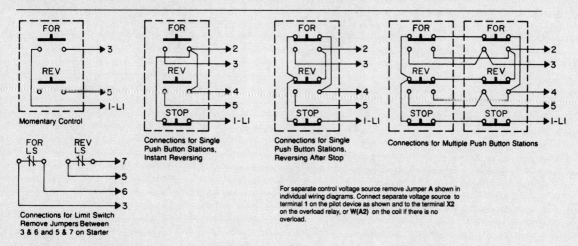

Courtesy of Furnas Electric Company

APPENDIX B

FORMULAS

DC CIRCUITS

$R = E/I, E = I \times R, I = E/R, W = E \times I$

- R = Resistance a circuit has to current flow, measured in ohms
- E = Electromotive force applied to a circuit, measured in volts
- I = Current flow in an electric circuit, measured in amperes
- W = Watts: the amount of electrical work done, measured in watts

AC CIRCUITS

$$L = \frac{3.19\, N^2 A \mu}{10^8 l}$$

- L = Inductance: the property of a circuit that causes it to oppose a change in current, measured in Henrys
- 3.19 = Constant applied to English units
- N = Number of turns or loops of a coiled conductor
- A = Area of the cross section of the core of the coil, measured in inches
- μ = mu: the permeability of the core material
- 10^8 = Constant for coils
- l = Length of the coil, measured in inches

(OR)

$$L = \frac{1.26\, \mu A N^2}{lm \times 10^8}$$

- A = Square centimeters for this formula
- lm = Mean length of the coil

$$C = \frac{.224\, AKN}{10^6\, t}$$

- C = Capacitance: the property of a circuit that causes it to oppose a change in voltage, measured in farads

313

.224 = Constant applied for measurement in English units

A = Area of the capacitive plates measured in square inches

K = Dielectric constant dependent on dielectric type

N = Number of capacitive actions, depends on the number of plates

10^6 = Constant applied to English unit measurements

t = Thickness of the dielectric in mil inches

X_L = $2\pi FL$

X_L = Inductive Reactance: the opposition to AC current caused by induction, measured in ohms

2π = 2 · 3.1416 — Constant for formula

F = Frequency: the frequency of the applied voltage, measured in Hertz

L = Inductance: measured in Henrys

$X_C = \dfrac{1}{2\pi FC}$

X_C = Capacitive reactance: the opposition to AC current caused by capacitance, measured in ohms

2π = (2) (3.1416) constant for formula

F = Frequency: the frequency of the applied voltage, measured in ohms

C = Capacitance: measured in farads

Z = $\sqrt{R^2 + (X_L - X_C)^2}$

Z = Impedance: the total opposition to current flow in an AC circuit caused by (R) resistance, (X_L) inductive reactance, (X_C) capacitive reactance

SINGLE-PHASE POWER

$VA = E \times I$

VA = Volt amp: the apparent power used in an AC system

$W = VA \times PF$ or $W/VA = PF$

W = Watts: the true power consumed in an AC circuit

PF = Power factor: a factor applied to the apparent power that will yield the true power consumed, usually expressed as a percentage

SINGLE-PHASE MOTORS

$HP = \dfrac{33{,}000 \text{ ft-lbs}}{\min}$

or

$HP = \dfrac{FRN}{5252}$

$HP = \dfrac{\text{output watts}}{746}$

HP = Horsepower: the measurement of the mechanical output work done by a motor

F = Force: in pounds, measured on a scale

R = Radius: the radius of the circular arc of motion of the rotation of the motor shaft

N = Number of revolutions of the motor shaft measured in RPM

746 = Constant used to convert watts to horsepower

% efficiency = $\dfrac{\text{watts output}}{\text{watts input}}$

% efficiency = the percentage of the input watts that are converted to useful output work

$N_S = \dfrac{120 \cdot F}{\text{No. of poles}}$

N_S = Synchronous speed of an AC rotating magnetic field

120 = Constant conversion factor for formula
F = Frequency: the frequency of the AC voltage
No. of poles = The number of magnetic poles established in the motor

% slip = $\dfrac{N_S - N_R}{N_S}$

% slip = The percent slip is the percentage of N_S that the rotor of an induction motor slips behind N_S

N_S = Synchronous speed

N_R = Actual rotor speed at time of measurement

% Regulation = $\dfrac{N_{(NL)} - N_{(FL)}}{N_{FL}}$

% Regulation = The percentage of full-load speed that rotor will fluctuate from no load to full load

N_{NL} = Speed measured with no load, measured in RPM

N_{FL} = Speed measured with full load, measured in RPM

MOTOR TORQUE RELATIONSHIP

$Td^2 = K\Phi I_R$

Td^2 = Torque developed is dependent on the motor design and the relation between the stator and rotor. It is usually measured in foot-pounds

K = A constant applied, based on the motor design

Φ = Magnetic field strength of stator

I_R = Magnetic field produced by the rotor current

THREE-PHASE POWER FORMULAS

$3\Phi VA = E \times I \times 1.73$

$3\Phi VA$ = Three-phase VA or apparent power, measured in volt-amps

1.73 = A constant ($\sqrt{3}$) applied to three-phase circuits to calculate total power

$3\Phi W = E \times I \times 1.73 \times PF$

$3\Phi W$ = Three-phase true power measured in watts

Phase (coil) voltage = .58 × line voltage for wye connection

Phase current = line current for wye connection

Phase (coil) current = .58 × line current for delta connection

Phase voltage = line voltage for delta connection

TRANSFORMERS

$E_p/E_s = N_p/N_s = I_s/I_p,$ $E_p \cdot I_p = E_s \cdot I_s$
assume 100% efficient

E_p = Voltage of the transformer primary

E_s = Voltage of the transformer secondary

N_p = Number of turns of the transformer primary

N_s = Number of turns of the transformer secondary

I_s = Current leaving the transformer secondary

I_p = Current entering the transformer primary

VOLTAGE DROPS

$$V_d = \frac{2 \times K \times L \times I}{D}$$

$$3\Phi V_d = \frac{2 \times K \times L \times I}{D} \times .886$$

V_d = Volt drop of the entire circuit caused by wire resistance

K = Resistivity figure of the wire: use K of 12 for copper conductors loaded more than 50 percent of rated ampacity

Use 11 for K if copper conductors are loaded less than 50 percent of rated ampacity

Use 18 for K if aluminum conductors are used

L = Length of the circuit one way, measured in feet

I = Current in the circuit conductors, measured in amperes

D = Area of the cross section of wire, measured in circular mils

.866 = Constant factor used in three-phase circuits

APPENDIX C

SYMBOLS

FUSE

POWER OR CONTROL

STANDARD-DUTY SELECTOR

2-POSITION   3-POSITION

HEAVY-DUTY SELECTOR

2-POSITION
1-CONTACT CLOSED

3-POSITION
1-CONTACT CLOSED

2-POSITION SELECTOR PUSH BUTTON
1-CONTACT CLOSED

CONTACTS	SELECTOR POSITION			
	A		B	
	BUTTON		BUTTON	
A B 1○ ○2 3○ ○4	FREE	DEPRES'D	FREE	DEPRES'D
1–2	1			
3–4		1	1	1

PUSH BUTTONS
MOMENTARY CONTACT

SINGLE CIRCUIT
NO   NC

DOUBLE CIRCUIT
NO & NC

MUSHROOM HEAD   WOBBLE STICK

ILLUMINATED

MAINTAINED CONTACT

TWO SINGLE CONTACT

ONE DOUBLE CONTACT

PILOT LIGHTS
INDICATE COLOR BY LETTER

NON-PUSH-TO-TEST

PUSH-TO-TEST

CONTACTS
INSTANT OPERATING

WITH BLOWOUT (ARC SUPRESSION COIL)
NO   NC

WITHOUT BLOWOUT
NO   NC

TIMED CONTACTS—CONTACT ACTION RETARDED WHEN COIL IS

ENERGIZED
NO   NC

DEENERGIZED
NO   NO

COILS

SHUNT   SERIES

OVERLOAD RELAYS

THERMAL   MAGNETIC

Symbols

INDUCTORS

IRON CORE

AIR CORE

TRANSFORMERS

AUTO

IRON CORE

AIR CORE

CURRENT

DUAL VOLTAGE

AC MOTORS

SINGLE-PHASE

3-PHASE SQUIRREL CAGE

WOUND ROTOR

DC MOTORS

ARMATURE

SHUNT FIELD
(SHOW 4 LOOPS)

SERIES FIELD
(SHOW 3 LOOPS)

COMM OR COMPENS FIELD
(SHOW 2 LOOPS)

WIRING

NOT CONNECTED

CONNECTED

POWER

CONTROL

WIRING TERMINAL

GROUND

CONNECTIONS

MECHANICAL

MECHANICAL INTERLOCK

RESISTORS

FIXED
-[RES]-

HEATING ELEMENT
-[H]-

ADJUSTED BY FIXED TAPS
-[RES]-

RHEOSTAT POT OR ADJUSTMENT TAP
-[RH]-

CAPACITORS

FIXED

ADJUSTABLE

SPEED (PLUGGING)

ANTIPLUG

BELL

BUZZER

HORN SIREN, ETC.

METER

INDICATE TYPE BY LETTER

METER SHUNT

HALF-WAVE RECTIFIER

FULL-WAVE RECTIFIER

BATTERY

–|·|⊢

SEMICONDUCTORS

DIODE

SILICONE-CONTROLLED RECTIFIER (SCR)

LIGHT-EMITTING DIODE

SPST, NO

| SINGLE BREAK | DOUBLE BREAK |

SPST, NC

| SINGLE BREAK | DOUBLE BREAK |

SPDT

| SINGLE BREAK | DOUBLE BREAK |

Symbols

DPST, 2 NO

SINGLE BREAK

DOUBLE BREAK

DPST, 2 NC

SINGLE BREAK

DOUBLE BREAK

DPDT

SINGLE BREAK

DOUBLE BREAK

SYMBOLS FOR SOLID-STATE CONTROL DEVICES

LIMIT SW NC

LIMIT SW NO

INPUT "COIL"

OUTPUT NC

METHOD OF SWITCHING ELECTRICAL CIRCUITS WITHOUT THE USE OF CONTACTS = STATIC CONTROL. THE SYMBOLS SHOWN ARE USED BY ENCLOSING THEM IN A DIAMOND SHAPE.

TERMS

SPST—SINGLE POLE SINGLE THROW

SPDT—SINGLE POLE DOUBLE THROW

DPST—DOUBLE POLE SINGLE THROW

DPDT—DOUBLE POLE DOUBLE THROW

NO—NORMALLY OPEN

NC—NORMALLY CLOSED

Courtesy of Walter Alerich. *Electric Motor Control,* 5th ed., Clifton Park, NY: Delmar (1993).

APPENDIX

TROUBLESHOOTING

INTRODUCTION

Troubleshooting is the process used to help track down problems in malfunctioning equipment or to test and modify new equipment during installation. Electrical troubleshooting employs the use of test equipment as well as your knowledge and skill in dealing with electrical circuits. In this book, you have gained skill and knowledge in the operations of motors and controls. You must also rely on your senses of sight, sound, smell, and feel. As you gain experience in electrical systems, you will become accustomed to the usual heat produced by a normally operating motor. While running, the motor should not be so hot that you cannot put your hand on it. The physical appearance of the motor and its controller will also be helpful in diagnosing the problems. If the motor is dirty or covered with grease, the likelyhood of overheating is great. If the motor pulley wobbles or has a great deal of end play in the shaft, the bearings or bushings may be worn out and in need of replacing. If the motor smells like burned electrical insulation, it is probably ready for replacement. Likewise, the motor controls may be too hot, or may chatter or be burned. Use your senses to help you in the problem solving even before you work on the electrical system.

As part of the troubleshooting procedure, a thorough inspection is necessary to try to determine the cause of the malfunction. Check the motor and the control for signs of excessive heat, bad bearings, or gritty noise as the motor shaft spins slowly. Check for loose connections or loose parts on the motor and the drive mechanisms and check for clear air circulation paths for motor cooling.

Inspection during troubleshooting may reveal poor maintenance. Regular maintenance should include a cleaning of the motor. Clean the external surface of the motor to help in the heat dissipation. Clean the air vents to provide cooling air flow. Check for proper tension on the drive parts; then tighten the motor mountings. Check all electrical connections in the motor and to the controller for good solid connections. Make sure the bearings are lubricated with the proper lubricant and that the lubricant is clean and dry.

Be sure the motor and the controller are suited for the environment. Make sure the motor is designed for damp locations if it is to be exposed to fluids. Be sure the motor has the properly rated housing if it is to be used in hazardous locations. The enclosure must also be rated for the location where it is to be installed.

TROUBLESHOOTING PROCEDURES

After all the mechanical causes for malfunction have been checked, the electrical troubleshooting can begin. Be sure to follow OSHA rules for safe work practices. Be sure to lock out and tag your work according to policies and procedures that are established. If at all possible, work with the circuit deenergized to test for problems.

When using electrical test equipment, be sure the test equipment is rated for the voltage and current of the circuit you are working on. Be sure the test equipment is functioning properly and that you are sure of the test equipment indications. If the circuit you are testing has an unknown voltage, start with the highest scale possible. If the current values are unknown, likewise start with the highest scale possible.

As in any skilled craft, knowledge of the system and the proper function of the system or control is essential if you are to diagnose and repair a malfunctioning system. Your first task is to determine what the system is supposed to do. If there are documents or schematic diagrams, study them to determine the normal operating sequence. Consult the equipment operators, if available, to determine what the symptoms of the malfunction were and what the normal operation should be.

After gathering as much information as possible, begin the electrical troubleshooting procedure by checking for the most simple problems first. Be sure that you have power at the proper points in the circuit. Check for blown fuses, tripped circuit breakers, and tripped overloads. If the control circuit uses a control transformer, check to be sure the fuse in the secondary circuit is good. Verify that the transformer is functioning and that the control voltage is present at the control circuit points. After you have established that the proper voltage is delivered to the circuit and that all safety lockouts or door access switches are in place, then begin isolating the problem. (See Figure D-1.)

First determine if the problem is in the power circuit or in the control circuit. To do this, disconnect the power to the motor at the bottom of the motor starter. Attempt to operate the control circuit to close the circuit to the controller magnetic coil. Does the controller energize and provide power at the bottom of the motor controller? (See Figure D-2.) If the control coil on the first rung does energize, the problem is in the control circuit leading up to the control coil. To determine if the problem is in the stop stations or emergency circuits or in the starting control devices, use a voltmeter to determine if there is a complete circuit through the stop stations. Move the voltmeter one step at a time to determine where the control voltage is lost. (See Figure D-3.) If the voltage is available at the right side of the stop stations, then the problem is in the start controls or in the magnetic coil itself. Check with a voltmeter at the right side of each start control to determine if the control voltage is present when the start station is pressed. If voltage is present, then the problem is in the magnetic coil or the mechanical action of the controller. Double-check to be sure the overload contacts are closed. The last step to determine the control problems is to measure directly across the magnetic coil. If there is voltage present and the armature assembly is free to move, then the

Troubleshooting

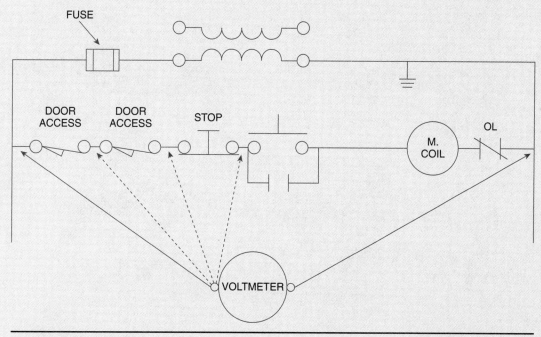

FIGURE D-1

coil must be replaced. It is good practice to try to determine why the coil failed, if it burned out. See the section on controllers in this book to determine possible causes of failure.

If the control circuit is functioning properly and the motor controllers are being energized properly, then the problem can be traced to the power circuit. Be sure to check the motor and the driven machinery to find if there are physical reasons the motor may not function. The power circuit is usually easier to troubleshoot. If the proper power is delivered to the bottom of the controller and the motor still does not function, the problem is between the controller and the motor, or the motor itself is not functioning. Check the motor lead connection housing to determine if the leads have been disconnected. Check for power at the motor leads with the motor controller energized. Did a connection between the controller and the motor fail? Disconnect the motor and test for proper voltage at the motor again. Sometimes when testing the motor power, the motor windings act as a source of voltage through induction and the readings at the motor can be misleading.

When checking the power circuit, remember to test the incoming power and test the fuses in the disconnect switch. The best way to test the fuses is to disconnect power, use a fuse puller, and pull the fuses out of the circuit to test them with an ohmmeter. This out-of-the-circuit testing avoids confusion when testing them in the circuit and reading through other circuit components. Reinstall the fuses and test them again under load by reading across them with a voltmeter. (See Figure D-4.) There should be no voltage drop across a good fuse. If all the phases are delivered properly, the motor is the problem.

Testing the actual motor for faults can lead you to the decision to replace the motor. If the motor is a specialized motor, the decision may be to repair a minor fault. Usually the

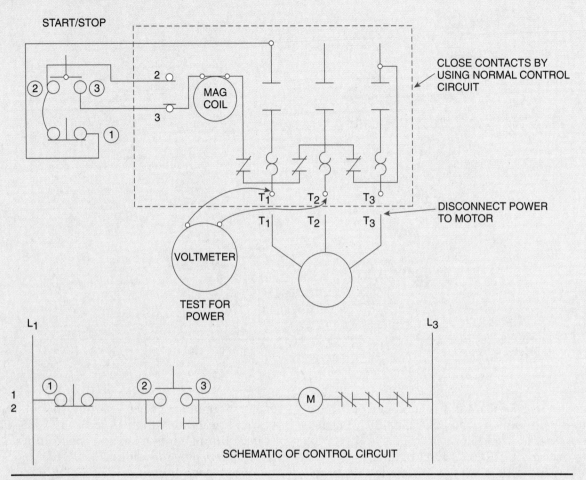

FIGURE D-2

maintenance electrician will replace the motor in question to lessen down time and get the system back in operation. Often in a maintenance organization, the motor can have minor parts replaced to make the motor usable for future needs.

When testing the motor for faults, be sure to gather as much information as possible about the motor condition before replacing the motor. Does the motor run hotter than normal? Does the motor fail to start? Does a two-speed motor only run in one speed?

If the control system is operating properly and the motor still does not operate properly, there are several items to check. Check the voltage at the motor leads for proper values. The voltage delivered to the motor should be within 10 percent of the rated nameplate rating. If the voltage is too high, the motor will have excessive current and run hot. If the voltage is too low, the motor will not run at rated speed or torque and will also run hot and eventually ruin the motor. The motor voltage rating should match the applied voltage. Either change the motor or find the problem with the supply voltage.

If the motor voltage is correct for the motor connected, test the motor bearings for

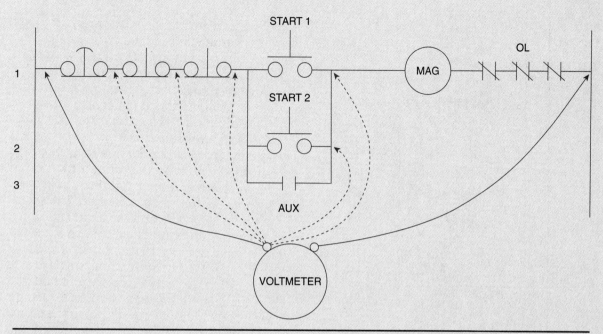

FIGURE D-3

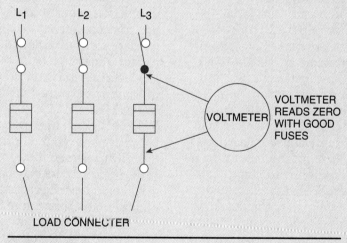

FIGURE D-4

free rotation. Lock off the motor disconnect and spin the motor shaft by hand. It should spin freely and easily without any bearing drag. If the bearings present a drag to the motor or if the bearings are loose in the bearing race, the motor will have extra load added to the mechanical load and will draw too much current and overheat. Replace the motor or replace the motor bearings.

If the motor voltage and the bearings are satisfactory, test the motor under load. Connect the motor back to the source of power. Be sure

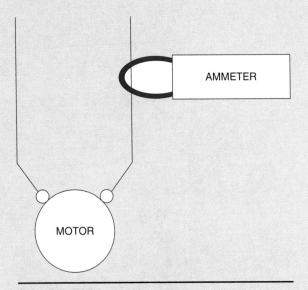

FIGURE D-5

to double check the wiring and the connectors from the motor to the line. Operate the motor under the normal load conditions. Use a clamp-on ammeter such as in Figure D-5. Test all the leads to the motor and verify that the motor is not drawing more that the nameplate current. If you are measuring the current to a DC motor, be sure a use a clamp-on meter that will measure DC current. The standard clamp-on meter does not measure DC current. If the current draw is too large, check the mechanical load. Has additional load been added? The motor may have been undersized for the application if the current draw is consistently too high.

Other things to check if there is premature motor failure are determined by the operating conditions of the motor. Be certain that a motor that is started and stopped frequently is designed for the use. Frequent starting, stopping, or jogging duty is not recommended for standard-duty motors as the motor does not get a chance to cool off. Motors that operate in severe conditions, such as extreme heat or cold, should be designed for the application. These conditions can change the operating characteristics of the motor. Higher than normal temperatures will prevent the motor from cooling properly. Extreme cold can affect the motor bearings and add drag to the motor during starting. Damp or wet locations will also help deteriorate the windings of a motor if it is not designed to keep the windings dry. Motors that are totally enclosed fan cooled (TEFC) especially should be kept clean. The outside of the motor has cooling fins to help dissipate heat when an external fan blows air over the outside of the motor. The fan must be free to blow air over the motor, and the motor housing should be free of dirt and dust.

If the motor fails to come up to the proper speed, verify that the control system is operating and not leaving the motor in a reduced voltage starting mode. Also on single phase motors, check to be sure the starting switches are operating or if the capacitor is still good on capacitor start motors. If the load on the motor is too heavy or requires more staring torque than the motor can deliver, a different type of motor may be required.

To test capacitors used in capacitor motors, use the following procedure. Check the capacitors for burn marks, leaking electrolyte, punctures, or signs of ruptured container. If the capacitor appears to be in good condition, then an electrical test can be completed. Remove the capacitor from the circuit. CAUTION: THE CAPACITOR MAY STILL CONTAIN STORED ENERGY. Discharge the capacitor after the power is removed by shorting across the terminals with a wire. If there is still energy stored in the capacitor, there may be a large spark during discharge. After the capacitor is discharged, use an ohmmeter to measure across the terminals. It is best to use an ohmmeter with an analog face and a needle indicator. Use the R times 10 scale if available. If the capacitor has failed in the shorted position, the ohmmeter scale will indicate zero ohms and stay in that condition. If the capacitor has failed in the open condition, then the ohmmeter will stay in the infinite ohms condition. The capacitor may also be leaky, which would be indicated by the ohmmeter moving to the zero ohms position and then slowly moving back to the middle

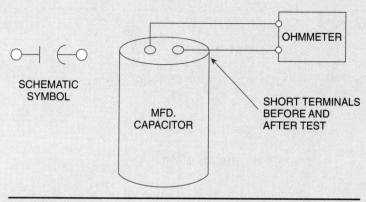

FIGURE D-6

ohms range and staying there. All three of these conditions would require replacement of the capacitor. An indication for a good capacitor is when the ohmmeter first swings to the shorted, or zero ohms, condition and then charges the capacitor so that the ohms eventually read infinite again. This indicates that the capacitor is charged and there is no further flow of DC through the capacitor. Be sure to discharge the capacitor after the test. (See Figure D-6.)

Basic items to check on electronic drives can lead you to simple corrections. Usually the newer drives have a lot of self-diagnostics. The block approach to troubleshooting can be used here. Find the section that is not functioning and replace that section if possible. In variable-frequency AC electronic drives there are several blocks to consider. The general blocks are shown in Figure D-7. The AC to DC converter section converts incoming AC to DC. Check at the output bus of the converter section to determine if the converter is operating. Measure with the appropriate voltmeter. The DC bus voltage should be about 1.4 times the AC RMS voltage if all the rectifiers are working. There is a controller board where the control decisions and adjustable control parameters are set. The only practical way to test this block is to make some changes on the control settings. Remember to mark or register the original settings. Because of the variety of options available on the variable-speed drives, the best source to verify correct settings is the operating manual. The last section of the electronic drive that can be generally tested is the inverter section. The DC to AC inverter changes the DC bus voltage back to an AC voltage output. The output board is made up of electronic drivers for large power applications. These can be SCRs, bipolar transistors, or insulated gate bipolar transistors (IGBTs). If these devices are working, there should be a voltage output at the terminals that can be measured with an oscilloscope. The frequency and amplitude should vary as the controls are adjusted.

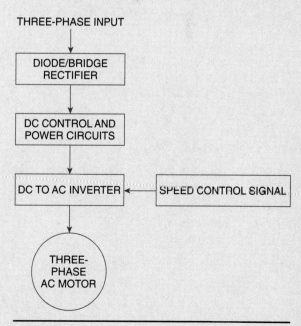

FIGURE D-7

TROUBLESHOOTING FLOWCHARTS

PROBLEM: CONTACT WELDING OR PREMATURE WEAR

IS THE CONTACTOR PROPERLY RATED FOR THE MOTOR HORSEPOWER AND VOLTAGE APPLIED?

    YES | NO
             └──→ REPLACE WITH PROPERLY RATED CONTACTOR.

IS THERE JOGGING DUTY OR RAPID STARTING AND STOPPING?

    NO | YES
             └──→ INCREASE THE SIZE OR RATING OF THE CONTACTOR.

IS THE PROPER VOLTAGE RATED COIL USED IN THE CONTACTOR?

    YES | NO
             └──→ REPLACE WITH THE PROPERLY RATED COIL SO THAT THE CONTACTS DON'T SLAM SHUT OR FAIL TO CLOSE.

IS THERE NONCONDUCTING MATERIAL PRESENT ON THE CONTACT SURFACE?

    NO | YES
             └──→ REMOVE OIL, GREASE, DIRT, OR SAND THAT PREVENTS FULL CONTACT SURFACE CLOSURE.

IS THERE ENOUGH CONTACT PRESSURE (SLIDE PAPER BETWEEN CLOSED CONTACTS)?

    YES | NO
             └──→ REPLACE CONTACTS AND SPRING ASSEMBLY.

IS THERE ARC CONTROL APPLIED TO THE CONTACTS?

    NO | YES
             └──→ VERIFY THAT ARC QUENCHING IS WORKING.

CHECK THE LOAD FOR EXCESSIVE CURRENT OR CONSTANT OVERLOAD.

Troubleshooting

PROBLEM: CONTACTOR FAILS TO HOLD

IS THE CONTACTOR COIL ENERGIZING AND MOVING THE CONTACT ASSEMBLY?

YES | NO
→ CHECK THE CONTROL CIRCUIT AND THE COIL FOR AN OPEN CIRCUIT.

DOES THE HOLDING OR SEALING CIRCUIT CLOSE?

YES | NO
→ CHECK THE CONTROL CIRCUIT FOR OPERATIONS OR CHECK THE WIRING TO THE CONTACTOR AUXILIARY CONTACTS.

IS THERE A PHYSICAL STOP THAT KEEPS THE CONTACTS FROM CLOSING?

NO | YES
→ REMOVE THE PHYSICAL BARRIER, AS LONG AS IT IS NOT A SAFETY FEATURE.

DOES THE VOLTAGE TO THE MAGNETIC COIL STAY CONSTANT?

YES | NO
→ CHECK BRANCH CIRCUIT FOR VOLTAGE VARIATIONS.

IS THE MAGNETIC ASSEMBLY DIRTY OR STICKY?

NO | YES
→ CLEAN AND LUBRICATE WITH A NONCONDUCTING CLEANER.

ARE THE OVERLOAD RELAYS STAYING CLOSED?

YES | NO
→ CHECK THE OVERLOAD RELAYS FOR INTERMITTENT OPERATION.

REPLACE THE CONTACTOR IF NECESSARY.

PROBLEM: TIMER MALFUNCTION

IS THE DEFECTIVE TIMER A MECHANICAL TIMER?

NO | YES → CHECK FOR FREE MOVEMENT OF THE MECHANISM.

IS THE TIMER A DASHPOT TIMER?

NO | YES → CHECK FOR PROPER AIR FLOW OR OIL FLOW THROUGH THE ORIFICES AND CHECK TO BE SURE THE DIAPHRAGM IS NOT RUPTURED.

IS THE TIMER AN ELECTRONIC TIMER?

NO | YES → CHECK FOR THE PROPER VOLTAGE APPLIED, PROPER CONTACT SEQUENCE, PROPER SETTING OR APPLICATION.

CHECK FOR PROPER INPUT TIMER INITIATION SIGNAL OR REPLACE TIMER IF NECESSARY.

PROBLEM: OVERLOAD RELAY TRIPPED

IS THERE A MECHANICAL OVERLOAD ON THE MOTOR?

    NO | YES
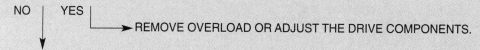
    ↓   → REMOVE OVERLOAD OR ADJUST THE DRIVE COMPONENTS.

IS THE PROPER VOLTAGE AVAILABLE AT THE MOTOR?

    YES | NO
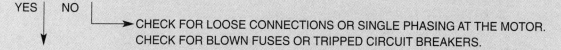
    ↓   → CHECK FOR LOOSE CONNECTIONS OR SINGLE PHASING AT THE MOTOR. CHECK FOR BLOWN FUSES OR TRIPPED CIRCUIT BREAKERS.

IS THE OVERLOAD RELAY SET OR RATED FOR THE MOTOR INSTALLED?

    YES | NO
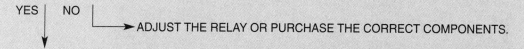
    ↓   → ADJUST THE RELAY OR PURCHASE THE CORRECT COMPONENTS.

ARE THERE LOOSE CONNECTIONS OR ARE THE MOTOR CURRENTS IMBALANCED?

    NO | YES
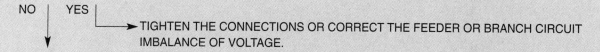
    ↓   → TIGHTEN THE CONNECTIONS OR CORRECT THE FEEDER OR BRANCH CIRCUIT IMBALANCE OF VOLTAGE.

TEST THE ACTUAL TRIP POINT OF THE RELAY AND REPLACE IF NECESSARY.

PROBLEM: BURNED MAGNETIC COIL

REMOVE POWER FOR THE FIRST TEST!

IS THE ARMATURE ASSEMBLY FREE TO MOVE?

    YES | NO
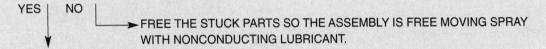
→ FREE THE STUCK PARTS SO THE ASSEMBLY IS FREE MOVING SPRAY WITH NONCONDUCTING LUBRICANT.

IS THE COIL RATING USED CORRECT FOR THE AVAILABLE VOLTAGE?

    YES | NO
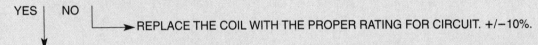
→ REPLACE THE COIL WITH THE PROPER RATING FOR CIRCUIT. +/−10%.

IS THE PROPER VOLTAGE AVAILABLE AT THE COIL DURING OPERATION?

    YES | NO
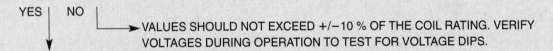
→ VALUES SHOULD NOT EXCEED +/−10 % OF THE COIL RATING. VERIFY VOLTAGES DURING OPERATION TO TEST FOR VOLTAGE DIPS.

IS THE CONTROLLER IN A HIGH AMBIENT TEMPERATURE?

    NO | YES
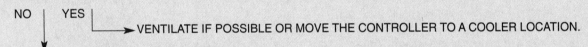
→ VENTILATE IF POSSIBLE OR MOVE THE CONTROLLER TO A COOLER LOCATION.

IS THERE POOR CONTACT TO THE COIL CAUSED BY LOOSE TERMINALS OR CORROSION?

    NO | YES
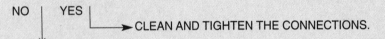
→ CLEAN AND TIGHTEN THE CONNECTIONS.

ARE THE SHADING RINGS INTACT, AND IS THERE GOOD CLOSURE ON THE MAGNETIC PIECES?

    YES | NO
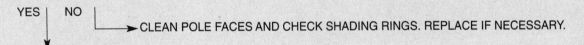
→ CLEAN POLE FACES AND CHECK SHADING RINGS. REPLACE IF NECESSARY.

REPLACE DEFECTIVE COIL WITH NEW CORRECTLY RATED COIL.

GLOSSARY

Accelerating Relay Any type of relay used to aid in starting a motor or to accelerate a motor from one speed to another. Accelerating relays may function by motor armature current (current limit acceleration), armature voltage (CEMF acceleration), or definite time (definite time acceleration).

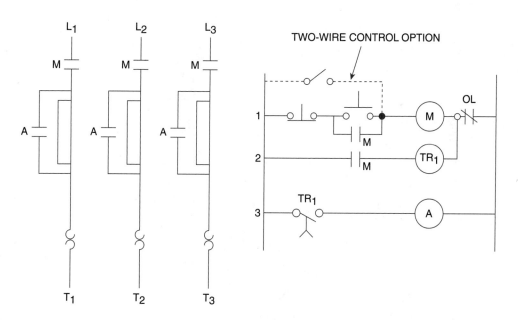

ELECTRIC MOTORS AND MOTOR CONTROLS

Accessory (control use) A device that controls the operation of magnetic motor control. (Also see *Master Switch, Pilot Device,* and *Push Button*)

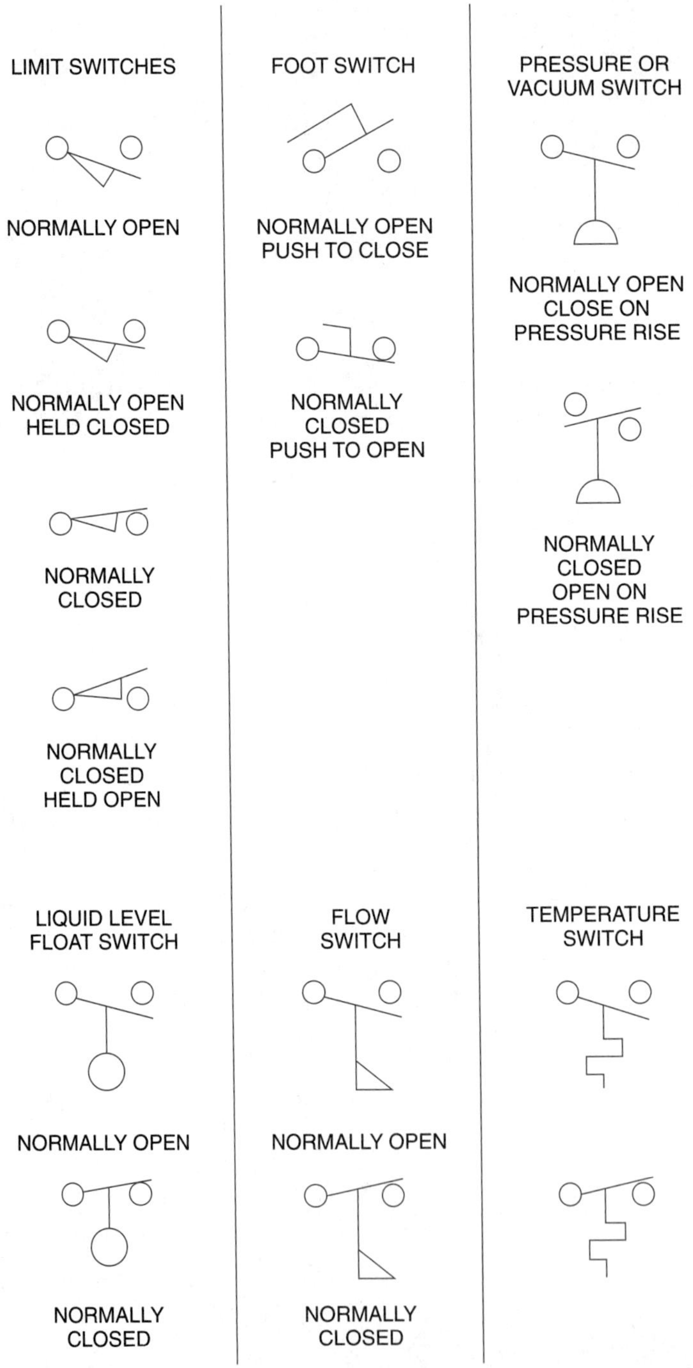

LIMIT SWITCHES

NORMALLY OPEN

NORMALLY OPEN HELD CLOSED

NORMALLY CLOSED

NORMALLY CLOSED HELD OPEN

FOOT SWITCH

NORMALLY OPEN PUSH TO CLOSE

NORMALLY CLOSED PUSH TO OPEN

PRESSURE OR VACUUM SWITCH

NORMALLY OPEN CLOSE ON PRESSURE RISE

NORMALLY CLOSED OPEN ON PRESSURE RISE

LIQUID LEVEL FLOAT SWITCH

NORMALLY OPEN

NORMALLY CLOSED

FLOW SWITCH

NORMALLY OPEN

NORMALLY CLOSED

TEMPERATURE SWITCH

Across-the-Line Method of motor starting that connects the motor directly to the supply line on starting or running. (Also called *Full-Voltage Control*)

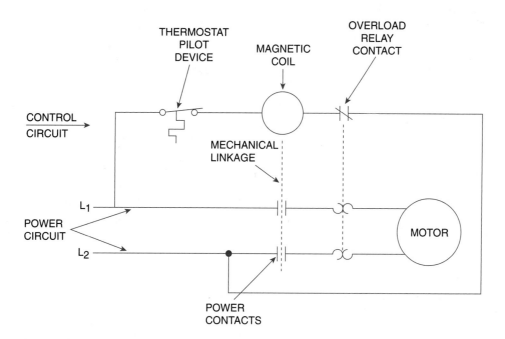

AC Synchronous Motor An AC motor that has a special rotor that is excited by DC, creating a magnetic field that allows the rotor to follow the stator at the same speed of the stator (synchronous speed).
Alternating Current (AC) Current changing both in magnitude and direction; most commonly used current.
Ambient Temperature The temperature surrounding a device.
Ampacity The maximum current rating of a wire or cable.
Ampere Unit of electrical current.
Analog Signal Having the characteristic of being continuous and changing smoothly over a given range, rather than switching on and off between certain levels.
Armature The part of an AC apparatus where a magnetic field is induced to create its own magnetic field.
ASA American Standards Association.
Automatic Self-acting, operating by its own mechanism when actuated by some triggering signal; for example, a change in current strength, pressure, temperature, or mechanical configuration.

Automatic Starter A self-acting starter that is completely controlled by master or pilot switches or other sensing devices; designed to control automatically the acceleration of a motor during the acceleration period.

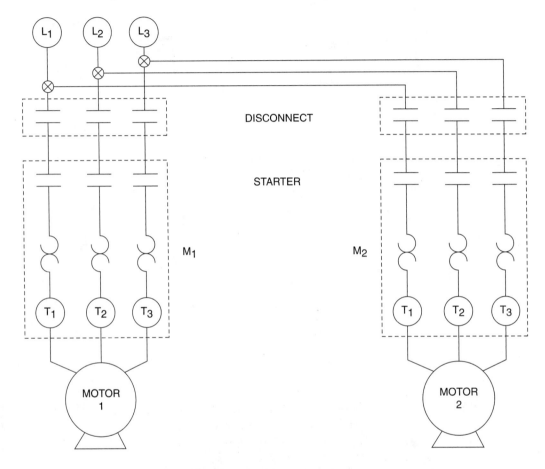

Autotransformer A transformer with one continuous winding that is used to change the voltage of a circuit. It uses one common lead between the primary and the secondary.

Auxiliary Contacts Contacts of a switching device in addition to the main circuit contacts; auxiliary contacts operate with the movement of the main contacts.

Blowout Coil Electromagnetic coil used in contactors and starters to deflect an arc when a circuit is interrupted.

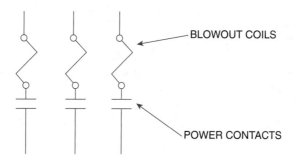

Branch Circuit That portion of a wiring system that extends beyond the final overcurrent device protecting the circuit.

Brake An electromechanical friction device to stop and hold a load. Generally electric release spring applied—coupled to motor shaft.

Breakdown Torque (of a motor) The maximum torque that will develop with the rated voltage applied at the rated frequency, without an abrupt drop in speed. (ASA)

Busway A system of enclosed power transmission that is current- and voltage rated.

Capacitor-Start Motor A single-phase induction motor with a main winding arranged for direct connection to the power source and an auxiliary winding connected in series with a capacitor. The capacitor phase is in the circuit only during starting. (NEMA)

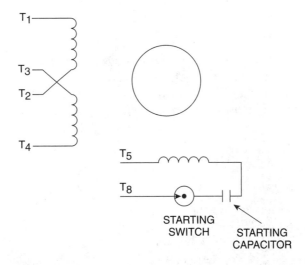

CEMF Counter electromotive force is a counter (opposite) voltage that is produced from mutual induction that opposes the applied voltage.

Circuit Breaker Automatic device that opens under abnormal current in carrying circuit; circuit breaker is not damaged on current interruption; device is ampere, volt, and horsepower rated.

Closed Transition A term given to the time when a starter transfers(transitions) from a start to a run position. Closed transition has no voltage interruption during the transition.

Commutator The segmentation of the rotor connection to the rotating member of a DC machine. This allows brushes to connect to the correct segment of the rotor.

Conduction Motor A motor that relies on a conductive path to allow current flow to the rotor of a motor rather than inducing a voltage and current to the rotor.

Contact A conducting part that acts with another conducting part to complete or to interrupt a circuit.

Contactor A device to establish or interrupt an electric power circuit repeatedly.

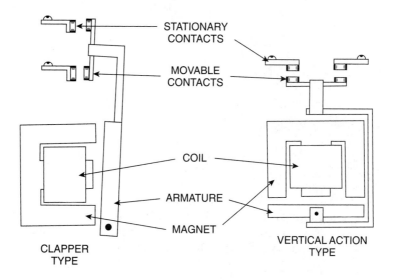

Controller A device or group of devices that governs, in a predetermined manner, the delivery of electric power to apparatus connected to it.

Controller Function Regulate, accelerate, decelerate, start, stop, reverse, or protect devices connected to an electric controller.

Controller Service Specific application of controller. General purpose: standard or usual service. Definite purpose: service condition for specific application other than usual.

Copper Loss The watts lost in an apparatus due to the copper resistance and the current flow in the circuit. This is considered a loss of power that is dissipated as heat.

Cumulative Compounding In DC motors or generators, the compounding effect of the series and the shunt field aid each other, thus creating an additive or (cumulative) effect.

Current Relay A relay that functions at a predetermined value of current. A current relay may be either an overcurrent or undercurrent relay.

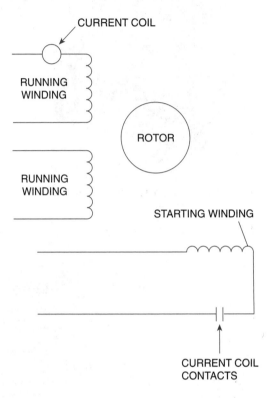

Dashpot Consists of a piston moving inside a cylinder filled with air, oil, mercury, silicon, or other fluid. Time delay is caused by allowing the air or fluid to escape through a small orifice in the piston. Moving contacts actuated by the piston close the electrical circuit.

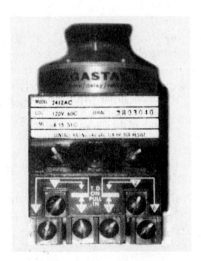

DC Braking The braking action of a motor that is accomplished by adding DC to the stator, which tries to hold the rotor in the stationary position.

Definite Time (or time limit) Definite time is a qualifying term indicating that a delay in action if purposely introduced. This delay remains substantially constant regardless of the magnitude of the quantity that causes the action.

Definite-Purpose Motor Any motor designed, listed, and offered in standard ratings with standard operating characteristics or mechanical construction for use under service conditions other than usual or for use on a particular type of application. (NEMA)

Delta A connection pattern for three-phase motors and generators that is vectorally represented by the Greek letter Δ.

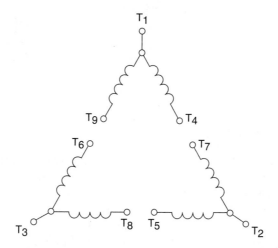

Delta-Connected Motor A three-phase AC motor that has the winding connected in a delta pattern rather than a wye connection.

Device A unit of an electrical system that is intended to carry but not utilize electrical energy.

Differential Compounding In DC motors and generators, the compounding effect of the series and shunt field oppose each other, creating a differential of the two fields.
Differential Voltage The difference between two voltages (two opposite voltages) is the differential value of the voltages.
Digital The representation of data in the form of pieces, bits, or digits.
Diode A two-element electronic device that permits current to flow through it in only one direction.
Direct Current (DC) A continuous nonvarying current in one direction.
Disconnecting Means (Disconnect) A device, or group of devices, or other means whereby the conductors of a circuit can be disconnected from their source of supply.

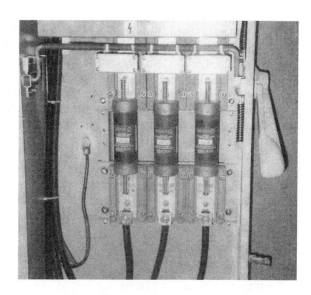

Dropout Voltage The voltage value or percentage of the applied voltage that allows an armature of a magnetic coil to release, or drop open, from an energized "held in" position.

Drum Controller Electrical contacts made on surface of rotating cylinder or sector; contacts made also by operation of a rotating cam.

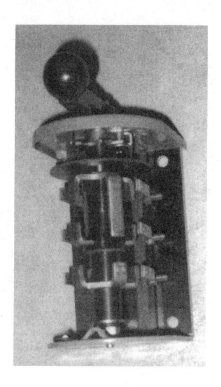

Drum Switch A drum switch is a switch that has electrical connecting parts in the form of fingers held by spring pressure against contact segments or surfaces on the periphery of a rotating cylinder or sector.

Duty Specific controller functions. Continuous (time) Duty: constant load, indefinite long time period. Short-Time Duty: constant load, short or specified time period. Intermittent Duty: varying load, alternate intervals, specified time periods. Periodic Duty: intermittent duty with recurring load conditions. Varying Duty: varying loads, varying time intervals, wide variations.

Dynamic Braking Using the motor as a generator, taking it off the line, and applying an energy-dissipating resistor to the armature.

Eddy Currents Circulating induced currents contrary to the main currents; a loss of energy that shows up in the form of heat.

Electrical Interlocking Accomplished by control circuits in which the contacts in one circuit control another circuit.

Electric Controller A device, or group of devices, that governs, in some predetermined manner, the electric power delivered to the apparatus to which it is connected.

Electronic Control Control system using gas and/or vacuum tubes or solid-state devices.

Glossary

Enclosure Mechanical, electrical, and environmental protection for control devices.

**NEMA Type 1
General Purpose
Surface Mounting**

Type 1 enclosures are intended for indoor use primarily to provide a degree of protection against contact with the enclosed equipment in locations where unusual service conditions do not exist. The enclosures are designed to meet the rod entry and rust-resistance design tests. Enclosure is sheet steel, treated to resist corrosion.

**NEMA Type 1
Flush Mounting**

Flush mounted enclosures for installation in machine frames and plaster wall. These enclosures are for similar applications and are designed to meet the same tests as NEMA Type 1 surface mounting.

NEMA Type 3
Type 3 enclosures are intended for outdoor use primarily to provide a degree of protection against windblown dust, rain, and sleet; and to be undamaged by the formation of ice on the enclosure. They are designed to meet rain ■, external icing ■, dust, and rust-resistance design tests. They are not intended to provide protection against conditions such as internal condensation or internal icing.

NEMA Type 3R

Type 3R enclosures are intended for outdoor use primarily to provide a degree of protection against falling rain, and to be undamaged by the formation of ice on the enclosure. They are designed to meet rod entry, rain ■, external icing ■, and rust-resistance design tests. They are not intended to provide protection against conditions such as dust, internal condensation, or internal icing.

NEMA Type 4

Type 4 enclosures are intended for indoor or outdoor use primarily to provide a degree of protection against windblown dust and rain, splashing water, and hose-directed water; and to be undamaged by the formation of ice on the enclosure. They are designed to meet hosedown, dust, external icing ■, and rust-resistance design tests. They are not intended to provide protection against conditions such as internal condensation or internal icing. Enclosures are made of heavy-gauge stainless steel, cast aluminum, or heavy-gauge sheet steel, depending on the type of unit and size. Cover has a synthetic rubber gasket.

**NEMA Type 3R, 7 & 9
Unilock Enclosure
For Hazardous
Locations**

This enclosure is cast from "copper-free" (less than 0.1%) aluminum and the entire enclosure (including interior and flange areas) is bronze chromated. The exterior surfaces are also primed with a special epoxy primer and finished with an aliphatic urethane paint for extra corrosion resistance. The V-Band permits easy removal of the cover for inspection and for making field modifications. This enclosure meets the same tests as separate NEMA Type 3R, and NEMA Type 7 and 9 enclosures. For NEMA Type 3R application, it is necessary that a drain be added.

■ Evaluation criteria: No water has entered enclosure during specified test.
■ Evaluation criteria: Undamaged after ice that built up during specified test has melted (Note: **Not** required to be operable while ice laden).
■ Evaluation criteria: No water shall have reached live parts, insulation, or mechanisms.

NEMA Type 4X Nonmetallic, Corrosion-Resistant Fiberglass-Reinforced Polyester

Type 4X enclosures are intended for indoor or outdoor use primarily to provide a degree of protection against corrosion, windblown dust and rain, splashing water, and hose-directed water; and to be undamaged by the formation of ice on the enclosure. They are designed to meet the hosedown, dust, external icing [2], and corrosion-resistance design tests. They are not intended to provide protection against conditions such as internal condensation or internal icing. Enclosure is fiberglass reinforced polyester with a synthetic rubber gasket between cover and base. Ideal for such industries as chemical plants and paper mills.

NEMA Type 6P

Type 6P enclosures are intended for indoor or outdoor use primarily to provide a degree of protection against the entry of water during prolonged submersion at a limited depth; and to be undamaged by the formation of ice on the enclosure. They are designed to meet air pressure, external icing [2], hosedown, and corrosion-resistance design tests. They are not intended to provide protection against conditions such as internal condensation or internal icing.

NEMA Type 7 For Hazardous Gas Locations Bolted Enclosure

Type 7 enclosures are for indoor use in locations classified as Class I, Groups C or D, as defined in the National Electrical Code. Type 7 enclosures are designed to be capable of withstanding the pressures resulting from an internal explosion of specified gases, and contain such an explosion sufficiently that an explosive gas-air mixture existing in the atmosphere surrounding the enclosure will not be ignited. Enclosed heat-generating devices are designed not to cause external surfaces to reach temperatures capable of igniting explosive gas-air mixtures in the surrounding atmosphere. Enclosures are designed to meet explosion, hydrostatic, and temperature design tests. Finish is a special corrosion-resistant, gray enamel.

NEMA Type 9 For Hazardous Dust Locations

Type 9 enclosures are intended for indoor use in locations classified as Class II, Groups E, F or G, as defined in the National Electrical Code. Type 9 enclosures are designed to be capable of preventing the entrance of dust. Enclosed heat generating devices are designed not to cause external surfaces to reach temperatures capable of igniting or discoloring dust on the enclosure or igniting dust-air mixtures in the surrounding atmosphere. Enclosures are designed to meet dust penetration and temperature design tests, and aging of gaskets. The outside finish is a special corrosion-resistant gray enamel.

NEMA Type 12

Type 12 enclosures are intended for indoor use primarily to provide a degree of protection against dust, falling dirt, and dripping noncorrosive liquids. They are designed to meet drip [1], dust, and rust-resistance tests. They are not intended to provide protection against conditions such as internal condensation.

NEMA Type 13

Type 13 enclosures are intended for indoor use primarily to provide a degree of protection against dust, spraying of water, oil, and noncorrosive coolant. They are designed to meet oil exclusion and rust-resistance design tests. They are not intended to provide protection against conditions such as internal condensation.

---

[1] Evaluation criteria: No water has entered enclosure during specified test.
[2] Evaluation criteria: Undamaged after ice that built up during specified test has melted (Note: **Not** required to be operable while ice laden).

Eutectic Alloy Metal with low and definite melting point; used in thermal overload relays; converts from a solid to a liquid state at a specific temperature; commonly called solder pot.

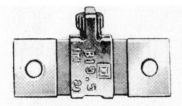

ONE-PIECE THERMAL UNIT

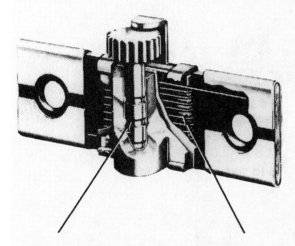

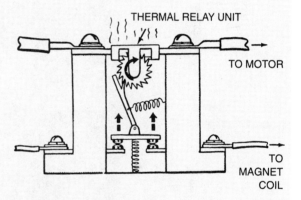

Solder pot (heat-sensitive element) is an integral part of the thermal unit. It provides accurate response to overload current, yet prevents nuisance tripping.

Heater winding (heat-producing element) is permanently joined to the solder pot, so proper heat transfer is always insured. No chance of misalignment in the field.

Drawing shows operation of melting alloy overload relay. As heat melts alloy, ratchet wheel is free to turn—spring then pushes contacts open.

Feeder The circuit conductor between the service equipment, or the generator switchboard of an isolated plant, and the branch circuit overcurrent device.

Feeler Gauge A precision instrument with blades in thicknesses of thousandths of an inch for measuring clearances.

Filter A device used to remove the voltage/current ripple produced by a rectifier.

Frequency Number of complete variations made by an alternating current per second; expressed in Hertz. (See *Hertz*)

Full-Load Torque (of a motor) The torque necessary to produce the rated horsepower of a motor at full-load speed.

Full-Voltage Control (Across-the-line) Connects equipment directly to the line supply on starting.

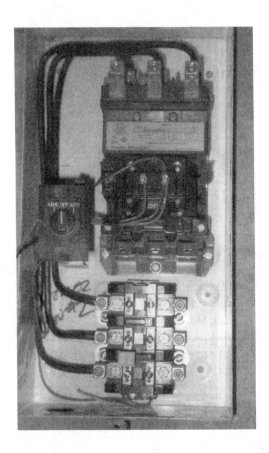

Fuse An overcurrent protective device with a fusible member that is heated directly and destroyed by the current passing through it to open a circuit.

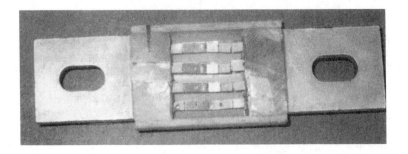

General-Purpose Motor Any open motor having a continuous 40C rating and designed, listed, and offered in standard ratings with standard operating characteristics and mechanical construction for use under usual service conditions without restrictions to a particular application or type of application. (NEMA)

Growler A tool used in motor analysis and repair to check the rotor or stator for shorts or opens. The name reflects the growling noise it makes as it vibrates and indicates condition of the motor pieces.

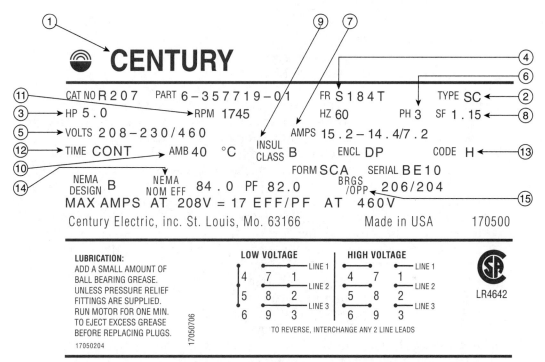

1. MANUFACTURER'S NAME
2. TYPE
3. HORSEPOWER
4. FRAME
5. VOLTAGE RATINGS
6. PHASE—SINGLE-PHASE, 3-PHASE, OR DC
7. AMPERAGE
8. SERVICE FACTOR
9. INSULATION CLASS
10. AMBIENT TEMPERATURE
11. RPM
12. DUTY OR TIME
13. CODE LETTER
14. NOMINAL EFFICIENCY
15. POWER FACTOR

Heat Sink   Metal used to dissipate the heat of solid-state components mounted on it; usually finned in form.

Hertz   International unit of frequency, equal to one cycle per second of alternating current.

High-Voltage Control   Formerly, all control above 600 volts. Now, all control above 5000 volts. See *Medium Voltage* for 600- to 5000-volt equipment.

Horsepower   Measure of work done. The measurement includes weight moved, distance moved and time (see Chapter 2).

Motor HP = $\dfrac{F \times R \times N}{5252}$

F = Force or scale in pounds
R = Radius of motor pulley in feet
N = Speed of motor shaft in RPM

Hysteresis   The effects in a magnetic material that restricts the realignment of the magnetic fields within the metal. The hysteresis causes a lag of magnetic effect after the magneto motive force (MMF) is removed.

IEC   International Electrotechnical Commission, a European standards association that publishes standards for electrical equipment that is used throughout the world.

Impedance   Total opposition to current flow in an electrical circuit.

Induction Motor   A motor that relies on the principles of induction to transfer energy to the rotor and therefore needs no electrical connection to the rotor. A pure induction motor relies on slip to create a inductive effect.

Input   Power delivered to an electrical device.

Inrush Current   The initial surge of current to an AC apparatus that occurs as the counter EMF has not produced an effective opposition to current. As the inductive effects or the CEMF from a rotating rotor build up, the inrush current diminishes to normal levels.

Instantaneous   Instantaneous is a qualifying term indicating that no delay is purposely introduced in the action of the device.

Integral   Whole or complete; not fractional.

Interface   A circuit permitting communication between the central processing unit (of a programmable controller) and a field input or output.

Interlock   To interrelate with other controllers; an auxiliary contact. A device is connected in such a way that the motion of one part is held back by another part.

Inverse Time A qualifying term indicating that a delayed action is introduced purposely. This delay decreases as the operating force increases.

Jogging (Inching) Momentary operations; the quickly repeated closure of the circuit to start a motor from rest for the purpose of accomplishing small movements of the driven machine.

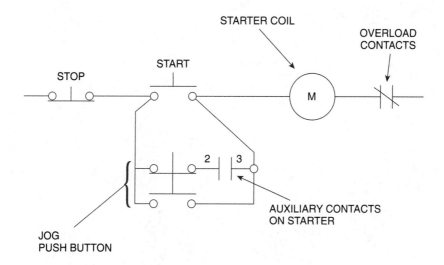

Jumper A short length of conductor used to make a connection between terminals or around a break in a circuit.

Ladder Diagram An electrical operations diagram that resembles the construction of a ladder. The diagram has two runners that represent the supply voltage and has rungs of the ladder crossing to represent each operation of a control circuit.

Lenz's Law Paraphrased, his law states that the results of induction oppose the cause of induction. For example, as current flows to create a magnetic field, the field induces a voltage into the same coil that opposes the original current.

Limit Switch A mechanically operated device that stops a motor from revolving or reverses it when certain limits have been reached.

Load Center Service entrance; controls distribution; provides protection of power; generally of the circuit breaker type.

Local Control Control function, initiation, or change accomplished at the same location as the electric controller.

Locked Rotor Current (of a motor) The steady-state current taken from the line with the rotor locked (stopped) and with the rated voltage and frequency applied to the motor.

Locked Rotor Torque (of a motor) The minimum torque that a motor will develop at rest for all angular positions of the rotor with the rated voltage applied at a rated frequency. (ASA)

Lockout A mechanical device that may be set to prevent the operation of a push button.

Logic A means of solving complex problems through the repeated use of simple functions which define basic concepts. Three basic logic functions are and, or, and not.

Low-Voltage Protection (LVP) Magnetic control only; nonautomatic restarting; three-wire control; power failure disconnects service; power restored by manual restart.

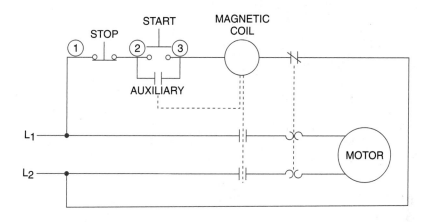

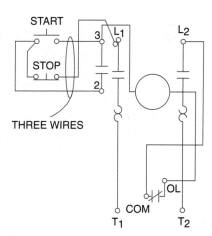

Low-Voltage Release (LVR) Manual and magnetic control; automatic restarting; two-wire control; power failure disconnects service; when power is restored, the controller automatically restarts motor.

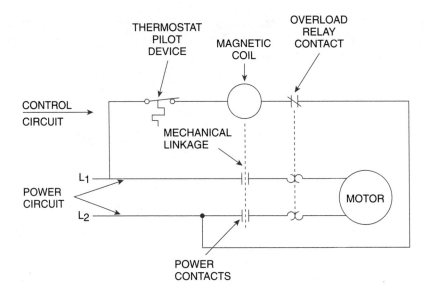

Magnet Brake Friction brake controlled by electromagnetic means.
Magnetic Contactor A contactor that is operated electromagnetically.
Magnetic Controller An electric controller; device functions operated by electromagnets.
Magnet Wire Wire that is used expressly for producing a magnetic field. Motors are wound with magnet wire to create the needed magnetic field.
Maintaining Contact A small control contact used to keep a coil energized, actuated by the same coil usually. Holding contact; pallet switch.

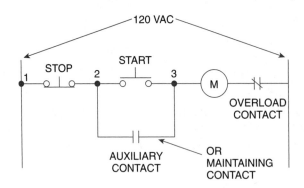

Manual Controller An electric controller; device functions operated by mechanical means or manually.

Master Switch A main switch to operate contactors, relays, or other remotely controlled electrical devices.

Megohmeter A name created by the manufacturer of ohmmeters that are designed to measure megohms or resistance in for magnet wire. Normally there should be no connection between the magnet wire turns in a motor. If the insulation deteriorates or breaks down, the effects are measured in the reduction in megohms of resistance.

Memory Part of a programmable controller where instructions and information are stored.

Medium Voltage Control Formerly known as high voltage; includes 600- to 5000-volt apparatus; air break or oil-immersed main contactors; high interrupting capacity fuses; 150,000 kVA at 2300 volts; 250,000 kVA at 4000 to 5000 volts.

Microprocessor A small computer.

Momentary Contact The type of contact that is not designed to stay closed or open but only change condition momentarily.

Motor Device for converting electrical energy to mechanical work through rotary motion; rated in horsepower.

Motor Bearings The mechanical bearings on the rotor shaft of a motor designed to have minimum friction for the turning shaft.

Motor Circuit Switch Motor branch circuit switch rated in horsepower; capable of interrupting overload motor current.

Motor Control Center An enclosure that is designed to house several motor starters and their associated controls, disconnects, protection, and so on.

Motor Controller The device that controls the power delivered to the motor. This may be a manual control, a magnetic control, or an electronic control.

Motor-Driven Timer A device in which a small pilot motor causes contacts to close after a predetermined time.

Motor Starter The term "starter" implies that there is more than just an on–off control. A starter may have more complex systems that reduce the voltage or change the starting characteristics of the motor. A starter could be a simple magnetic controller with thermal overload protection added.

Multifunction-Function Timer A style of electric timer that is programmable to operate in many different functional modes. This makes one timing module field adjustable to fit many situations.

Multispeed Starter An electric controller with two or more speeds; reversing or nonreversing; full or reduced voltage starting.

NEMA National Electrical Manufacturers Association.

NEMA Size Electric controller device rating; specific standards for horsepower, voltage, current, and interrupting characteristics.

NEMA Size	Continuous Amp. Rating	600 VOLTS MAXIMUM				
			Maximum Horsepower Rating [2] Full-load current must not exceed the "Continuous Ampere Rating"		Maximum Horsepower Rating For Plugging Service [1]	
		Volts	Single Phase	3 or 2 Phase	Single Phase	Three Phase
00	9	120 208 240 480 600	⅓ — 1 — —	¾ 1½ 1½ 2 2	— — — — —	— — — — —
0	18	120 208 240 480 600	1 — 2 — —	2 3 3 5 5	½ — 1 — —	1 1½ 1½ 2 2
1	27	120 208 240 480 600	2 — 3 — —	3 7½ 7½ 10 10	1 — 2 — —	2 3 3 5 5
2	45	120 208 240 480 600	3 — 7½ — —	— 15 15 25 25	2 — 5 — —	— 10 10 15 15
3	90	120 208 240 480 600	— — — — —	— 30 30 50 50	— — — — —	— 20 20 30 30
4	135	120 208 240 480 600	— — — — —	— 50 50 100 100	— — — — —	— 30 30 60 60
5	270	120 208 240 480 600	— — — — —	— 100 100 200 200	— — — — —	— 75 75 150 150
6	540	208 240 480 600	— — — —	200 200 400 400	— — — —	150 150 300 300
7	810	208 240 480 600	— — — —	300 300 600 600	— — — —	— — — —
8	1215	208 240 480 600	— — — —	450 450 900 900	— — — —	— — — —
9	2250	208 240 480 600	— — — —	800 800 1600 1600	— — — —	— — — —

Noise A condition that interferes with the desired voltage, or signal, in a circuit. Noise can produce erratic operation.

Nonautomatic Controller Requires direct operation to perform function; not necessarily a manual controller.

Nonreversing Operation in one direction only.

Normally Open and Normally Closed When applied to a magnetically operated switching device, such as a contactor or relay, or to the contacts of these devices, these terms signify the position taken when the operating magnet is deenergized. The terms apply only to nonlatching types of devices.

One-Shot Timer A timer that is programmed to create one timing cycle per sequence, until reset. This timer does continue producing timing cycles after the first sequence.

Open Delta This term refers to the method of creating three-phase power from only two sets of coils as opposed to three. This method yields true three-phase power but at a reduced capacity compared to a full closed delta system.

Open Transition Opposite of closed transition; as a motor is started and transfers from starting mode to running mode, there is a short period when the electric circuit is open or disconnected. The motor coasts for that very short time and then reconnects to complete the transition.

Optoisolator A device used to connect sections of a circuit by means of a light beam.

Output Devices Elements such as solenoids, motor starters- and contactors that receive input.

Overload Protection Overload protection is the result of a device that operates on excessive current, but not necessarily on short circuit, to cause and maintain the interruption of current flow to the device governed. *Note:* Operating overload means a current that is not in excess of six times the rated current for AC motors, not in excess of four times the rated current for DC motors.

Overload Relay Running overcurrent protection; operates on excessive current; not necessarily protection for short circuit; causes and maintains interruption of device from power supply. Overload relay heater coil: Coil used in thermal overload relays; provides heat to melt eutectic alloy. Overload relay reset: Push button used to reset thermal overload relay after relay has operated.

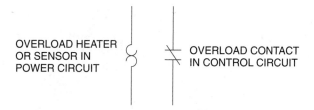

Panelboard Panel, group of panels, or units; an assembly that mounts in a single panel; includes buses, with or without switches and/or automatic overcurrent protection devices; provides control of light, heat, power circuits; placed in or against wall or partition; accessible from front only.

Part Winding Starter This method of starting a motor is used to reduce the starting current drawn from the line, and also reduce the starting torque. Voltage is first applied to part of the motors windings, then as the speed of the rotor increases the motor's remaining windings are connected.

Permanent Magnet A permanent magnet is a magnetic material that retains strong magnetic properties after the MMF has been removed. As opposed to temporary magnets that stay magnetized with the presence of the electrical current flow, a permanent magnet stays magnetized.

Permanent-Split Capacitor Motor A single-phase induction motor similar to the capacitor-start motor except that it uses the same capacitance that remains in the circuit for both starting and running. (NEMA)

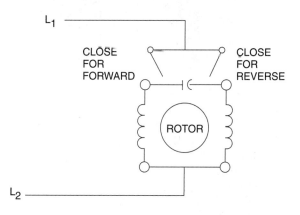

Permeability The ease with which a material will conduct magnetic lines of force.

Phase Relation of current to voltage at a particular time in an AC circuit. Single-phase: A single voltage and current in the supply. Three-phase: Three electrically related (120-degree electrical separation) single-phase supplies.

Phase-Failure Protection Phase-failure protection is provided by a device that operates when the power fails in one wire of a polyphase circuit to cause and maintain the interruption of power in all the wires of the circuit.

Phase-Reversal Protection Phase-reversal protection is provided by a device that operates when the phase rotation in a polyphase circuit reverses to cause and maintain the interruption of power in all the wires of the circuit.

Phase Rotation Relay A phase rotation relay is a relay that functions in accordance with the direction of phase rotation.

Pickup Voltage The amount of voltage needed to pick up, or energize, a magnetic coil and allow it to attract a moveable armature. This voltage allows enough current to flow in a coil to pull or pick up a movable piece by magnetic attraction.

Glossary

Pilot Device Directs operation of another device. Float switch: A pilot device that responds to liquid levels. Foot switch: A pilot device operated by the foot of an operator. Limit switch: A pilot device operated by the motion of a power-drive machine; alters the electrical circuit with the machine or equipment.

LIMIT SWITCHES	FOOT SWITCH	PRESSURE OR VACUUM SWITCH
NORMALLY OPEN	NORMALLY OPEN PUSH TO CLOSE	NORMALLY OPEN CLOSE ON PRESSURE RISE
NORMALLY OPEN HELD CLOSED	NORMALLY CLOSED PUSH TO OPEN	NORMALLY CLOSED OPEN ON PRESSURE RISE
NORMALLY CLOSED		
NORMALLY CLOSED HELD OPEN		
LIQUID LEVEL FLOAT SWITCH	FLOW SWITCH	TEMPERATURE SWITCH
NORMALLY OPEN	NORMALLY OPEN	NORMALLY OPEN
NORMALLY CLOSED	NORMALLY CLOSED	NORMALLY CLOSED

Plugging Braking by reversing the line voltage or phase sequence; motor develops retarding force.

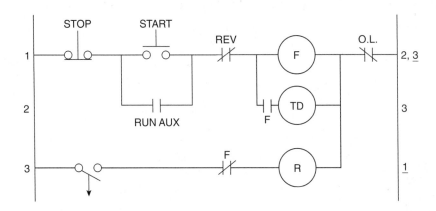

Pole The north or south magnetic end of a magnet; a terminal of a switch; one set of contacts for one circuit of main power.

Potentiometer A variable resistor with two outside fixed terminals and one terminal on the center movable arm.

Power Factor A factor applied to the apparent power is an AC system that will yield true power. The formula for true power when you know apparent power is VA × Power Factor = Watts of True Power.

Pressure Switch A pilot device operated in response to pressure levels.

Primary Resistor Starter A type of reduced voltage motor starter that uses a resistor in series with the motor to drop a series voltage and therefore reduce the voltage to the motor. As the inrush current drops, the series drop of voltage diminishes and the motor voltage increases.

Primary Winding The winding of a transformer that is connected to the power source. This is the primary supply of power for the transformer which transfers energy to the secondary winding by electromagnetism.

Printed Circuit A board on which a predetermined pattern of printed connections has been formed.

Programmable Logic Controller A type of electronic control that can be locally programmed to provide the same functions as discrete components such as relays, timers, counters, and so on.

Programmable Relay A component that performs the relay function but is programmable to be a standard relay or a timing relay or various combinations of functions.

Programmable Timer A timing style relay that is field programmable to perform many discrete functions.

Proximity Sensor The sensor that makes no direct contact with the sensed object but can "sense" by being in the same proximity. Typically the sensor is an inductive sensor or a capacitive sensor.

Pull-Up Torque (of AC motor) The minimum torque developed by the motor during the period of acceleration from rest to the speed at which breakdown occurs. (ASA)

Push Button A master switch; manually operable plunger or button for an actuating device; assembled into push-button stations.

Rectifier A device that converts AC into DC.

Relay Operated by a change in one electrical circuit to control a device in the same circuit or another circuit; rated in amperes; used in control circuits.

Control Relays

Product Selection

AC RELAYS

	Relay Contacts		Overlapping Side Mounted Contacts		Relay Arrangement	Auxiliary	Cat. No.
	NO	NC	NO	NC			
With 1 or 2 sets of overlapping contacts 4 Pole Control Relays IP20 660V Max.	4	0	1	1			700–FZ1510
	3	1	1	1			700–FZ1420
	2	2	1	1			700–FZ1330
	4	0	2	2			700–FZ2620
	3	1	2	2			700–FZ2530
	2	2	2	2			700–FZ2440

Remote Control Controls the function initiation or change of an electrical device from some remote point or location.

Remote Control Circuit Any electrical circuit that controls any other circuit through a relay or an equivalent device.

Repulsion-Induction Motor A motor that starts as a repulsion motor with a commutator and brushes but is in the running mode as an induction motor.

Residual Magnetism The retained or small amount of remaining magnetism in the magnetic material of an electromagnet after the current flow has stopped.

Resistance The opposition offered by a substance or body to the passage through it of an electric current; resistance converts electrical energy into heat; resistance is the reciprocal of conductance.

Resistor A device used primarily because it possesses the property of electrical resistance. A resistor is used in electrical circuits for purposes of operation, protection, or control; commonly consists of an aggregation of units.

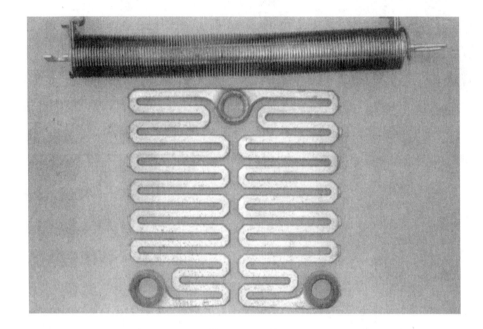

Armature Regulating Resistors Used to regulate the speed or torque of a loaded motor by resistance in the armature or power circuit.

Dynamic Braking Resistors Used to control the current and dissipate the energy when a motor is decelerated by making it act as a generator to convert its mechanical energy to electrical energy and then to heat in the resistor.

Field Discharge Resistors Used to limit the value of voltage which appears at the terminals of a motor field (or any highly inductive circuit) when the circuit is opened.

Plugging Resistors Used to control the current and torque of a motor when deceleration is forced by electrically reversing the motor while it is still running in the forward direction.

Starting Resistors Used to accelerate a motor from rest to its normal running speed without damage to the motor and connected load from excessive currents and torques, or without drawing undesirable inrush current from the power system.

Rheostat   A resistor that can be adjusted to vary its resistance without opening the circuit in which it may be connected.

Ripple   An AC component in the output of a DC power supply; improper filtering.

Rotor   The rotating portion of a motor that is free to spin and delivers the twisting effort of the motor to the load.

Safety Switch   Enclosed manually operated disconnecting switch; horsepower and current rated; disconnects all power lines.

Schematic Diagram   A drawing of the electrical circuits that show the electrical relationships of one component to each other. It is designed to make it easy to follow the path of current through the circuit and determine the sequence of events. The parts are not drawn in the real physical location.

SCR   Silicon-controlled rectifier; a semiconductor device that must be triggered on by a pulse applied to the gate before it will conduct.

Seal-In Voltage   The voltage value on a magnetic coil when the movable armature holds tight against the magnetized coil. This voltage is less than the voltage required to pull in but more than the dropout value.

Secondary Winding   The winding on a transformer that receives its energy by induction, or magnetic transfer, from the primary coil. This winding then has an output voltage and current.

Selector Switch   A master switch that is manually operated; rotating motion for actuating device; assembled into push-button master stations.

Self-Excited   A term given to a generator when no external voltage or current needs to be applied to start the generation. The small amount of voltage required is actually produced by spinning an armature in a residual magnetic field already in the generator core.

Semiautomatic Starter   Part of the operation of this type of starter is nonautomatic while selected portions are automatically controlled.

Semiconductors   See *Solid-State Devices*.

Semimagnetic Control   An electric controller whose functions are partly controlled by electromagnets.

Sensing Device   A pilot device that measures, compares, or recognizes a change or variation in the system that it is monitoring; provides a controlled signal to operate or control other devices.

Separately Excited   As opposed to self-excited, a separately excited generator *does* need an external voltage source to begin generation. A separately excited generator cannot start without an external source of power.

Service   The conductors and equipment necessary to deliver energy from the electrical supply system to the premises served.

Service Equipment   Necessary equipment, circuit breakers or switches and fuses, with accessories mounted near the entry of the electrical supply; constitutes the main control or cutoff for supply.

Service Factor (of a general-purpose motor) An allowable overload; the amount of allowable overload is indicated by a multiplier that, when applied to normal horsepower rating, indicates the permissible loading.

Shaded Pole Motor A single-phase induction motor provided with auxiliary short-circuited winding or windings displaced in magnetic position from the main winding. (NEMA)

Shading Ring A small copper ring on a induction coil that allows the magnetic field of the shading ring to lag behind the main magnetic field. The effect is to shade or delay the magnetic field for a very short time, thus evening out the expansion and contraction of the main field.

Single Phasing A condition that occurs in a three-phase system when one of the phases fails. There are only two phases left, which create a single-phase operation.

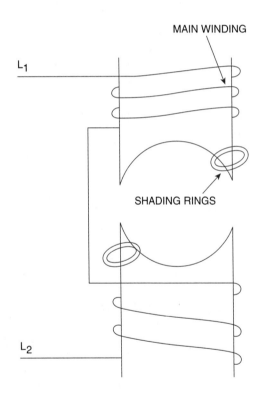

Slip Difference between rotor RPM and the rotating magnetic field of an AC motor.

Signal The event, phenomenon, or electrical quantity that conveys information from one point to another.

Snubber A circuit that suppresses transient spikes.

Solenoid A tubular, current-carrying coil that provides magnetic action to perform various work functions.

Solenoid-and-Plunger A solenoid-and-plunger is a solenoid provided with a bar of soft iron or steel called a plunger.

Solder Pot See *Eutectic Alloy*.

Solid-State Devices Electronic components that control electron flow through solid materials such as crystals; for example, transistors, diodes, integrated circuits.

Solid-State Relay A relay is a device used to cause a change in one circuit, in response to a change in the original circuit. A solid state relay uses electronics and not moving parts to accomplish this relay action.

Special-Purpose Motor A motor with special operating characteristics or special mechanical construction, or both, designed for a particular application and not falling within the definition of a general-purpose or definite-purpose motor. (NEMA)

Split Phase A single-phase induction motor with auxiliary winding, displaced in magnetic position from, and connected in parallel with, the main winding. (NEMA)

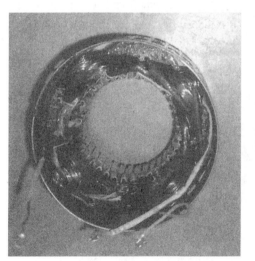

Starter A starter is a controller designed for accelerating a motor to normal speed in one direction of rotation. *Note:* A device designed for starting a motor in either direction of rotation includes the additional function of reversing and should be designated as a controller.

Starting Torque The amount of torque developed by a motor as it starts from a stationary, or stopped, condition. Starting torque is higher than normal running torque because of the high starting currents present in the rotor and the stator.

Startup The time between equipment installation and the full operation of the system.

Static Control Control system in which solid-state devices perform the functions. Refers to no moving parts or without motion.

Stator The stationary part of a motor.

Stepper Motor A motor designed to operate in incremental steps. It is controlled by a stepper motor controller and is pulsed to cause the motor to rotate in definite measured steps.

Surge A transient variation in the current and/or potential at a point in the circuit, unwanted, temporary.

Switch A switch is a device for making, breaking, or changing the connections in an electric circuit.

Switchboard A large, single panel with a frame or assembly of panels; devices may be mounted on the face of the panels, on the back, or both; contains switches, overcurrent, or protective devices; instruments accessible from the rear and front; not installed in wall-type cabinets. (See *Panelboard*)

Synchronous Speed Motor rotor and AC rotating magnetic field in step or unison.

Tachometer Generator Used for counting revolutions per minute. Electrical magnitude or impulses are calibrated with a dial gauge reading in RPM.

Temperature Relay A temperature relay is a relay that functions at a predetermined temperature in the apparatus protected. This relay is intended to protect some other apparatus such as a motor or controller and does not necessarily protect itself.

Temperature Switch A pilot device operated in response to temperature values.

Terminal A fitting attached to a circuit or device for convenience in making electrical connections.

Thermal Protector (as applied to motors) An inherent overheating protective device that is responsive to motor current and temperature. When properly applied to a motor, this device protects the motor against dangerous overheating due to overload or failure to start.

Three-Wire Control A method of motor control that utilizes three conductors to operate a magnetic starter. Using a momentary start/stop pushbutton station and three wires to the controller, the motor has low-voltage release and low-voltage protection. The motor will not automatically restart if the power returns to normal.

Time Limit See *Definite Time*.

Timer A pilot device that also is considered a timing relay; provides adjustable time period to perform function; motor driven; solenoid actuated; electronic.

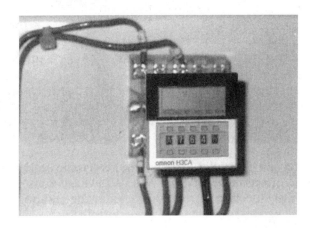

Torque The torque of a motor is the twisting or turning force which tends to produce rotation.

Transducer A device that transforms power from one system to power of a second system: example, heat to electrical.

Transformation Ratio The ratio of total line voltage on the primary side of a transformer system to the total line voltage on the secondary side of a transformer system.

Transformer Converts voltages for use in power transmission and operation of control devices, an electromagnetic device.

Transistor An active semiconductor device that can be used for rectification and amplification by placing the proper voltage on its electrodes.

Transient See *Surge*.

Transient Temporary voltage or current that occurs randomly and rides an AC sine wave.

Trip Free Refers to a circuit breaker that cannot be held in the on position by the handle on a sustained overload.

Troubleshoot To locate and eliminate the source of trouble in any flow of work.

Two-Wire Control A method of motor control using two wires to connect the pilot device to the motor controller. With two wires, an open or closed contact on the pilot device operates the magnetic starter. This system has low-voltage release but not "no voltage protection." If the power fails, the contactor opens; when the supply voltage returns, the motor may restart if the pilot device is closed.

Undervoltage Protection Undervoltage protection is the result when a device operates on the reduction or failure of voltage to cause and maintain the interruption of power to the main circuit.

Undervoltage Release Occurs when a device operates on the reduction or failure of voltage to cause the interruption of power to the main circuit but does not prevent reestablishment of the main circuit on the return of voltage.

Variable-Frequency Drive An electronic controller that is designed to re-create a set frequency of the AC output to drive AC motors at variable speeds. The frequency of the AC can be set from a few hertz to over 60 Hz on some models.

Voltage Relay A relay that functions at a predetermined value of voltage. A voltage relay may be either an overvoltage or an undervoltage relay.

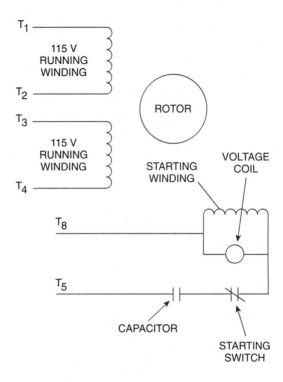

VOM Volt-ohm-milliammeter; a test instrument designed to measure voltage, resistance, and milliamperes.

Wild Leg On a four-wire delta system, one phase has a grounded center tapped winding and the voltage from the grounded conductor to the opposite phase conductor is higher than to the other two-phase conductors. The phase with the higher voltage to ground is known as the "wild leg," the "high leg," the "stinger leg," and so on.

Wiring Diagram A diagram of the motor and its controller, where the physical location of the components is approximated on the drawing. This allows the electrician to see where to look for a particular component and how many wires are connected to it.

Wound Rotor Motor A type of induction motor that was used for variable-speed requirements and is still used for heavy industrial loads. The rotor is actually wound with coils, as opposed to cast squirrel cage windings. The rotor has slip rings that allow the user to control the amount of rotor current and therefore the operational speed and torque of the motor.

Wye-Connected Motor A three-phase motor that has its internal windings connected in a wye pattern. The coils are designed to be the correct voltage when connected in a wye pattern.

Wye–Delta Starter A motor starter that is designed to start a motor by first connecting the motor windings in a wye pattern during starting and then reconnecting the same windings into a delta pattern for running mode.

This process requires that the motor have all coil ends available and that it designed to run in delta connection. This starter reduces the inrush current to the motor and also softens the initial twisting effort.

Wye/Star A connection pattern for three-phase motors and transformers that is vectorally represented by the letter Y. This also is known as a star pattern.

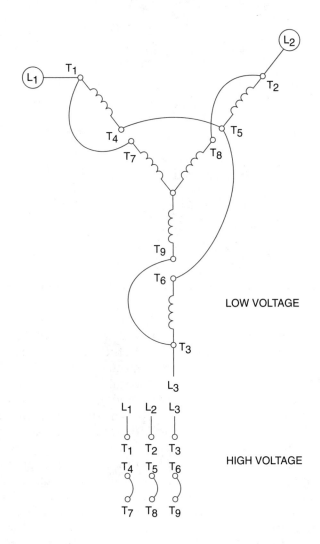

INDEX

Accelerating relays, 335
Accessory control, 336
AC chatter, 85
AC circuits, 313–314
Across-the-line starting, 158, 337
AC synchronous motors, 169, 182, 337
Additive polarity, 251f, 252
Air pressure switches, 96
Alternate pole method, 53
Alternating current (AC), 337
Alternators
   loss of output of, 42–43
   parallel operation of, 38–41,
     38–42, 39f, 41f
Altitude, 5
Ambient temperature, 5, 337
American National Standards Institute
   (ANSI), 250, 252
Ammeters, clamp-on, 259–260, 259f, 260f
Ampacity, 337
Amperage, 2–3
Ampere, 337
Ampere turns, 241
Analog signal, 337
ANSI standards, 250, 252
Antiplugging, 188, 189f
Arc horns, 127–128, 128f
Armature reaction, 283, 292–293, 298
Armature regulating resistors, 366
Armatures, 7, 286, 337

Armature windings, 287, 287f
Armortisseur winding, 182
Automatic, 337
Automatic starters, 338
Autotransformers
   explanation of, 242–243, 338
   function of, 255–256, 255f
Autotransformer starters
   control diagram for, 178f
   explanation of, 169, 176–177, 177f, 179
Auxiliary contacts, 338

Bearings. *See* Motor bearings
Bifurcated cable, 101, 101f
Blowout coils, 339
Brakes/braking
   DC, 169, 186, 188–190
   drum, 191f
   dynamic, 170, 186, 190–191, 366
   electrical, 186
   explanation of, 339
   magnet, 356
   mechanical, 191–192, 192f
Branch circuits, 216, 339
Breakdown torque, 21, 340
Brushless DC motors, 301, 301f
Busway, 340

Cadmium sulfide cells, 98
Capacitive sensors, 102, 103, 103f

Capacitor motors
　explanation of, 21–23
　permanent split, 24–25, 25f
　two-value, 21, 24, 24f
Capacitor-start motors
　explanation of, 22f, 23, 340
　use of, 21
CEMF (counter electromotive force), 283–284, 295, 296f, 340
Circuit breakers, 340
Circulating current, 42
Clamp-on ammeters, 259–260, 259f, 260f
Closed transition, 132, 162, 340
Code letters, 56
Cogeneration, 43
Coil connections, 311
Coils, 91–93
Combination starters, 140
Commutator, 284, 286, 340
Compound field generators, 290–292, 290f
Compound fields, 284
Compound motors, 300–301, 300f
Conduction motors, 1, 6, 7, 340
Consequent poles, 53
Contactors
　explanation of, 61, 341
　magnetic, 86, 87f, 88, 90, 92f
　NEMA, 86, 89–90
　size of, 139–140
Contacts
　convertible, 111–112
　explanation of, 340
　operation of, 127–128, 127f
Contact welding, 330
Continuity tests, 204
Contractors
　explanation of, 66
　magnetic, 66, 67f, 356
　troubleshooting, 331
Control circuits
　diagram of, 309, 310
　explanation of, 134
　schematic of, 326f
　troubleshooting, 143–144
Controller function, 341

Controllers
　drum, 65, 65f
　explanation of, 341
　manual, 62–63, 63f
　multispeed, 146, 147f, 148–149, 148f, 149f, 150f
　programmable logic, 150–151, 151f
　soft start, 170
　solid-state, 125, 126f
　toggle-switch-style, 64f, 65
Controller service, 341
Control relays, 113, 113f, 114f
Control transformers, 246, 256, 256f
Convergent beams, 100, 100f
Copper loss, 241, 252, 341
Core-type transformers, 246, 248, 248f
Cumulative compound, 284, 290–291, 341
Current relays, 342
Current source input (CSI), 152, 154–155, 155f, 156f
Current transformers Q, 257–259, 258f, 259f

Dashpot, 120, 342
Dashpot timers, 120
DC braking, 169, 186, 188–190, 343
DC circuits, 313
DC generators
　explanation of, 284, 285f, 286, 286f, 287f
　parallel operation of, 293–294, 293f
DC motors, 294–296, 294f, 295f
　brushless, 301, 301f
　manual control of, 303, 304f
　NEC and, 303–305
Definite-purpose motors, 343
Definite time, 343
Delta-connected motors
　explanation of, 46, 47f, 195, 343
　lead identification and, 207, 207f, 209–210, 210f, 211f
　motor connections and, 52
Delta-delta-connected transformers, 263, 265f, 266–267
Delta system
　explanation of, 31, 343
　three-phase true power formula for, 46, 47f
　wye vs., 49

Demand meters, 277–278
Design letters
   code letters vs., 56
   explanation of, 5–6
Devices, 109–110, 344
Differential compound, 284, 291, 344
Differential voltage, 284, 295, 344
Diffuse scan, 99–100, 100f
Digital, 344
Diode, 344
Direct current (DC), 344
Direct scan, 101, 101f
Disconnecting means, 215–217, 217f, 344
Disk brakes, 192, 192f
Distribution transformers, 246, 246f
Dropout voltage, 85, 93, 345
Drum controllers
   explanation of, 133, 134f, 345
   use of, 65, 65f, 74
Drum switch, 346
Dual-ramp system, 180
Duty, 5, 346
Dynamic braking
   explanation of, 170, 186, 346
   use of, 190–191
Dynamic braking resistors, 366

Eddy currents
   explanation of, 46, 47f, 195, 213,
      242, 343, 346
   laminations and, 246
Efficiency
   explanation of, 5
   formula for, 58, 214
   motor, 58, 214–215
Electrical brakes, 186
Electrical interlocking, 346
Electrical symbols, 308
Electric controllers, 346
Electromechanical relays (EMR), 110–113,
      111f, 112f, 113f, 114f, 115–116,
      115f, 116f
Electromechanical timers, 119
Electron flow theory, 32
Electronic control, 346
Electronic control devices, 139–140, 140f

Electronic drives, 158–159
Electronic overload, 77, 78
Electronic speed control
   alternate full-voltage starting methods
      and, 158–159
   current source input and, 154–155, 155f
   DC motors and, 302–303, 303f
   electronic drives and NEC and, 157–158
   explanation of, 151–152, 152f
   flux vector drive and, 156, 158f
   pulse width modulation and, 153–154,
      154f, 155f
   variable-fequency drive considerations
      and, 156–157
   variable voltage input and,
      152–153, 153f
Enclosures
   explanation of, 104, 346
   IEC requirements, 107f–108f
   NEC requirements, 104, 105f–106f
   types of, 347–348
Energy management, 278, 280
Eutectic alloy, 75, 76f, 349
Exciters, 37–38, 38f

Faraday's law, 242, 249
Feeders, 215, 349
Feeler gauge, 349
Fiber optics, 101
Fieldbus, 136–137, 137f
Field discharge resistors, 366
Filter, 349
Flashing the fields, 293
Flow switches, 96
Flux vector controller, 158f
Flux vector drive, 156
Foot switches, 96
Formulas
   AC circuits, 313–314
   DC circuits, 313
   motor torque relationship, 315
   single-phase motors, 314–315
   single-phase power, 314
   three-phase power, 315
   transformers, 315
   voltage drops, 316

Frequency
    calculation of generated, 34–35, 35f
    explanation of, 349
Frog-leg-wound armatures, 287
Full-load torque, 349
Full-voltage control, 350
Fuses
    explanation of, 350
    ratings of, 80–81
    types of, 81–82, 81f
    use of, 80

General-purpose motors, 351
Generating plants, 242
Generators. *See also specific types of generators*
    compound field, 290–292
    DC, 284, 285f, 286, 286f, 287f
    left-hand rule for, 43, 43f
    motorization of, 42–43
    output power of, 36–37
    output voltage of, 35–36
    self-excited, 367
    series, 288–289, 288f, 289f
    shunt-wound, 289–290, 289f, 290f
    standby, 43, 43f
Growlers, 202–203, 203f, 351

Hazardous locations
    determination of, 109f
    explanation of, 104, 106, 109
H-core-type transformers, 248, 248f
Heat sink, 352
Hertz, 352
High leg. *See* Wild leg
High-voltage controls, 352
Horsepower
    explanation of, 55, 352
    testing, 211–214, 212f
    torque and, 55
Horsepower watts output, 58
Human machine interface (HMI), 137, 138f
Hybrid motors, 238, 238f
Hysteresis, 195, 213, 352
Hysteresis loss, 242

Ice cube relays, 113
IEC (International Electrotechnical Commission)
    contactors, 139
    enclosures, 107f–108f
    explanation of, 86, 352
    relays, 116
IEC standards, 93–94
Impedance, 352
Impedance protected, 5
Inch stations. *See* Jog stations
Induction motors
    explanation of, 2, 6–7, 352
    power factor of, 57–58
    three-phase, 50
    troubleshooting, 58
Inductive sensors, 102, 103f
Input, 352
Input minus loss method, 213–214
Inrush current, 86, 91, 352
Instantaneous, 352
Instrument transformers
    current, 257–259, 258f, 259f
    explanation of, 246, 256
    potential, 256–257, 257f
Insulation classes, 3, 5, 272–273
Integrals, 352
Interface, 352
Interlocks, 142, 352
Inverse time, 353
Isolation transformers, 242

Jogging, 353
Jogging control, 61
Jog stations, 73, 73f
Jumpers, 353

Ladder diagrams
    explanation of, 70, 71f, 86, 353
    two-speed motor, 146, 148, 148f
    use of, 115
Laminations, 246
Lap-wound armatures, 287
Left-hand rule for generators, 43, 43f
Lenz's law, 2, 25, 353
Limit switches, 96, 35496

Liquid-level switches, 96
Load center, 354
Local control, 354
Locked rotor, 17, 171
Locked rotor current
   explanation of, 19, 158, 354
   formula for, 21
Locked rotor torque, 354
Lockout, 354
Lockout relays, 187–188
Logic, 354
Low-voltage protection (LVP)
   explanation of, 62, 355
   three-wire control and, 69, 69f
   two-wire control and, 68
Low-voltage release (LVR), 62, 355

Magnet brake, 356
Magnetic coils, 91, 334
Magnetic contactors, 66, 67f, 356
Magnetic controllers, 356
Magnetic fields
   explanation of, 10
   rotating, 10–13, 11f
   transformers and, 244
Magnetic field windings, 287–288
Magnet wires, 196, 201, 356
Maintaining contact, 62, 356
Manual control
   of DC motors, 303, 304f
   explanation of, 132–134, 133f, 134f
Manual controllers, 62–63, 63f, 357
Master switches, 357
Mechanical brakes, 191–192, 192f
Medium voltage control, 357
Mcgohmmeters
   explanation of, 196, 357
   function of, 201–202, 202f
Memory, 357
Microprocessors, 357
Momentary contact, 62, 70, 357
Motor bearings
   explanation of, 196, 357
   function of, 196–198, 197f, 198f
   lubrication of, 200–201, 200f
   replacement of, 198–200, 199f

Motor circuit switch, 357
Motor control centers (MCC), 132, 166, 166f, 358
Motor controllers
   explanation of, 86, 196, 358
   function of, 216–217
Motor controls
   critical nature of, 136
   fuse protection and, 80–82, 81f, 82f
   jog and reversing, 73–75, 73f, 74f
   manual, 62–63, 63f
   overload protection and, 63, 64f, 65–67, 67f
   overload selection and, 75–78, 76f, 77f, 78f, 79f, 80
   requirements for, 86
   single-phase motors and, 82–83
   solid-state, 125, 126f, 127
   three-wire automatic, 69–73, 69f, 70f, 71f, 72f
   two-wire automatic, 67–68, 68f
Motor-driven timers, 358
Motorization of generator, 42–43
Motor rotation tester, 52, 52f
Motors. *See also* Three-phase motors; *specific types of motors*
   armature reaction in, 298
   compound, 300–301, 300f
   DC, 294–296, 294f, 295f
   delta-connected, 46, 47f, 195, 207, 207f, 209–210, 210f, 211f, 343
   efficiency of, 58, 214–215
   explanation of, 62, 357
   hybrid, 238, 238f
   installation and maintenance of, 62–63, 67, 72, 75, 77, 78, 80, 82–84, 215–217
   lead identification of, 203–210, 204f, 205f, 206f, 207f, 208f, 209f, 210f, 311f
   maintenance checks on, 201, 201f
   multispeed, 31, 53, 54f, 55–56
   nameplate information for, 2–3, 3f, 4f, 5–6, 6f
   permanent magnet, 223, 235, 299, 299f
   replacement of, 196
   repulsion, 223, 228–231, 229f, 230f
   repulsion-induction, 223

Motors (*continued*)
   right-hand rule for, 37, 37f
   series, 299–301, 299f
   shunt, 296–298, 297f
   single-phase, 6–28
   speed control and direction of, 298
   split-phase, 7–9, 9f
   stepper, 223, 235–238, 236f, 237f
   synchronous, 231, 233–235, 233f, 234f, 235f
   torque, 223, 238
   wound rotor, 223–226, 224f, 225f, 226f, 227f, 228, 373
Motor starters. *See also* Starters
   explanation of, 62, 358
   operation of, 86
   sizes of, 88f
   synchronous, 182–183, 184f, 185, 185f
   voltage and, 70, 71f
Motor torque relationship formulas, 315
Multifunction timers, 121–122, 122f, 124, 358
Multispeed controllers, 146, 147f, 148–149, 148f, 149f, 150f
Multispeed motors, 31
Multispeed starters, 358

Nameplate information, 2–3, 3f, 4f, 5–6, 6f
National Electrical Manufacturers Association (NEMA). *See* NEMA (National Electrical Manufacturers Association)
NEC requirements
   DC motors, 303–305
   electronic drives, 158–159
   motor installation and maintenance, 62–63, 67, 72, 75, 77, 78, 80, 82–84, 215
   overload and short protection, 149
   for permanent magnet motors, 235
   resistance starters, 176
   synchronous motors, 185
   transformer, 273–274
   wound rotor motors, 226, 338
   wye-delta and, 162–163
NEMA (National Electrical Manufacturers Association), 86
NEMA size, 104, 349

NEMA standards
   coil, 93
   enclosures, 105f–106f
   explanation of, 86
Neutral current, 262
Noise
   explanation of, 360
   solid-state controls and, 127
Nonautomatic controllers, 360
Nonreversing, 360
Normally closed, 70, 360
Normally open, 70, 360
No-voltage protection
   explanation of, 62
   two-wire control and, 68
No-voltage release, 62

Off-delay timers, 119, 120
On-delay timers, 119–120
One-shot timers, 121, 360
Open delta, 242, 360
Open-delta transformers, 267–268
Open transition, 132, 162, 360
Optoisolators, 360
Output devices, 360
Output voltage
   explanation of, 37
   of generators, 35–36, 36f
Overcurrent protection, 217–218
Overload protection
   explanation of, 360–361
   function of, 63, 64f, 65–67, 67f
   overload heaters and, 79f
   types of, 75–77, 76f
Overload relays
   explanation of, 361
   troubleshooting, 333
Overload selection, 75–78, 76f, 77f, 78f, 79f, 80
Overload selection chart, 79f

Panelboard, 361
Part winding starters
   diagrams for, 164f
   explanation of, 132, 163, 165–166, 361
   function of, 159
   options for, 165, 165f

Peak switching, 118
Percent impedance, 242, 270
Permanent magnet motors, 223, 235, 299, 299f
Permanent magnets, 361
Permanent-split capacitor motors, 24–25, 25f, 361
Permeability, 362
Phase-failure protection, 362
Phase-failure relays, 166
Phase-reversal protection, 362
Phase rotation meters, 39–40, 39f
Phase rotation relays, 362
Phases, 362
Phase sequence, 31
Photodiode sensors, 96, 98, 98f
Photoelectric controls, 96, 98–102, 98f, 99f, 100f, 101f
Photoresistive cells, 96
Photo sensors, 99
Photovoltaic cells, 99, 99f
Pickup voltage, 86, 92, 362
Pilot control, 311
Pilot devices
    explanation of, 62, 94, 96, 96f, 363
    momentary contact, 70
    types of, 96
    use of, 68
Pilot switches, 96, 97f
Plugging
    explanation of, 170, 364
    lockout relay for, 187–188
    use of, 186–187, 186f
    using time delay, 188, 188f
Plugging resistors, 366
Pneumatic timers, 120–121, 120f
Polarized field frequency relay (PFFR), 183, 184f, 185
Poles
    consequent, 53
    explanation of, 364
Potential transformers, 256–257, 257f
Potentiometers, 364
Power circuits
    explanation of, 134
    for resistance starter, 172f

Power factor (PF)
    correction of, 277, 279f
    explanation of, 5, 31–32, 36, 364
Power transformers, 246, 246f
Premature wear, 330
Pressure switches, 364
Primary resistor starters
    autotransformer, 176–177, 177f, 178f, 179
    explanation of, 170, 171, 172f, 173, 174f–175f, 176, 364
    NEC and, 176
Primary winding, 242–244, 243f, 364
Prime mover, 32, 41–42
Printed circuits, 364
Profibus, 136
PROFINET, 137
Programmable logic controllers (PLC), 132, 139, 150–151, 151f, 364
Programmable relays, 124, 124f, 364
Programmable timers, 121, 364
Proximity sensors
    explanation of, 86, 364
    types of, 102–104, 102f, 103f
Pull-up torque, 365
Pulse width modulation (PWM), 153–154, 154f, 155f
Push buttons, 365

Radial loads, 197, 197f
Rectifiers, 365
Recycle timers, 121
Reduced voltage starters, 170–171, 179
Regenerative braking, 186
Relays. *See also specific relays*
    electromechanical, 110–113, 111f, 112f, 113f, 114f, 115–116, 115f, 116f
    explanation of, 86, 365
    IEC, 116
    phase-failure, 166
    polarized field frequency, 183, 184f, 185
    programmable, 124, 124f
    solid-state, 116–118, 117f
    types and styles of, 109–110, 116
Remote control, 365
Remote control circuits, 365
Repulsion-induction motors, 223, 234f, 365

Repulsion motors
   applications for, 230
   explanation of, 223, 228–229, 229f
   operation of, 229–230, 230f
   styles of, 231, 232f, 233f
Repulsion start-induction run (RSIR)
       motors, 231, 234f
Residual magnetism, 365
Resistance, 365
Resistance starters, 176
Resistors
   explanation of, 171, 173, 365
   types of, 366–367
   use of, 171
Retroreflective scanning, 99, 99f
Reversing three-phase motor starters,
       140–144, 141f, 142f
Rheostats, 367
Right-hand rule for motors, 37, 37f
Ripples, 367
Rotating magnetic fields
   explanation of, 10–13, 11f
   function of, 34, 34f
   in three-phase motors, 49, 49f
Rotors
   explanation of, 2, 7, 8f, 367
   locked, 17, 21
   reversing direction of, 17–18
   squirrel cage, 13–14, 15f, 55
   testing of, 203, 203f
   types of, 20f
   variable reluctance, 238

Safety switches, 367
Schematic diagrams
   explanation of, 70–73, 72f, 132, 367
   multispeed controllers and, 146
   wiring vs., 70, 71f
Sealing circuit, 132, 141–142
Seal-in voltage, 86, 92–93, 367
Secondary winding, 242–244, 243f, 367
Selector switches, 367
Self-excited generators, 288, 289f, 367
Semiautomatic starters, 367
Semiconductors. *See* Solid-state devices
Semimagnetic control, 367

Sensing devices, 367
Sensors
   proximity, 102–104, 102f, 103f
   types of, 96, 98–102
Separately excited, 290, 367
Sequence control, 144–146, 144f, 145f, 146f
Series motors, 299–301, 299f
Service equipment, 367
Service factors
   explanation of, 368
   overload protection and, 77
Services, 367
Servo motors, 237–238
Shaded pole motors
   explanation of, 25, 26f, 27, 368
   identification of, 205, 205f
Shading ring, 86, 89, 90f, 368
Shell-type transformers, 247, 248f
Shunt motors, 296–298, 297f
Shunt-wound generators, 289–290, 289f, 290f
Signals, 369
Single-phase generators, 33
Single-phase motors, 6–28
   capacitor-start, 21–23
   explanation of, 2, 6–7
   formulas for, 314–315
   magnetic fields and, 10–13
   permanent split capacitor, 24–25, 25f
   principles of, 7–10, 8f, 9f, 10f
   reverse of, 65, 65f, 73–74
   shaded pole, 25, 26f, 27
   slip and, 15–18
   starting winding switches and, 18–19,
      19f, 20f, 21
   two-value capacitor, 24, 24f
   types of, 28, 203
   universal, 27, 28f
   use of, 82–83
Single-phase power formulas, 314
Single-phase transformers, 242
Single phasing, 58, 132, 166, 368
Slip
   explanation of, 2, 369
   split-phase motors and, 15–18
Snubber, 369
Soft start controllers, 170

Soft start starters, 179
Solder pot, 75, 76f. *See also* Eutectic alloy
Solenoid, 369
Solenoid-and-plunger, 369
Solid-state devices, 369
Solid-state motor control, 125, 126f, 127
Solid-state reduced voltage starters, 179–182, 180f
Solid-state relays (SSR)
 electromechanical vs., 118
 explanation of, 116–118, 117f, 118f, 369
Special-purpose motors, 369
Specular scan, 100, 101f
Speed control
 electronic, 151–159, 152f, 153f, 154f, 155f, 156f, 157f, 158f, 302–303, 303f
 explanation of, 2
 formula for percent, 16, 16f
 method of, 53
 for shaded pole motor, 205, 205f
 in universal motors, 26, 27f
Split phase, 2, 369
Split-phase motors
 explanation of, 7–9, 9f
 slip and, 15–18
Squirrel cage rotors
 function of, 55
 principles of, 13–14, 15f
 testing of, 203
Starters. *See also* Motor starters
 autotransformer, 169, 176–177, 177f, 178f, 179
 combination, 140
 explanation of, 134–135, 370
 part winding, 163, 164f, 165–166, 165f
 primary resistor, 170, 171, 172f, 173, 174f, 175f, 176
 reduced voltage, 170–171, 174f–175f, 179–182, 180f
 resistance, 176
 selection of, 93–94, 95f
 solid-state, 127, 179–182, 180f
 styles of, 90–93, 92f
 synchronous motor, 182, 183, 184f, 185, 185f
Starting resistors, 366–367

Starting torque, 132, 370
Startup, 370
Static control, 370
Stators, 2, 370
Step-down transformers, 245
Stepper motors
 explanation of, 223, 235–237, 236f, 237f, 238f, 370
 servo, 237–238
Step-up transformers, 245
Subtractive polarity, 250, 251f
Surge, 370
Switchboard, 370
Switches
 explanation of, 370
 starting winding, 18–19, 19f, 20f, 21
Symbols list, 317–321
Synchronous clock timers, 118–119, 119f
Synchronous motors
 explanation of, 231, 233–234
 NEC requirements for, 185
 Warren motor and, 234–235, 235f
Synchronous motor starters, 182, 183, 184f, 185, 185f
Synchronous speed, 2, 11–13, 371
Synchroscope
 explanation of, 32, 40, 40f
 use of, 41, 41f

Tachometer generators, 371
Temperature-activated switches, 96
Temperature relays, 371
Temperature switches, 371
Terminals, 371
Thermal protectors
 explanation of, 5, 371
 use of, 63, 64f, 78
Three-phase generation
 alternators and, 38–43
 exciters and, 37–38
 explanation of, 32–34
 frequency and, 34–35
 output power and, 36–37
 output voltage and current, 35–36
 standby and cogeneration systems and, 43–49

Three-phase motors
　efficiency and, 58
　explanation of, 49–50
　input minus loss method for, 213–214
　lead identification and, 207
　motor design letters and, 56–57, 57f
　multispeed, 53, 54f, 55–56
　power factor and, 57–58
　rotation checkers and, 52
　rotation fields and, 50–52
　troubleshooting, 58, 59f
　wye- and delta-connected, 52–53
Three-phase motor starters, reversing, 140–144, 141f, 142f
Three-phase power formulas, 315
Three-phase transformers
　delta-delta-connected transformers, 263, 265f, 266–267
　explanation of, 261
　wye-wye-connected, 261–263, 261f, 264f
Three-phase true power formula, 46, 47f
Three-phase watts
　calculation of, 46, 47f
　explanation of, 46, 48–49
Three-wire control
　automatic, 69–73, 69f, 70f, 71f, 72f
　explanation of, 62, 371
Time delays, 188
Time limit, 371
Timers
　electronic, 121–122, 122f, 123f, 124–125, 124f
　explanation of, 371
　troubleshooting, 332
　types of, 118–121, 119f, 120f, 121f
Toroidal-core-type transformers, 248–249, 248f
Torque
　breakdown, 21, 340
　counter, 37, 37f
　explanation of, 32, 37, 372
　formula for, 55, 56
　horsepower and, 55
　wiring combinations and, 55, 56f
Torque motors, 223, 238
Transducers, 372

Transformation ratio, 268–269
　explanation of, 242, 372
Transformers
　classes of, 246
　connection of, 253, 254f, 255–256, 268–269
　control, 256, 256f
　design of, 245–249, 248f
　explanation of, 247f, 372
　formulas for, 315
　function of, 242–245
　instrument, 256–260, 258f, 259f, 260f
　losses of, 252
　multivoltage, 253, 253f
　nameplates for, 270–273, 271f, 272f
　NEC requirements for, 273–274
　open-delta- or V-connected, 267–268
　operation of, 249–250
　polarity of, 250, 251f, 252
　regulation of, 252
　three-phase, 261–263, 263f, 264f, 265f, 266–267, 266f
　as three-phase units, 269–270
Transient, 372
Transistors, 372
Trip free, 372
Troubleshooting
　control circuits, 143–144
　explanation of, 372
　flowcharts for, 330–334
　induction motors, 58
　overview of, 323–324
　procedures for, 324–329
　three-phase motors, 58, 59f
Two-value capacitor motors, 24, 24f
Two-wire control
　automatic, 67–68, 68f
　explanation of, 62, 68, 372

Undervoltage protection, 372
Undervoltage release, 372
Universal motors
　explanation of, 27, 28f
　testing of, 206–207
Universal switching, 118

Vacuum switches, 96
Variable-frequency drives, 132, 156–157, 158f, 372
Variable reluctance rotors, 238
Variable voltage input (VVI), 152–153, 153f
V-connected transformers, 267–268
Voltage
   differential, 284, 295, 344
   dropout, 85, 93, 345
   effect on coils of, 91, 93
   generator output, 292
   output, 35, 36f, 37
   transformer connections for, 253, 254, 254f, 255
Voltage relays, 372
Voltage transformation, 244–245, 245f
Volt drop formulas, 274–277, 316
VOM (volt-ohm-milliammeter), 373

Ward-Leonard system, 284
Warren motor, 234–235, 235f
Wave-mound armatures, 287
Wild leg, 242, 267, 373

Wiring diagrams
   coil connections and pilot control, 311
   control circuits, 309, 310
   electrical symbols, 308
   explanation of, 132, 373
   multispeed controllers and, 146, 147f
   schematic vs., 70, 71f
Wound rotor motors
   explanation of, 223, 224, 224f, 373
   operation of, 225–226, 225f, 226f, 227f, 228
Wye-connected motors
   explanation of, 45, 45f, 46f, 196, 373
   lead identification and, 52, 207, 207f, 208f
   motor connections and, 52
Wye-delta starters
   explanation of, 132, 373–374
   function of, 159–161, 161f, 162f
   NEC and, 162–163
Wye/star, 374
Wye systems, 32, 49
Wye-wye-connected transformers, 261–263, 261f, 264f